Energy: Supply and Demand

Focusing on trends in energy supply and demand, this text provides students with a comprehensive account of the subject and an understanding of how to use data analysis and modeling to make future projections and study climate impacts.

Developments in technology and policy are discussed in depth, including the role of coal, the fracking revolutions for oil and gas, the electricity grid, wind and solar power, battery storage, and biofuels. Trends in demand are also detailed, with analysis of electrical demands such as LEDs, air conditioning, heat pumps, and information technology, and the transportation demands of railroads, ships, and cars (including electric vehicles). The environmental impacts of the energy industry are considered throughout, and a full chapter is dedicated to climate change. Real-life case studies and examples add context, and over 400 full-color figures illustrate key concepts.

Accompanied by a package of online resources including solutions, video examples, sample data, and PowerPoint slides, this is an ideal text for courses on energy and is accessible to a range of students from engineering and related disciplines.

David B. Rutledge is the Tomiyasu Professor of Engineering, emeritus, at Caltech. He is a founder of the Wavestream Corporation (a manufacturer of transmitters for satellite uplinks) and his recent research has focused on modeling for projections of energy supply. He is a Fellow of the IEEE and a recipient of the Teaching Award of the Associated Students at Caltech.

Energy: Supply and Demand

David B. Rutledge

California Institute of Technology

CAMBRIDGE
UNIVERSITY PRESS

University Printing House, Cambridge CB2 8BS, United Kingdom

One Liberty Plaza, 20th Floor, New York, NY 10006, USA

477 Williamstown Road, Port Melbourne, VIC 3207, Australia

314–321, 3rd Floor, Plot 3, Splendor Forum, Jasola District Centre, New Delhi – 110025, India

79 Anson Road, #06–04/06, Singapore 079906

Cambridge University Press is part of the University of Cambridge.

It furthers the University's mission by disseminating knowledge in the pursuit of education, learning, and research at the highest international levels of excellence.

www.cambridge.org
Information on this title: www.cambridge.org/9781107031074
DOI: 10.1017/9781139381208

First published 2020

Printed in Singapore by Markono Print Media Pte Ltd

A catalog record for this publication is available from the British Library.

Library of Congress Cataloging-in-Publication Data
Names: Rutledge, David B., 1952– author.
Title: Energy : supply and demand / David B. Rutledge, California Institute of Technology.
Description: Cambridge, United Kingdom ; New York, NY : Cambridge University Press, [2020] |
Includes bibliographical references and index.
Identifiers: LCCN 2019015702 | ISBN 9781107031074 (hardback : alk. paper)
Subjects: LCSH: Fossil fuels.
Classification: LCC TP318 .R88 2020 | DDC 553.2–dc23
LC record available at https://lccn.loc.gov/2019015702

ISBN 978-1-107-03107-4 Hardback

Additional resources for this publication at www.cambridge.org/rutledge

To my wife
Dale

Contents

Preface

Where does our energy come from? How do we use it? This book is an introduction to the fossil fuels and the alternatives, the electrical grid, energy use in buildings and transportation, agriculture, and climate-change policy for fossil fuels. Modeling is emphasized for understanding trends and for making projections. It is important to appreciate what models can tell us about how energy systems are evolving. Students should learn the distinction between physical laws that allow precise predictions and model projections that can be wrong. Models did not predict the shale gas revolution.

The material evolved from years of teaching classes to Caltech students at all levels. The students in the courses complete homework and laboratory exercises and they take tours of a natural gas power plant and a solar power station. At the end of the term, each student selects a topic to investigate and makes a presentation to the other students. After completing the course, students should be comfortable making energy calculations and developing models for energy systems. In addition, students should be able to critically assess articles, books, and films on energy. They should be familiar with the strengths and weaknesses of the arguments made by early writers like Stanley Jevons and King Hubbert. They should appreciate the potential of new energy technology. Finally, students should be able to recognize when government policies are working and when they are not.

Some homework problems involve locating energy databases online and downloading and analyzing the data. The online databases often provide their information in Excel format. Some students lack experience in Excel, and it is helpful for an instructor to demonstrate the functions that are used in the homework.

It is a pleasure to acknowledge people who have helped me in writing this book. Dale Yee took photographs and drew figures. Professor Joseph Shepherd at Caltech has been a collaborator in the classes. I have not found a question on combustion that he could not answer. Kent Potter at Caltech developed exercises and critiqued many of the ideas. Dr. Romeo Flores, formerly of the United States Geological Survey and the nation's foremost expert on coal, encouraged the work at the early stages. Jean Laherrere, formerly of the Total oil company, graciously provided production statistics. Jean taught many of us how to look at data. Several people read early drafts and made thoughtful suggestions, including the late Tom Tombrello, professor of physics at Caltech and erstwhile director of research at Schlumberger, Dr. Euan Mearns of Aberdeen, Scotland, founder of the blog Energy Matters, and my brother John Rutledge, Vice-President and Water Resources Group Manager at Freese and Nichols, Inc., consulting civil engineers in Fort Worth, Texas. At the Cambridge University Press, I would like to thank Julie Lancashire, who has supported the project from the beginning, Nicola Chapman, who managed a challenging production process, and my wonderful editor, Heather Brolly.

Please let me know of errors by email at rutledge@caltech.edu.

1 Preliminaries

Chop your own wood,
and it will warm you twice.

Henry Ford

Let us begin by describing the sort of energy we are interested in. In this book, our emphasis will be on useful energy, energy that people apply to their purposes, like heat from a wood stove (Figure 1.1) or the energy derived from eating rice (Figure 1.2). Nowadays most of us buy our energy, although some people do split their own firewood. The energy *supply* is the amount that producers sell, like the 562 million tonnes[1] of oil that flowed from Saudi wells in 2017. The energy *demand* is the amount that consumers buy, like gasoline for a car and electricity to light a home. Energy sales are an important part of the economy, and for this reason most countries keep good energy statistics, although they often miss the energy that people generate for themselves, like the firewood they split and the electricity from home solar panels (Figure 1.3). For energy statistics, a useful publication is the BP *Statistical Review of World Energy*, which has been published annually by the oil company since 1951. In recent years, the *Statistical Review* has been a collaborative effort between the economics group at BP and the Heriot-Watt University Centre for Energy Economics Research and Policy. The *Statistical Review* is the most current, most consistent, and most accessible of the energy data sources. Measured in terms of the quantity of data, the US Department of Energy's Energy Information Administration (EIA) comes first. Often one can find arcane numbers in an obscure part of the EIA website that appear nowhere else. For agriculture, the best reference is FAOSTAT, the online database of the United Nations Food and Agriculture Organization, abbreviated FAO. Other energy data sources are listed in the Further Reading section at the end of this chapter. Studying these statistics can help one identify trends and gain perspective on energy issues. In energy, the past is often a good guide to the future and the trends may indicate whether a policy proposal is realistic or not. However, there can be unpredictable shocks because of wars, economic collapse, and new technology. For example,

---

[1] A tonne is a thousand kilograms. Technical language varies between countries, and this can be confusing. Much of the American energy literature uses the short ton, which is 2,000 pounds. Americans commonly write and say "metric ton" instead of tonne to distinguish it from the short ton.

Figure 1.1 A wood stove for heating a home. This stove, manufactured by the Morso Company in Denmark, produces 6 kilowatts of heat. It has an efficiency of 75%, that is, only a quarter of the heat is lost up the chimney. There is a secondary pre-heated air supply at the top of the fire box that reduces air pollution by making the burn more complete. A wood stove can save money. A 6-kilowatt electric heater might cost a dollar an hour to run. However, it is hard work to saw and split the wood. Photograph by the author.

Figure 1.2 A rice field in Japan. Rice provides one-fifth of the world's food energy. Like wood, rice is primarily made up of carbohydrates, and this gives it a similar energy density. The world food supply was reckoned by the United Nations Food and Agriculture Organization (FAO) in 2013 to be 2,884 food calories per person per day. Credit: Jason Hickey/CC BY 2.0, https://creativecommons.org/licenses/by/2.0/

Figure 1.3 A solar photovoltaic array for a home in the state of New Mexico. There are six panels with a capacity of 190 W each. The array is connected to the electrical grid, which supplies electricity when the sun is not shining. When the array produces more electricity than needed, the electric cooperative buys the surplus. Solar arrays provide electricity without noise or air pollution. The solar panels themselves have become inexpensive, but there are substantial costs associated with the installation and integration into the grid. Germany has been the world leader in solar panel installations, with a capacity of 500 watts per person in 2017. Photograph by the author.

the combination of hydraulic fracturing and horizontal wells, aided by the unusual American system of private mineral rights, has resulted in an astonishing increase in US oil and gas production.

Engineering calculations often start from the principle of energy conservation that is expressed by the First Law of thermodynamics. This law states that when energy is converted from one form to another, it is neither created nor destroyed. So what do we mean when we talk about producing or consuming energy? We are not talking about violating the First Law. Rather, it is an accounting convention where we track useful energy. Consider a computer that is supplied with 100 watts of electricity during an 8-hour workday. During that time electrical energy is converted in the computer to heat in the processor, light in the monitor, and sound in the speakers. That heat, light, and sound energy are no longer available for practical purposes, and we say that the computer consumed 800 watt-hours of electricity. For a production example, consider coal. Coal is a rock that formed over millions of years from plants that were buried by sediment. The energy in sunlight enabled the plants to grow, and heat and pressure in the earth drove the chemical processes that converted the plant remains to coal. When the coal is mined, it is available for burning, and we say that the energy produced is the combustion heat of the coal.

1.1 Plan of the Book

This chapter introduces ideas that we will encounter repeatedly. In the following chapter, we turn to historical energy sources that were important before the transition to coal, considering horses, whale oil, and wood. Then we study the major fossil fuels, coal and hydrocarbons. It is important to get a perspective on the factors that affect fossil-fuel resources and to appreciate the potential of new technology for fossil fuels. We will develop projections for the ultimate production of the fossil fuels.

Next we consider the food supply. Agriculture is both a significant consumer of energy in fertilizer production and a significant producer of energy through biofuels. Its most important impact is the enormous land area needed for agriculture. In some places, like Africa, the yield increase has not been sufficient to prevent significant deforestation. On the oceans, commercial fishing has put severe pressure on many fisheries. However, a transition to aquaculture is underway.

We will consider the role of each of the components of the electricity supply, starting with the fossil fuels and the traditional alternatives, hydroelectric and nuclear power. Many countries are now emphasizing the new alternatives: wind, solar, geothermal, and biofuels for new electricity capacity, so it is important to investigate how the new alternatives will affect the grid.

Stationary demand is industrial energy use, and lighting, heating, and cooling for buildings. The emphasis in many countries has been in making buildings more efficient. Usually the goal has been to reduce energy consumption, but the policy can also make homes more comfortable and make offices better places to work.

In transportation, oil is king, although electric vehicle sales are beginning. It is important to understand the differences between highway and rail networks as infrastructure, and between passenger and freight transportation systems.

We conclude with a discussion of climate change. In contrast to the technical solutions developed for air pollution, there is no quick way to reduce carbon-dioxide emissions from burning fossil fuels. Our discussion will consider the carbon cycle, sea level rise, and the temperature indexes. We will develop an estimate of the long-run temperature sensitivity to ultimate fossil-fuel production.

1.2   Units

The basic energy unit is the *joule*, abbreviated J. It is named after James Joule, an English physicist who in 1845 characterized the conversion of mechanical energy to heat energy. In his measurements, Joule spun a paddle wheel in a can of water. He found that this warmed the water and he measured the temperature rise. Expressed in modern units, the ratio was 4,400 J per kilogram of water per kelvin.[2] This amount of energy is the basis of the *food calorie*, abbreviated Cal. Food calories are used in food content labels in the United States, Canada, and Japan. The modern value of the food calorie is 4,184 J.

Many systems are characterized by how fast energy is produced or consumed. This quantity is the *power*. The unit of power is the watt (W), named after James Watt, a Scottish mechanical engineer who invented the first practical steam engine in 1765. A watt is one joule per second. The maximum transmitter power from cell phones is about a watt. There is also an energy unit that includes "watt" in the name. This is the watt-hour, abbreviated Wh. A watt-hour is the energy delivered by a 1-W source in an hour, or 3,600 J. It is common in media reports to confound watts with watt-hours. This is an easy error to make, but it marks the journalist and the editor as having a weak understanding of energy. We will use the watt-hour only for electricity.

For many energy discussions, both the joule and the watt are small units, and prefixes are applied to indicate larger amounts. The standard SI[3] unit prefixes are given in Table 1.1. Be careful to distinguish the "k" for "kilo" from the "K" for kelvin, the temperature unit. The larger units starting with "M" for "mega" are capitalized. One warning is that in energy writing, you will often see "M" or "m" for one thousand, and "MM" or "mm" for one million. Older dictionaries often show the pronunciation of the first "g" in "giga" as soft, like the word gigantic. However, most people now pronounce both g's hard, as in giggle, and newer dictionaries reflect this. Usually we choose the prefix that leaves the number between 1 and 1,000. For example, we would write 10 kW rather than 10,000 W or 0.01 MW. By convention 1 km² is not one thousand square meters but rather a square kilometer. In this book, we freely put prefixes in front of most units but not another prefix. So we write 1 G\$ for a billion dollars and 1 Tm³ for a trillion cubic meters, but 1.5 million km² for one million, five hundred thousand square kilometers.

The units we use are shown in Table 1.2. Standard SI units will be used for many calculations in this book. However, energy markets use many non-SI units. International oil prices are commonly quoted in US dollars per barrel. The barrel

---

[2] Joule expressed the result in British units: 817 foot-pounds per degree Fahrenheit for a pound of water. This amount is the basis for the Btu (British thermal unit), which is equivalent to 1,055 J. Joule's measurement was within 5% of the correct value.

[3] From the French, *Système International d'unités*, based on combinations of the meter, kilogram, and second and the prefixes.

Table 1.1 The standard SI unit prefixes.

Prefix	Pronounced	Meaning
y	yocto	10^{-24}
z	zepto	10^{-21}
a	atto	10^{-18}
f	femto	10^{-15}
p	pico	10^{-12}
n	nano	10^{-9}
μ	micro	10^{-6}
m	milli	10^{-3}
k	kilo	10^{3}
M	mega	10^{6}
G	giga	10^{9}
T	tera	10^{12}
P	peta	10^{15}
E	exa	10^{18}
Z	zetta	10^{21}
Y	yotta	10^{24}

Table 1.2 Units in this book. The abbreviation "p" for person is not standard, but it is convenient.

Symbol	Name	Type	Representative use	Equivalent
$	US dollar	currency	prices	
p	person	population	per-person demand	
H	hash	calculation	transaction validation	
lm	lumen	luminous flux	lamp ratings	
lx	lux	flux density	illumination	$1\,lm/m^2$
°C	degrees Celsius	temperature	reference is freezing water	
°F	degrees Fahrenheit	temperature	annual cooling degree days	
K	kelvin	temperature	temperature changes	
J	joule	energy	basic energy unit	kgm^2/s^2
W	watt	power	basic power unit	J/s
N	newton	force	vehicle drag	J/m
Nm	newton-meter	torque	motor output	J/radian
Wh	watt-hour	energy	electricity	3,600 J/Wh
Btu	British thermal unit	energy	American energy statistics	1,055 J/Btu
Cal	food calorie	energy	food energy	4,184 J/Cal

Table 1.2 (cont.)

Symbol	Name	Type	Representative use	Equivalent
toe	tonne-of-oil equivalent	energy	fuel comparisons	42 GJ/toe
boe	barrel-of-oil equivalent	energy	fuel comparisons	7.33 boe/toe
hp	horsepower	power	motor and engine output	746 W/hp
rpm	revolutions per minute	frequency	motor and engine output	
l	liter	volume	gasoline consumption	1,000 l/m^3
m	meter	length	fundamental SI length unit	
km^2	square kilometer	area	land	10^6 m^2/km^2
m^3	cubic meter	volume	natural gas production	900 toe/Mm^3 (gas)
b	barrel	volume	oil production	7.33 b/t (oil)
kg	kilogram	mass	fundamental SI mass unit	
t	tonne, metric ton	mass	coal production	2 t/toe (world coal)
tC	metric tons of carbon	mass	CO_2 emissions	12/44 tC/tCO$_2$
Pa	pascal	pressure	atmosphere	N/m^2
V	volt	voltage	electrical circuits	
A	ampere	current	electrical circuits	
Ω	ohm	resistance	electrical circuits	V/A
F	farad	capacitance	electrical circuits	J/V^2
T	tesla	magnetic field	electrical machines	Vs/m^2
Hz	hertz	frequency	AC circuits	cycle per second
s	second	time	fundamental SI time unit	
h	hour	time	electricity production	3,600 s
d	day	time	food supply	24 h
y	year	time	energy production	8,760 h

unit goes back to early oil production in the state of Pennsylvania in the 1800s. It is a volume unit, equal to 42 gallons, or 159 liters. The barrel for whale oil was ten gallons smaller. Oil production may also be quoted in metric tons. The tonne is a unit of mass, and in practice, the density of different grades of oil vary enough that it may take anywhere from six to eight 42-gallon barrels to make up a tonne. We will follow the lead of the BP *Statistical Review* and convert at 7.33 barrels per tonne. The toe, or tonne-of-oil equivalent, is an energy unit equal to 42 GJ. It is useful for comparisons involving different energy sources. The energy density of mined coal varies considerably from mine to mine and from country to country. However, at the world level, the average energy density of mined coal has been relatively stable at 2 t/toe (tonnes per tonne-of-oil equivalent). There are several definitions for horsepower that differ slightly. When it matters, we will use the electrical horsepower that appears in electrical motor and generator ratings for calculations. It has the advantage that the conversion to watts is a whole number, 746 W/hp. The number of hours in a year that is not a leap year, 8,760, is a useful number to remember for converting annual electricity production to average power. In a leap year there are 24 more hours.

1.2.1   Capacity Factors

The power rating for a source is called the *capacity*, or often *nameplate capacity*, because it is common for a generator capacity to be given on an attached metal plate. Capacities vary widely. Each solar panel in the array in Figure 1.3 has a capacity of 190 W, while the Diablo Canyon Nuclear Power Plant in California has a capacity of 2.2 GW. The capacity is specified at standard testing conditions. For example, solar panels are tested in a strong light of 1 kW/m^2 with the panel cooled to 25 °C. However, a panel outdoors will usually be hotter than 25 °C when the sun is shining in the summer, and this reduces the output. For this reason, a solar panel may never actually achieve its power capacity in practice. The capacity of solar farms is often specified as the capacity of the electronic inverters that produce AC electricity for the grid, rather than the total nameplate capacity of the panels. Capacity is what a power company pays for when it buys a generator. However, in making comparisons, one should be aware that the average power from a source may be much less than the capacity. For example, a solar panel will have no output at night and reduced output on cloudy days. We define the *capacity factor* CF as

$$CF \equiv P_a / P_c \qquad\qquad 1.1$$

where $\equiv$ indicates a definition, P_a is the average power and P_c is the nameplate capacity. As an example we will calculate the capacity factor for the six solar panels in Figure 1.3. These have a capacity of 190 W each, or 1.14 kW for the entire array. In one year of production, the array produced 1.52 MWh. The average power is given by

$$P_a = 1.52\,\text{MWh} / 8,760\,\text{h} = 174\,\text{W} \qquad\qquad 1.2$$

and the capacity factor is given by

$$CF = 174\,\text{W} / 1.14\,\text{kW} = 15\%. \qquad\qquad 1.3$$

In practice the capacity factor of solar panels varies considerably with the location. A solar array in cloudy England may have a 10% CF, while the panels at the Kayenta Solar Facility shown on the cover of this book have a 23% CF. The solar panel CFs are low compared to most other generators. For comparison, the worldwide average CF for nuclear power plants is 80%.

The low capacity factor of solar panels is a fundamental limitation. We cannot command the sun to shine at night or push away clouds over a solar farm. The electrical utility will need to have capacity available to connect to the grid when the solar output drops. This extra capacity is often provided by natural gas generators. In this situation the natural gas generators are not competing with solar farms, but instead are complementary. This means that it may not be appropriate to compare the prices of generation from different sources calculated separately. Consumers are not looking for access to a particular solar farm when the sun is shining. They want continuous access to electricity. The high capacity factor for nuclear plants can also be limiting. The electricity demand varies during the day, and with their enormous steam boilers, nuclear plants are not able to follow this changing load. Generators sell electricity into markets where the prices vary dramatically during the day, even going negative at times. When prices are low, it is difficult for the owners of the nuclear plants who cannot turn them off.

1.2.2   Payback Times

In the United States, when a residential solar array is connected to the grid, the owner receives a federal income tax credit of 30% of the installation price. The utilities charge for the electricity supplied to the residence after the solar generation to the grid has been subtracted. This is called *net metering*. It is equivalent to allowing the electricity meter to run backwards. In net metering, people effectively receive the marginal retail price for the electricity their panels generate. The marginal retail price varies widely so that it may make sense to buy solar panels in some states and not in others. California has price tiers with marginal rates up to 40¢/kWh. However, in other American states customers pay flat rates with marginal prices closer to 10¢/kWh. If an array produces more electricity during a month than the house uses, the utilities may pay a much lower price based on their accounting of the savings in accepting the solar power. These *avoided costs* may be as low as 3¢/kWh.

In Europe the system is different. The utility pays the solar panel owners for the electricity they generate. The price is called the *feed-in-tariff*, or FIT. The FITs are set by the government. The FITs have generally been larger than the retail price, but they have been dropping over time. A major difference between the two systems is who pays. In the US, the main subsidy is the income-tax tax credit, so the burden falls primarily on the high-income taxpayers who pay the majority of income taxes.[4]

---

[4] In the US, the most important source of tax revenue is the federal income tax. In other countries the value-added-tax may be as important. The US does not have a national value-added-tax. In 2017, the top 20% of households in income, those who reported $150,000 or more on their tax returns, received 52% of the income and paid 87% of the income taxes. For more discussion, see Laura Saunders, "Top 20% of earners pay 87% of income tax" in the *Wall Street Journal*, April 6, 2018.

The effects may not be at all apparent in the utility bills that an American homeowner pays. In Europe the burden effectively falls on the neighbors who do not have solar arrays, and who are likely to have lower incomes than people who do.

We can evaluate a capital investment like a solar array by its payback time T in years, given by

$$T = I / r \tag{1.4}$$

where I is the investment net of subsidies and r is the annual money return after the operating costs are subtracted. For a homeowner, the return r for a solar array might be the reduction in electricity bills. Longer payback times indicate poorer investments, but it is hard to be specific about how short a payback time should be. If the equipment has a limited lifetime, the payback time should certainly be shorter than that lifetime. For example, the manufacturer specifies the performance of the panels in Figure 1.3 only out to 25 years. One should also realize that to be attractive, the return should be larger than the distributions for an alternative investment like stocks. For energy systems, the original investment is lost forever. This is different from stocks, which can ordinarily be sold later, often at a profit. In our examples, we will assume a 10-year payback time.

1.3   Efficiency

When energy is converted from one form to another, the useful output energy will be less than the input energy. This is an informal statement of the Second Law of thermodynamics. We write the *efficiency* η (the Greek letter *eta*) of the conversion as

$$\eta = E_o / E_i \tag{1.5}$$

where E_o is the output energy and E_i is the input energy. In this formula η does not have units, and it is typically expressed as a percentage. For example, for the wood stove in Figure 1.1, the input energy is the combustion heat, and the output energy is the heat that goes into the home. Some heat is lost up the chimney, and we can write an equivalent efficiency formula in terms of the lost energy E_l as

$$\eta = 1 - E_l / E_i \tag{1.6}$$

In some situations, the word efficiency is used in a less precise way. For example, people talk about improving the efficiency of a house by adding insulation in the walls and roof. We define a *thermal resistance* R_t, given by

$$R_t \equiv T / P \tag{1.7}$$

where $\equiv$ denotes a definition. The parameter T is the temperature rise in the house from the stove and P is the stove power needed to maintain this rise. The units of R_t are K/W. In this case the input is power, that is, energy per unit time, rather than energy. The formula is dual to Ohm's law, with the temperature rise in the role of voltage and the heater power as current. Adding insulation increases the thermal resistance and reduces the power that is needed to heat the house. In characterizing insulation and windows, the power is normalized as a power per unit area, and the resistance is called an R-factor. In Section 10.4.3, the forcing, which is the radiative power imbalance of the earth due to greenhouse gases, is similarly normalized as a power per unit area.

1.3.1 Conservation and Efficiency

We distinguish between an efficiency improvement, like adding insulation, and conservation. In energy discussions, *conservation* means reducing energy use. A prominent example is a 1979 speech by US President Jimmy Carter, when he said,

> I am asking you for your good and for your nation's security to take no unnecessary trips, to use carpools or public transportation whenever you can, to park your car an extra day per week, to obey the speed limit, and to set your thermostats to save fuel. Every act of energy conservation like is this more than just common sense – I tell you it is an act of patriotism.[5]

Carter was referring to a national highway speed limit that had been set at 55 miles per hour in 1974 during the Nixon Administration. The limit was widely despised and widely violated. It was repealed in 1995 during the Clinton Administration.

This is conservation. It means using less. The meaning of the word conservation has evolved over time. A hundred years earlier the word had the sense of using a resource in a sustainable way and it was applied to logging and fishing. For a more recent example, Japan shut down all of its nuclear plants after the Fukushima reactor disasters in 2011, and for this reason, utilities had trouble meeting the electricity demand in the summer. As a conservation measure, many Japanese offices operated at 28 °C, a level that was uncomfortable for many people, given the high humidity in many Japanese cities.

Incidentally, determining an appropriate temperature range for a room is not as straightforward as it might appear. There are wide differences between countries and between individuals in the temperature range that is considered comfortable. Visitors to Finland often note that Finns like their homes quite warm during the arctic winter and take their guests to saunas that are even hotter. Visitors to Texas find that Texans like their homes cold when it is scorching outside. For some, it

---

[5] The full transcript of the speech is available from the Carter Library at
www.jimmycarterlibrary.gov/assets/documents/speeches/energy-crisis.phtml

is a matter of safety. Homes that are too hot or too cold are unhealthy for the elderly. In Europe, where residential air conditioning is not common, it has been estimated that 70,000 people died in the great heat wave of 2003.[6] Conservation and efficiency improvements both save energy. However, conservation reduces the standard of living, while increased efficiency improves the standard of living.

To a consumer, efficiency will usually be less important than cost in choosing a system to install. For example, an oil heater may be more efficient than a wood stove. However, if wood is freely available, a family might prefer a wood stove, even if the stove is less efficient than an oil furnace, and even if someone has to load the wood into the stove. In this situation, it is possible that in choosing a wood stove, the family will save money, consume more energy, and enjoy a warmer house all at the same time.

There are similar considerations at the national level. Lower-income countries may burn locally produced coal to generate electricity in preference to building more efficient natural gas generators. Coal can be mined with a pick and a shovel, and it can be transported in the back of a truck. Natural gas requires pipelines. Even though their mines are mostly closed now, the United Kingdom and France started with coal many years ago. China and India are following the same path, establishing their electricity grids primarily with coal that is mined within their own country. These two countries were the largest consumers of coal in 2017.

1.3.2   The Jevons Paradox

There is an important question that arises when the efficiency of energy systems is improved. How do people respond? Consider a family that adds insulation to their home. Previously they may have found that heating oil was so expensive that they only lit the furnace on the very coldest nights. The rest of the time they just shivered. However, with the insulation, the family might notice the heating oil bills dropping, and then decide that they can afford to put the furnace on a thermostat. Now the house is comfortable all of the time, but the energy savings are less than if they had followed their old routine. Economists call this reduced energy savings *rebound*. The family is more comfortable than before, so they are better off. It is even possible that the family might enjoy the comfort so much that their energy use goes up rather than down. This is called *backfire*. There can also be an *indirect rebound* effect. Let us assume that our family does not change its heating habits and its bill for heating oil drops. They might use the money they save to fly to visit relatives. The energy for the airplane fuel would offset the heating oil energy savings.

---

[6] Richard Keller tells the story of this horrific tragedy in his 2015 book, *Fatal Isolation*, University of Chicago Press.

The English economist Stanley Jevons is credited with noticing this effect. He wrote in his 1865 book, *The Coal Question*,

> It is wholly a confusion of ideas to suppose that the economical use of fuel is equivalent to a diminished consumption. The very contrary is the truth.

His book is also important because it was the first serious discussion of fossil-fuel supplies in the long run, and we will return to it in Section 4.5. Jevons only described the direct backfire effect. However, we will consider energy use more broadly, and call both rebound and backfire, direct and indirect, collectively the *Jevons Paradox*.

Jevons gave an example of efficiency improvements in steam engines. Early steam engines were only 1% efficient. This efficiency was so poor that the only important market for them was in pumping water out of coal mines, where coal was easier to get than anywhere else. However, in 1765 James Watt added a steam condenser that improved the efficiency by a factor of four. By Jevons' time, the efficiency of the best steam engines had increased to 14%. These improvements opened up new markets for steam engines in mills, factories, and locomotives. Even though the new steam engines were much more efficient than the old ones, coal consumption did not go down. Rather, it went up. Jevons calculated that British coal consumption was increasing at 3.5%/y. The increase in the number of new steam engines more than offset the effect of improved efficiency.

The Jevons Paradox presents a challenge. Government agencies often set efficiency standards for products with the intent of reducing energy consumption. It is common for the agencies to project energy savings under the assumption that the consumers make only small changes in their habits. However, in practice the energy savings may not actually take place. One example is mileage standards for cars in the United States. These were implemented in 1978. They go by the name of Corporate Average Fuel Economy standards, CAFE for short. Manufacturers cut the size of cars and the horsepower of the engines to meet the standards. However, consumers responded by buying sport-utility vehicles (SUVs) and light trucks. The result was that for thirty years from 1978 until the Great Recession began in 2008, US oil consumption did not budge. It was 870 Mt in 1978 and 875 Mt in 2008. It is difficult to call the program a success if the consumption does not actually go down. The only guaranteed way to reduce energy consumption at the national level has been economic collapse. The most important historical example is the breakup of the Soviet Union that followed the fall of the Berlin Wall in 1989. From 1989 to 1997, energy consumption in the countries of the former Soviet Union fell by 34%.

1.4   Energy Production

Figure 1.4 is a plot of world energy production. Production has risen dramatically since World War II ended in 1945. Hydrocarbons and coal made up 83% of the supply in 2017. By *hydrocarbons*, we mean crude oil, natural gas, which is predominantly methane, and the natural gas liquids ethane, propane, butane, and pentane. Hydrocarbons and coal are called *fossil fuels*, recognizing their origin millions of years ago in drifting plankton for hydrocarbons and in land plants for coal. *Alternatives* are everything else. This plot is for *primary energy*, the energy calculated before conversion to a secondary form like electricity. The distinction is important because a service provided directly by a fossil fuel may compete with one provided by electricity that is partly produced by alternatives. For example, a family may buy a natural gas heating system for their home or an electric one. Accounting for primary energy presents problems. Sometimes only the secondary electricity output is measured, as in solar farms and in hydroelectric power. Different sources have very different conversion efficiencies. For example, photovoltaic arrays typically convert light to electricity with an efficiency of 20%, while water turbines may be 90% efficient. This means that plotting actual

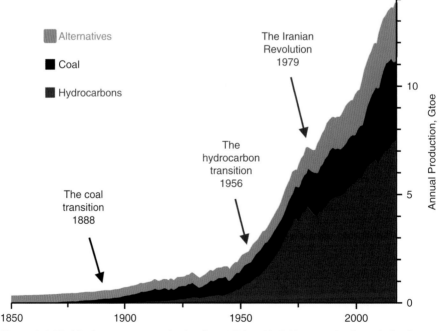

Figure 1.4 World primary energy production from 1850 to 2017. No correction is made for the non-fuel uses of coal and hydrocarbons. The data come from the BP *Statistical Review*, Brian Mitchell, 2007, *International Historical Statistics*, Palgrave Macmillan, Arnulf Grubler, 2003, *Technology and Global Change*, Cambridge University Press, and the online database of the United Nations Food and Agriculture Organization, FAOSTAT.

solar and water primary energies in a graph gives a misleading impression of their relative contributions. Because of this, it is conventional to use nominal conversion efficiencies for alternative sources. For example, the BP *Statistical Review* assumes a nominal plant conversion efficiency of $4.4\,\mathrm{MWh/toe} = 2.64/7 \approx 38\%$ for these sources and that convention is followed in this book. In 1850, the dominant alternative was firewood. Hydroelectric plants came into operation in the 1880s and nuclear power began in the 1950s. More recently biofuels, wind turbines, and solar power have become significant. Two transitions are noted in Figure 1.4. Coal passed alternatives in 1888. The reign of coal lasted 68 years until hydrocarbons passed coal in 1956. The hydrocarbon transition was close to the time that Norman Borlaug's Green Revolution in crop yield began. In 1958, Charles Keeling began his measurements of atmospheric carbon-dioxide levels in Hawaii. The hydrocarbon era that began in 1956 gives an appropriate time frame for considering the climate impacts of the carbon dioxide released by burning fossil fuels – 87% of the cumulative fossil-fuel carbon-dioxide emissions through 2017 occurred from 1956 on.

1.4.1 The Alternatives

We can get a better sense of how alternatives have fared from Figure 1.5, which shows their share over time. Starting in 1850 at 83%, there is a long decline because of increasing fossil-fuel consumption, finally bottoming out at 10% in 1973, the year of the Yom Kippur War and the Arab Oil Embargo. After 1973, the alternatives

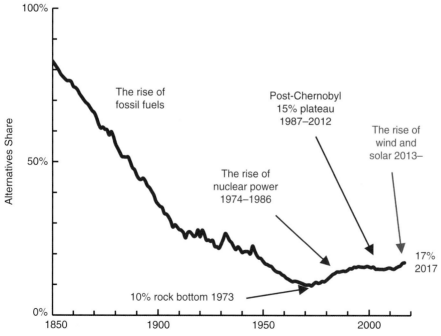

Figure 1.5 The alternatives share of primary energy over the years. This graph is based on the data in Figure 1.4.

share grew because many nuclear power plants were built. However, this slowed after the Chernobyl reactor disaster in the Ukraine in 1986. The alternatives share was flat at around 15% from 1987 through 2012. It started to pick up again in 2013 because of wind and solar power. The alternatives share was 17% in 2017.

1.4.2   Per-Person Consumption

We get a different perspective if we replot the energy on a per-person basis. This is shown in Figure 1.6. It makes the early years clearer. We will label the y-axis consumption this time to recognize the fact that in contrast to production, much energy consumption is on an individual basis. Wood and water are steady throughout at 0.2 toe/p. Coal and hydrocarbons each have an early growth phase, a war shock, and then a plateau. The hydrocarbon shock was 1979, the year of the Iranian Revolution, when the Shah was deposed. The following year Saddam Hussein led Iraq in an attack on Iran and oil production in both countries collapsed. Since that time, world hydrocarbon consumption has been around 1 toe/p. The coal shock was World War I that began in 1914. The plateau that followed at 0.4 toe/p lasted until 2002. In this century there has been a rise as lower-income countries built out

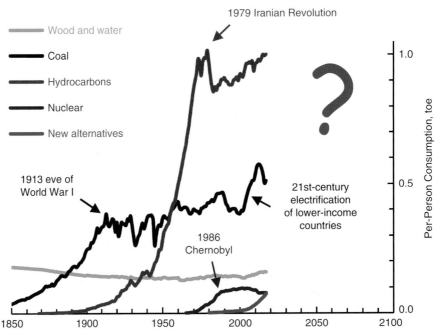

Figure 1.6 Annual world per-person energy consumption, based on the data in Figure 1.4. The population data from 1950 on are from the United Nations Population Division, "World Population Prospects: The 2017 Revision." The earlier population data are from the late Angus Maddison, formerly professor of economics at the University of Groningen. His data are available at www.ggdc.net/MADDISON/oriindex.htm

their electrical grids. The consumption stood at 0.5 toe/p in 2017. Nuclear power also shows a pattern of growth, shock, and plateau. For nuclear power, the shock was not war, but the Chernobyl disaster in the Soviet Union in 1986. The plateau was at 0.1 toe/p. In 2011, there was another nuclear disaster at Fukushima, Japan, that resulted from the Great Tohoku Tsunami. It is too early to tell whether this will change the nuclear energy trend again. New alternatives include wind, solar, geothermal, and biofuels. New alternatives are small, 0.08 toe/p in 2017. However, wind and solar generation are growing rapidly because of government subsidies.

It is interesting to compare the energy consumption in Figure 1.6 to the food supply noted in the caption of Figure 1.2 and to wood consumption. In energy terms, the world food supply of 2,884 Cal/p/d is 0.10 toe/p/y. Food crops equivalent to a tenth of the food supply are converted to biofuels. The most significant are in the US, where ethanol is made by fermenting corn, in Brazil, where ethanol is made from sugar cane, and in Germany where methane is produced from the anaerobic digestion of corn. World wood consumption is 0.07 toe/p, with half used as fuel.

Fossil fuels have allowed a ten-to-one increase in the per-person consumption of energy compared to wood and water alone. However, the curse of fossil fuels is that they run out. Thousands of sealed oil wells and ghost coal mining towns bear witness to this fact. Mankind has made a Faustian bargain.[7] When will the transition away from fossil fuels to alternatives happen? In Figure 1.6, the years through 2100 are filled with a giant question mark. However, we will see in Chapters 4 and 5 that we can get clues to the answer from the past production history.

1.5 Population

The per-person consumption plots in Figure 1.6 are flat for long periods. This shows that there is a strong association between consumption and population, and it makes it important to understand how populations evolve. Figure 1.7 shows the world population on a regional basis from the United Nations Population Division. Populations are determined by national censuses. The intervals between censuses vary and the census years differ even for countries that use the same interval. Thus the annual numbers in the graph include interpolations and extrapolations. The United States and the European Union do not have large enough populations to show up distinctly on a graph like this, so the population is given instead for the OECD countries. OECD stands for the Organization for Economic Cooperation and Development. The OECD members are the higher-income countries. In 2018, there were 36 members of the OECD, including most European countries, the US,

---

[7] Heinrich Faust was a character in German author Johann von Goethe's play *Faust*, written and rewritten over a sixty-year period from 1772 to 1831. In the play, Faust is a scholar who makes an agreement with the devil – in life, Faust's desires will be accommodated, but after death, Faust will serve the devil.

Canada, and Mexico in North America, and Japan, South Korea, Australia, New Zealand, Israel, and Chile. It will often be useful to compare the OECD countries to the non-OECD countries as a test of the effect of income. The OECD, China, India, and Africa have comparable populations, at least for now.

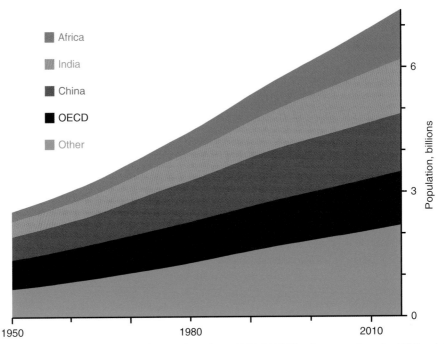

Figure 1.7 World population broken out by regions, 1950–2015. The data come from the UN Population Division, "World Population Prospects: The 2017 Revision." The data can be downloaded from https://population.un.org/wpp/Download/Standard/Population/

1.5.1 Doubling Times

Table 1.3 uses the data in Figure 1.7 in a different way. The annual growth rates in the table are calculated as

$$g = \frac{p_1}{p_0} - 1 \qquad\qquad 1.8$$

where p_1 is the population, p_0 is the population for the previous year, and g is the growth rate. If the growth rate is negative, we may drop the minus sign and call it a decline. The world growth rate peaked in 1968 at 2.1%/y. It has now fallen to 1.2%/y. China, with its compulsory One-Child Policy, went from the fastest-growing region in 1968 to the slowest. This policy has been associated with widespread selective abortion and infanticide of girls. An additional tragic result is that tens of millions of men are doomed never to marry and establish a family. The policy ended in 2016. Africa was the last region to peak, in 1983, and its current growth rate, 2.6%/y, is the highest.

Table 1.3 Regional population characteristics, based on the data in Figure 1.7.

Region	2015 population billions	2015 growth rate g, /y	Peak growth g, /y (peak growth year)
OECD	1.3	0.6%	1.4% (1956)
China	1.4	0.5%	2.8% (1968)
India	1.3	1.2%	2.4% (1974)
Africa	1.2	2.6%	2.9% (1983)
Other	2.2	1.2%	2.4% (1961)
World	7.4	1.2%	2.1% (1968)

One way to get a perspective on what these growth rates mean is to calculate how long it would take a population to double, assuming that the growth rates stay the same. We will derive a formula for this doubling time. The population in a given year is $1 + g$ times the population in the previous year. After t years, the population is $(1 + g)^t$ times the population in the first year. Thus for the doubling time t_2 we have

$$(1 + g)^{t_2} = 2. \qquad 1.9$$

We can then solve for t_2 as

$$t_2 = \ln(2) / \ln(1 + g). \qquad 1.10$$

In this formula it is not necessary for t_2 to be a whole number. For $g \ll 1$, we can use the first-order Taylor series approximation $\ln(1 + g) \approx g$ to write

$$t_2 \approx 69\% / g \text{ y} \qquad 1.11$$

This formula gives a simple way to estimate the doubling time. It is called the Rule of 69 in finance. Sometimes the numbers 70 or 72 are used instead of 69 for higher accuracy in particular interest-rate ranges. If there is any doubt, one can fall back on the exact expression, Equation 1.10. As an example we can calculate a doubling time for the population of Africa. Applying the rule, we divide the current growth rate, 2.6%/y, into 69% and get the doubling time, 27 y. This is significant because Africa has been the region with the slowest growth in agricultural yield.

1.5.2 The United Nations Projections

The UN Population Division also makes projections, and the current ones are shown in Figure 1.8. We can think of the medium projection as their best estimate, while the low and the high ones form a projection range. One way we can test the projections is to consider earlier UN projections for a date that has already passed, so that we know what actually happened. Figure 1.9 shows the historical projections starting 43 years earlier for the world population in 2000. The agreement is excellent. The actual population in 2000 is always between

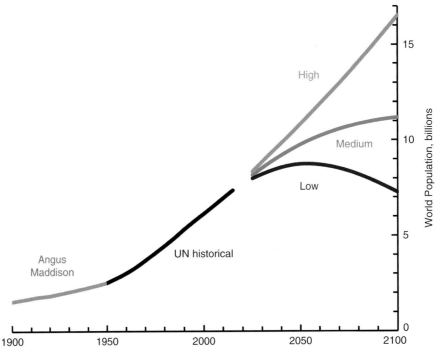

Figure 1.8 World population history and projections. The historical data from 1950 to 2015 and the projections are from the UN Population Division, "World Population Prospects: The 2017 Revision." The data before 1950 come from Angus Maddison.

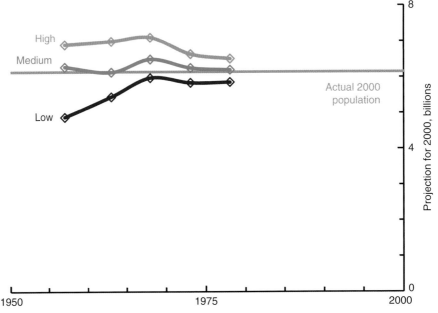

Figure 1.9 Historical United Nations projections for the world population in 2000, compared with the actual population.

the low and high projections, so the high and low projections form a range that captures the actual result. This is reassuring for the 2017 UN projection for 2060, because 2017 is 43 years before 2060, just as the 1957 projection was before 2000. The projection range for 2060 is 8.7 to 11.9 billion people. The African share of the world population in the medium projection for 2060 is 29%, compared with 16% in 2015.

1.5.3 Growth-Rate Plots

What can we say about the maximum population that the world might reach? For this we will develop our own set of projections. Figure 1.10 is called a growth-rate plot. The growth rate is plotted on the y-axis. On the x-axis, we plot the population. Plots like these for fossil-fuel production are called Hubbert linearizations, and we will return to them in Section 3.3.1. The plot starts on the left in 1950, and moves right as the population grows. Since the peak in growth rates in 1968, there has been a clear declining trend. To make a projection we assume that the growth rate will continue to decline in a linear fashion, eventually reaching the x-axis. At that point, the growth rate will be zero. A zero growth rate means that the population has stopped growing. In other words, there is a maximum. The x-intercept becomes our projection for the peak population.

Figure 1.11 shows the growth-rate plot again with fitted lines that start in different years. These lines minimize the sum of the squares of the differences between the y-coordinates of the line and the data. Finding these lines is called *regression*

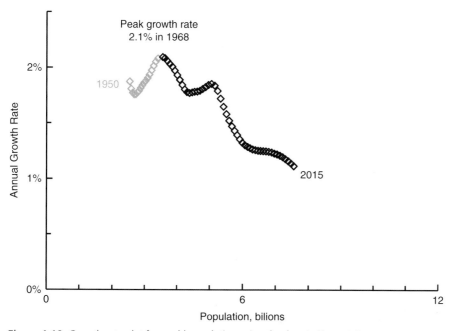

Figure 1.10 Growth-rate plot for world population using the data in Figure 1.8.

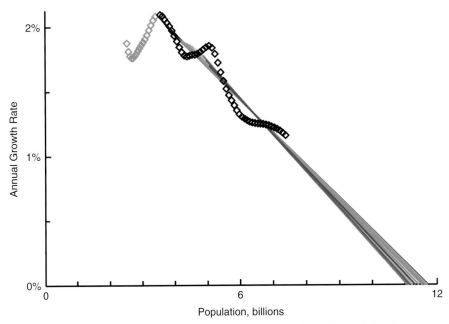

Figure 1.11 Using the growth-rate plot to make a projection for the peak population. The colors distinguish regression lines that start in different years.

analysis. The spreadsheet program Excel has functions for calculating the slope and the *y*-intercept of the regression line. The slope function is SLOPE(Y, X), where **Y** is the set of cells for the *y* variables, the growth rates in this case, and **X** is the set of cells for the *x* variables, the population data here. The *y*-intercept function is similarly given by INTERCEPT(Y, X). One thing that makes the regression analysis so useful is that it is a simple, standard procedure that is unaffected by personal bias. Anyone with the same list of *x*'s and *y*'s will get the same slope and intercept. There is also a standard test index for regression lines called r^2. This index varies between 0 and 1. The closer the relationship is to a straight line the closer the index will be to 1.[8] The Excel function for r^2 is RSQ(Y, X). In some situations, the slope will change due to a shock, and it may be appropriate to use the regression analysis to calculate an early slope and a late slope. We will see an example of this in Section 1.6.2 for income and electricity generation. There are also situations where the regression analysis is not appropriate. For example, if the *y*'s are nearly constant, then it is better simply to characterize them by an average. In addition, if the *y*'s are nearly constant except for a step in the middle of the data, then it may be appropriate to calculate an early average and a late average.

---

[8] In statistics, the r^2 index is the square of the correlation coefficient. This is a mouthful, so people usually just say, "*r* squared". More precisely, r^2 gives the fraction of the variance of *y* that is accounted for by the fitted line. There are problems with r^2 as an index when the slope is small. If *y* bounces around a constant value, the best fit line will be flat and r^2 will be close to zero. In this situation, it is better simply to characterize *y* in terms of its mean and variance.

For the growth rate plot, we actually need the x-intercept rather the y-intercept, and this is given by

$$p = -\frac{\text{INTERCEPT}\,(\mathbf{G}, \mathbf{P})}{\text{SLOPE}\,(\mathbf{G}, \mathbf{P})} \qquad\qquad 1.12$$

where p is the projection for the peak population, $\mathbf{G}$ is the set of cells for the growth rates, and $\mathbf{P}$ is the set of cells for the populations. One might think of simply swapping $\mathbf{X}$ and $\mathbf{Y}$ in the Excel INTERCEPT function to get the x-intercept, but this is not equivalent to Equation 1.12. In the derivation of the regression formulas, we minimize the square differences with the y variables rather than the x variables. Mathematically it is preferable to treat the variable with the largest relative fluctuations as the y variable for this reason. This would be the growth rates in this case. The statistical functions will be discussed further in Section 3.6.

The starting year is varied to give a set of projections. The first starting year is 1968, when the growth rate peaked. For the last projection, the 20th, the starting year is 1987. The range of the twenty population peak projections is 11.1 to 11.7 billion people. *Range* here is defined as the pair of minimum and maximum values. We will also express the range as a percentage by the formula

$$\text{Percentage range} = \pm\frac{p_{max} - p_{min}}{p_{max} + p_{min}} = \pm 3\% \qquad\qquad 1.13$$

where p_{min} to p_{max} is the range of the peak projections. This small $\pm 3\%$ range indicates that the projection is not sensitive to the starting year. For comparison, the population for the UN medium projection in 2100 is 11.2 billion people. Our projection range captures this value.

The growth-rate plot is our first example of a linearized model for making projections. Later on, we will use linearized plots to make projections of ultimate fossil-fuel production. The general procedure is to apply a transformation to the data so that the plot appears to be close to a straight line. Our eyes are good at telling whether a plot looks linear or not. The linearized model also can help us decide if a shock has occurred. In earlier plots in this chapter, the shocks were wars or nuclear disasters. In Figure 1.10, the curve was increasing before 1968, and we did not use these points in calculating the regression lines.

For population, one factor reducing growth rates was the oral contraceptive, informally, "the pill." The pill was developed by Dr. Gregory Pincus (Figure 1.12). Pincus had been a professor of biology at Harvard University. When he did not receive tenure there, he started the Worcester Foundation where he did his research. His research did not receive government funding. Instead he was supported almost entirely by one individual, Katherine McCormick. McCormick graduated from MIT, majoring in biology. She was an heiress to the International Harvester fortune. The pill was approved for sale by the Food and Drug Administration in 1960. It was marketed as Enovid-E by the Searle Corporation.

Figure 1.12 Gregory Pincus (1903–1967), the inventor of the birth-control pill. This photograph was taken when Pincus was a professor of biology at Harvard University. Credit: University of Massachusetts Medical School Archives, Lamar Soutter Library, University of Massachusetts Medical School, Worcester, Massachusetts.

A major advantage of a linearized plot for making projections is that the fitted line has only two parameters, the slope and the intercept. A paradox of projection modeling is that a model with more fitted parameters fits the existing data better, but the projections become unstable. This is shown in Figure 1.13 for a series of polynomial models of increasing order. The higher the order, the better the fit to the existing data, but the wilder the projection is. The reason for the wild behavior is that the higher-order polynomial curve fits balance large offsetting terms in the fitting range. However, outside the fitting range, the highest-power term dominates. If the coefficient of the highest power term is positive, as it is for the 3rd- and 4th-order polynomial, the plot goes off the top of the charts. If the coefficient is negative, as it is for the 5th and 6th orders, then the plot goes negative.

Models with large offsetting terms and high powers of the inputs can be accurate if they are based directly on physical law and accurate measurements. For example, cell

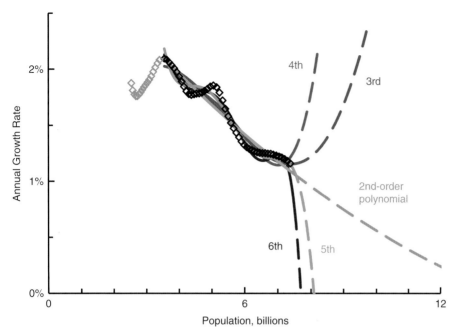

Figure 1.13 The growth rate plot again, this time with polynomial models fit by Excel.

phones use electromagnetic filters to block unwanted signals. These filters depend on Maxwell's equations for electric and magnetic fields for their operation. They operate through a delicate balance of electric and magnetic stored energy that is calculated as a high power of the frequency. If these energies are not calculated correctly to better than 1 part in 10^7, the cell phone would not be useable. The billions of cell phones in operation around the world testify that the models are that good.

One should be suspicious of projections from models with a large number of fitted parameters. These models have a tendency to predict catastrophes that result from the choice of the model rather than the data and they are hallmarks of a pseudoscience. Pseudoscience has much in common with *junk science* models, where a model is adjusted to give a desired result. *Pseudoscience* models can apparently explain existing data, but do not make good predictions. The late Professor Richard Feynman of Caltech, Nobel Laureate in Physics, spoke vividly about pseudoscience in a 1981 video interview for the BBC.

Because of the success of science, there is, I think, a kind of pseudoscience. Social science is an example of a science which is not a science; they don't do [things] scientifically, they follow the forms – or you gather data, you do so-and-so and so forth but they don't get any laws, they haven't found out anything. They haven't got anywhere yet – maybe someday they will, but it's not very well developed, but what happens is on an even more mundane level. We get experts on everything that sound like they're sort of scientific experts. They're not scientific, they sit at a typewriter and they make up something like, food grown with fertilizer that's organic is better for you than food grown with fertilizer that's inorganic – may be true, may not be true, but it hasn't been demonstrated one way or the other. But they'll sit there on the typewriter and make up all this stuff as if it's science and then become an expert on foods, organic foods and so on. There's all kinds of myths and pseudoscience all over the place. I may be quite wrong, maybe they do know all these things, but I don't think I'm wrong. You see, I have the advantage of having found out how hard it is to get to really know something, how careful you have to be about checking the experiments, how easy it is to make mistakes and fool yourself. I know what it means to know something, and therefore I see how they get their information and I can't believe that they know it, they haven't done the work necessary, haven't done the checks necessary, haven't done the care necessary. I have a great suspicion that they don't know that this stuff is [wrong] and they're intimidating people.

1.6   Energy and the Economy

Earlier we found that much of the rise in energy consumption was associated with the rise in population. There is also a strong association between the economy and energy consumption. Countries that have large energy consumption are likely to have large economies. Economics bureaus measure the size of an economy by the gross domestic product, or GDP. The *gross domestic product* is the money value of the goods and services produced by a country. The per-person GDP is a common income index. Generally, the higher the GDP is the better. Citizens in countries with a high per-person GDP buy homes and cars. They receive good education and medical care. The companies they work for provide safe work environments and they install effective pollution controls. The people live long and prosper.

To make GDP comparisons at different times and in different places, two adjustments are often made. One is for inflation. Few governments resist the temptation to expand the money supply faster than the increase of goods and services in the economy, and inflation is the result. A 1900 dollar was worth thirty times as much as a 2017 dollar. Inflation is widely resented as a hidden tax by people who are trying to save money. Economists often support inflation because it gives wage flexibility. People often resist declining nominal wages. Governments like inflation because it lets them wiggle out of their debts to bond holders. The inflation adjustment is done by multiplying the prices by a *deflator* factor. We will calculate deflator factors as the ratio of the inflation-adjusted and nominal annual crude-oil prices in the BP *Statistical Review*. It is not always appropriate to make an inflation adjustment. Consumers pay nominal prices, not inflation-adjusted ones, and some prices, like electricity in regulated markets, may not respond to inflation.

As an example, we consider the return for US government bonds. After World War II ended in 1945, the US government, like other combatants, had large debts to bond holders. There are markets where bond holders buy and sell the bonds.

According to the Ibbotson Associates, a group that tracks bond returns, people who held US government bonds in 1945 would have doubled their money by 1980. However, this was a nominal return. We can use a deflator to find the inflation-adjusted return. From the BP *Statistical Review*, the deflator in going from 1980 dollars to 1945 dollars is 4.6. This means that the inflation-adjusted value of the investment in 1980, instead of being worth twice as much as in 1946, was worth half as much. Taxes on the nominal income would have reduced this further. Thus a major portion of the war debt was foisted onto the bond holders.

The second adjustment involves currencies in different countries. The currency market exchange rates (MER) may not reflect local national prices appropriately. For example, when iron and coal prices are high, an Australian dollar might have the same value in currency markets as an American dollar. However, at those times, one has been able to buy more in the US with American dollars than one could buy in Australia with the same number of Australian dollars. Comparative GDP numbers are often adjusted for these price differences. The adjusted prices are indicated with the label PPP, for purchasing power parity. In some situations, particularly when a product is sold in an international market, like electricity in Europe, the PPP adjustment may not be appropriate, and it is better to work with the MER prices.

1.6.1   Oil Consumption and GDP

Figure 1.14 shows the historical relationship between world oil consumption and GDP in 2017 dollars. There was an extremely tight linear relationship between oil consumption and GDP from 1950 to 1973, the year of the Yom Kippur War and the Arab Oil Embargo The index r^2 is 0.995. The slope of the regression line is $6.49/l. In words, an increase of a liter of oil consumption is associated with an increase of $6.49 of GDP. It is important to note that we are not claiming a one-way cause and effect relationship. That is, we are not saying that an oil consumption increase causes a GDP increase exclusively, nor that a GDP increase causes an oil consumption increase exclusively. Causality clearly works in both directions here. Gasoline refined from oil allows a worker to drive a long distance to a job. The worker's pay increases the GDP. The worker's pay also allows the family to visit relatives by airplane, burning jet fuel that is refined from oil. The same considerations will apply to the relationship between electricity generation and GDP that we consider next. Moreover, oil production and electricity generation themselves are related because oil is burned to generate electricity, and electricity is used by the oil industry. It is interesting to extrapolate the line backwards to where oil consumption is zero. This gives $6.80/p/d. We interpret this as an estimate of the per-person GDP in a world without oil. For comparison, the economist Angus Maddison estimated the per-person GDP of the world in 1900 to be $6.47/p/d.

The 1973 and 1979 shocks broke the relationship between oil consumption and GDP. Since then oil consumption has been flat at 2 l/p/d, while income has

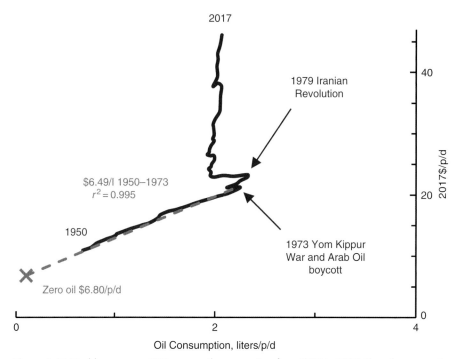

Figure 1.14 World per-person GDP versus oil consumption from 1950 to 2017. The oil consumption data from 1965 on are taken from the BP *Statistical Review*. The earlier oil data are for production rather than consumption and come from various editions of the UN *Yearbook of World Energy Statistics*. The GDP (PPP) data from 1980 on come from the World Bank, earlier data are from Angus Maddison. The GDP data are adjusted for inflation with 2017 as the reference year. The population data are from the United Nations Population Division.

doubled. The relationships at the level of individual countries are more complicated. Figure 1.15 shows oil consumption versus income for several countries, the European Union, and the world. There are two gray dashed lines. The higher one is defined by the points where the ratio of GDP to oil consumption is $22/l. This was the world average in 2017. For the lower line, the ratio is $11/l, the world average in 1980. In the figure, the gray curve for the world starts at the lower line in 1980 and wends upward to the higher line in 2017. For the higher income countries, like Japan, the US, and the EU, the curves start close to the initial $11/l lower line and then tilt backwards as they make their way up towards the current $22/l upper line. This means that they are reducing their oil consumption at the same time they are increasing their GDP. For the lower income countries, there is more variation. Initially the Russian Federation slid backwards down the lower line with both GDP and oil consumption dropping as the Soviet Union collapsed. Then it recovered and is now moving upwards along the $22/l line. India and Brazil start on the $22/l line and then move upward along it, with both oil consumption and GDP increasing. China is the most dramatic, starting well below the initial $11/l line and then moving well above the $22/l line. The improvement in Chinese

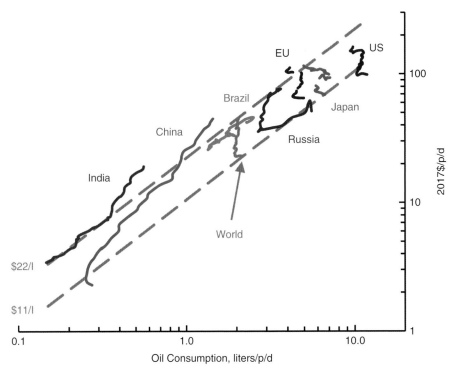

Figure 1.15 Per-person GDP versus oil consumption for selected countries, the EU, and the world, from 1980 through 2017, except for the Russian Federation, where the curve begins in 1985. The scales are logarithmic. The oil consumption data come from the BP *Statistical Review*. The GDP (PPP) data are from the World Bank, and are adjusted for inflation with 2017 as the reference year. The population data are from the United Nations Population Division.

per-person GDP is remarkable. In 1950, it was 10% of the world average. By 2017 it was 98%. Overall, oil consumption in the higher-income countries is decreasing, but this is offset by increases in the lower-income countries.

1.6.2   Electricity Generation and GDP

Figure 1.16 shows the historical relationship between world electricity generation and GDP. On a per-person basis, electricity generation has gone up by a factor of nine since 1950, reaching 386 W/p in 2017. Electricity is critical in economic development. With electricity, good things happen. A community can run a pump with a filter to provide clean water. This stops water-borne diseases. Meat can be refrigerated at the butcher shop and in the home. This prevents it from spoiling. There is light for reading and for socializing. A hospital can run an operating room. The graph shows a remarkable change in 1989. The slope almost doubles. Previously we have associated a change in an energy slope with a shock. The signal event of 1989 was the fall of the Berlin Wall. The collapse of the Soviet Union followed. The Communist Party that controlled the Soviet Union had favored strong central economic planning and control. After the collapse, many countries changed their policies to allow more economic freedom.

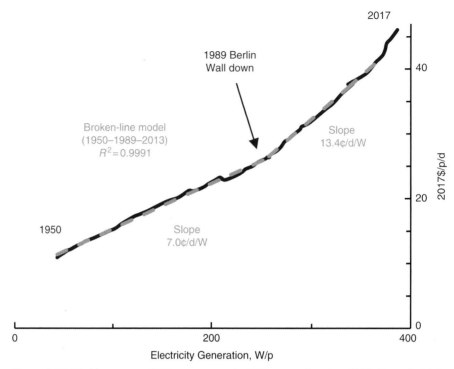

Figure 1.16 World per-person GDP versus gross electricity generation since 1950. Gross electricity generation includes power consumed by the generators; net generation does not. The electricity generation from 1985 on come from the BP *Statistical Review*, earlier generation through 1981 are from various UN *Yearbooks of World Energy Statistics*. The numbers from 1982 through 1984 are net generation from the EIA, with a 4% increase for comparison with the gross-generation data in the graph. The GDP (PPP) data from 1980 on come from the World Bank, earlier data are from Angus Maddison. The GDP data are adjusted for inflation with 2017 as the reference year. The population data are from the United Nations Population Division.

The broken dashed gray line in the figure is a model for the relationship between electricity and income from 1950 to 2013. The broken-line model is complicated by the change in slope. When a model is more complicated than a single straight line, people often capitalize and write R^2 instead of r^2. In Excel, we write RSQ(**D**, **M**) where **D** is the set of cells for the data and **M** is the set of cells for the model values. In this case $R^2 = 0.9991$. This is quite close to 1, indicating a good model. We can demonstrate how good the model is by plotting the income over time, and using the broken-line electricity model to predict the income. This is shown in Figure 1.17. Note that the model picks up even the small dips in the income curve. The relationship between electricity and income is so strong that people sometimes use electricity consumption to predict the GDP if there is a concern that a government might be manipulating the GDP data. There is a divergence from 2014 to 2017 where the model income does not keep up with the actual income gain. These years were not

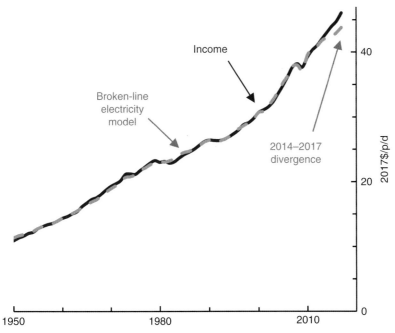

Figure 1.17 Comparing the predictions of the broken-line electricity model for income with the actual income from 1950 to 2017.

used for fitting the model, but they were left in the plot to make the discrepancy clear. The discrepancy amounted to 5% in 2017. Was it the introduction of alternatives?

Figure 1.18 shows electricity production versus income for selected countries, the European Union, and the world. The dashed line is a plot of electricity use against income at the 2017 world average of 11.9¢/d/W. Most countries are near this curve. Canada is the largest electricity user on a per-person basis. It has inexpensive hydroelectric power and it is common to use electricity for residential heating in its severe winters. The curves tend to move along the line, indicating steadily increasing income over time. The exception again is the Russia Federation with its initial retrograde movement in the aftermath of the collapse of the Soviet Union. China is well below the world average. This reflects the large industrial electricity use in China.

1.6.3 Electricity Access and Poverty

The average world citizen has done well since 1950. Inflation-adjusted income in 2017 was up by a factor of 4.2. At no other time in history has the average income increased remotely as quickly. But what about people with incomes that are less than this average? How have they done? The answer is, relatively speaking, they have also done well. There is a strong association between getting access to electricity and getting out of poverty. Figure 1.19 shows the fraction of the world population that the World Bank reckons to be above its poverty line from 1981 to 2015, plotted against the fraction that has electricity access. The fraction of the

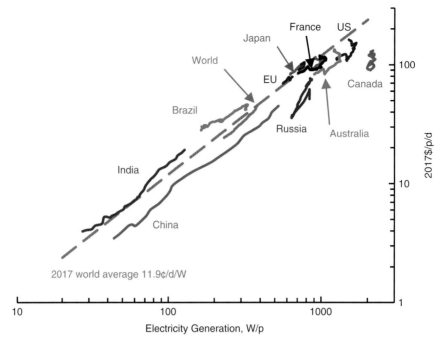

Figure 1.18 Per-person GDP versus gross electricity generation for selected countries, the EU, and the world, from 1985 through 2017. The scales are logarithmic. The GDP (PPP) data are from the World Bank, and are adjusted for inflation with 2017 as the reference year. Electricity generation is from the BP *Statistical Review*. Population is from the United Nations Population Division.

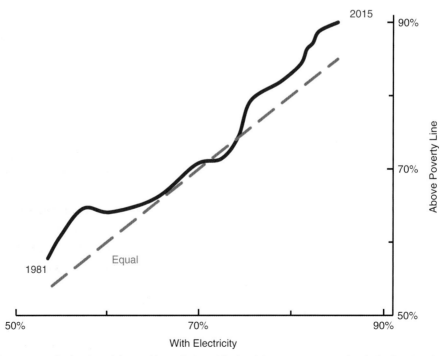

Figure 1.19 The fraction of the world population with electricity access compared with the fraction above the World Bank poverty line of $1.90/p/d (2011$ PPP). The electricity data come from various editions of the IEA *World Energy Outlook*. The dashed gray line shows equal fractions with electricity and above the poverty line.

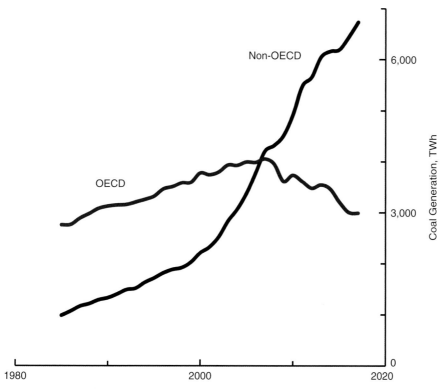

Figure 1.20 Electricity generation from coal for the OECD (higher-income) countries and the non-OECD (lower-income) countries from 1985 to 2017. The data are from the BP *Statistical Review*.

world with electricity access has risen from 54% to 85% during this time, while the share above the poverty line has risen from 58% to 90%.

How are people in lower-income countries gaining access to electricity? The answer is coal. Recall that in the discussion of Figure 1.6, we said that the recent increase in coal production was associated with lower-income countries building out their electrical grids. Figure 1.20 shows electricity generation from coal for the non-OECD countries. The non-OECD generation from coal was 2.4 times the OECD generation in 2017.

1.7 Agriculture

Agriculture is distinguished from any other human activity by its enormous land footprint. The FAO classifies 16,000,000 km² as "arable and permanent crops." *Arable* means land that can be plowed for planting, while *permanent crops* would include orchards and vineyards. Permanent meadows and pasture cover another 33,000,000 km². We define *agricultural land* as the sum, 49,000,000 km². This is 38% of the FAO world land area. Figure 1.21 shows the historical area devoted to cereals. Cereals include rice, wheat, and corn, as well as the less important barley,

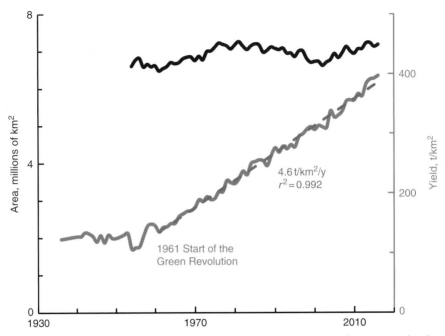

Figure 1.21 Land area and yield for world cereal production, from the United Nations Food and Agriculture Organization (FAO). From 1961 on, the data are from the FAO online database FAOSTAT. The earlier data come from various editions of the FAO *Yearbook of Food and Agricultural Statistics.*

rye, millet, and sorghum. It does not include sugar cane, grass harvested for hay, or oil seeds like soybeans and rape. The land area for cereals has been steady at around 7,000,000 km^2 from the beginning of the data series in 1950. The figure also shows the yield, which is the production per unit area. Of the hundreds of curves plotted in this book, this may be the most remarkable. The yield was flat at 130 t/km^2 before 1961. Then the Green Revolution began and the yield took off. This was a technical shock, utterly unpredictable. We can compare this with the war shock of the Iranian Revolution, the economic collapse shock of the Soviet Union, and the technical shock of shale gas and tight oil. The yield in 2016 was 2.9 times larger than in 1961. The population was 2.4 times larger, so cereal yield has kept ahead of population growth. In Section 6.5 we will see that without the yield improvement there would have been much pressure to convert forests to arable land.

The rise in yield is almost perfectly linear at 4.6 t/km^2/y. The r^2 value for the regression line that begins in 1961 is 0.992. There are many factors at work in the yield improvement. There has been a shift to corn, which has a higher yield than other cereals. Much credit was due in the beginning to Norman Borlaug's plant breeding program that was sponsored by the Rockefeller Foundation. In addition, farmers have increased the use of synthetic fertilizers. In recent years seed companies have been using genetic engineering techniques to develop new plant varieties that are resistant to herbicides and insect pests and that take advantage of longer growing seasons.[9] The enormous increase in agricultural yield has coincided with the vast majority of the fossil-fuel burn and its attendant carbon-dioxide emissions. In Section 1.4, it was noted that 87% of the fossil-fuel consumption has taken place from 1956 on. Plants use atmospheric carbon dioxide to synthesize glucose by photosynthesis. For a given set of conditions, plants have optimum CO_2 levels. Generally speaking, the optimum levels are higher than the current atmospheric levels, so rising CO_2 concentrations increase yield. This effect is called *CO_2 fertilization*, and it widely used in commercial greenhouses, typically by venting natural gas burners inside to give CO_2 levels that are double those outside. The nitrogen-oxide combustion products from burning fossil fuels are classified as pollutants because they contribute to the formation of smog. Smog is irritating and it is a health risk. For this reason, in higher-income countries strenuous efforts are made to reduce the emissions of nitrogen oxides from vehicles and power plants. However, these same nitrogen oxides act as fertilizers for plants when they are entrained in rain.

1.7.1 Life Expectancy and the Green Revolution

The increase in food supplies and electricity access has been accompanied by dramatic rises in life expectancy. Figure 1.22 shows the life expectancy at birth for the OECD

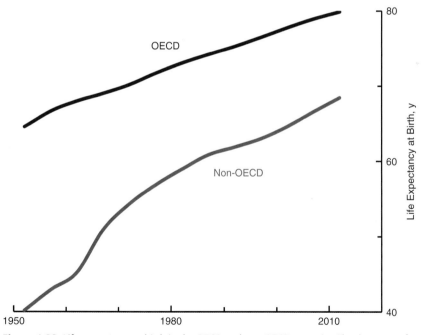

Figure 1.22 Life expectancy at birth in the OECD and non-OECD countries. The data come from the UN Population Division.

---

[9] People who are not farmers can be surprised to find out how subtle the issues are in understanding the effects of longer growing seasons. A good starting point is the paper by Ethan Butler, Nathaniel Mueller, and Peter Huybers, 2018, "Peculiarly pleasant weather for US maize," *Proceedings of the US Academy of Sciences*. Available at https://doi.org/10.1073/pnas.1808035115

and the non-OECD countries. For the higher-income OECD countries, life expectancy has increased from 65 years in the early 1950s to 80 years in the early 2010s. The increase is even larger in the non-OECD countries, from 40 years to 68 years.

This improvement was not predicted. On the contrary, many argued strenuously that it would not occur. The most prominent of these was Stanford professor Paul Ehrlich (Figure 1.23), who wrote in his popular 1968 book, *The Population Bomb*,

> The battle to feed all humanity is over. In the 1970s the world will undergo famines – hundreds of millions of people are going to starve to death in spite of any crash programs embarked upon now.

As we saw in Section 1.5, 1968 was high tide for the world population growth rate. The Green Revolution was already in progress at that time, and the man primarily responsible, Norman Borlaug, received the Nobel Prize two years later in 1970. A failed prediction has a cost. Many people decided not to have children because of Ehrlich's writing. This deprived them of the chance to have a family, which for many of us, is what gives our lives meaning.

Figure 1.23 Ecologist Paul Ehrlich, born 1932, author of *The Population Bomb*. This picture was taken in 1998. Credit: Linda Cicero/Stanford News Service.

1.7.2   Thomas Malthus on Population

Apocalyptic rhetoric aside, people have been thinking about population and famine for a long time. The most important contribution has been the *Essay on the Principle of Population* in 1798, by the Reverend Thomas Malthus (Figure 1.24). Even though it is an old book, it is well worth reading today. It is one of the most influential books ever written. Charles Darwin and Alfred Russel Wallace, who independently proposed the theory of evolution through natural selection, were both inspired by the book. Pierre Verhulst was influenced by Malthus to invent the logistic function for modeling population growth. We will use the logistic function in Section 3.3 for modeling fossil-fuel production. Stanley Jevons, basing his argument on Malthus' ideas, argued that coal production in the UK was likely to continue growing exponentially, and this needed to be taken into account in assessing how long British coal would last (Section 4.5.1). Malthus expressed the problem this way.

> Population, when unchecked, increases in a geometrical ratio. Subsistence increases only in an arithmetical ratio. A slight acquaintance with numbers will shew the immensity of the first power in comparison of the second.

In modern language, he is asserting that a population that increases exponentially in time will outrun a food supply that increases linearly in time. Malthus assumed that the food supply could only be increased by plowing more land, rather than through the yield increase of the Green Revolution.

Malthus' name is often invoked in resource discussions today. People who are pessimistic about humanity's prospects are sometimes called *Malthusians*. We will simply give Malthus credit for developing a framework for discussing the relationship between population and food. Malthus himself wrote

> If he [Malthus] should succeed in drawing the attention of more able men to what he conceives to be the principal difficulty in the way to the improvement of society and should, in consequence, see this difficulty removed, even in theory, he will gladly retract his present opinions and rejoice in a conviction of his error.

In this spirit, we can remove the "principal difficulty." The technology of the oral contraceptive makes it easy for couples to choose the number of children. Populations must no longer grow exponentially. All around the world, growth rates have been dropping. In several countries, notably Japan and Spain, the population may already have already peaked. For the past fifty years the food supply has progressed in Malthus' "arithmetical ratio," that is, linearly in time. The technology of the Green Revolution, funded like the pill by a private foundation, has enabled large increases in crop yield, allowing the food supply to increase without plowing more land. Measured by higher incomes and longer lives, "the improvement of society" that Malthus wished for has occurred.

Figure 1.24 Thomas Malthus (1766–1834), Anglican priest and political economist, and author of the *Essay on the Principle of Population*. Malthus also played an important role in the debates about England's Poor Laws that provided support for people who did not have jobs. Credit: Wellcome Collection CC BY 4.0, https://creativecommons.org/licenses/by/4.0/

1.8   Fossil-Fuel Independence

Many countries depend on imports of oil and gas for transportation fuel, heating, and electricity generation. The importing countries may feel vulnerable to supplies being cut off. In 1941, American President Franklin Roosevelt blocked US oil sales to Japan to protest its military actions in China. Japan has no oil fields of its own, and it responded by attacking the United States and the Dutch East Indies, now Indonesia, a large oil producer at the time. During the ensuing war, Japan suffered from severe shortages of bunker oil for its ships and gasoline for its airplanes. In 1973, Egypt and Syria attacked Israel on the Jewish holiday of Yom Kippur. The United States supported Israel in the conflict, and as a result Arab countries blocked oil exports to the United States. President Richard Nixon instituted price controls, resulting in long lines at gasoline stations. Later the United States developed the

Strategic Petroleum Reserve with storage for 700 Mb of oil, enough to replace imports for months at that time. As a final example, many countries blocked oil exports to South Africa during the 1960s and 1970s to protest the government's apartheid policies. However, the South African government had earlier investigated synthesizing oil from coal, based on the Fischer–Tropsch process developed in Germany. South Africa has excellent coal resources but limited oil resources. The result was the company SASOL (in Afrikaans, Suid-Afrikaanse Steenkool, Olie en Gasmaatskappy, in English the South African Coal, Oil, and Gas Corporation), founded in 1950. The Fischer–Tropsch process is technically difficult, but SASOL mastered it, and as a result, South Africa did not suffer oil shortages because of the boycotts.

One measure of a country's fossil-fuel independence is the ratio of production to consumption. We write

$$i = p / c \hspace{4cm} 1.14$$

where i is the fossil-fuel independence, p is the fossil-fuel production, and c is the consumption. A country with 0% independence imports all of its oil, gas, and coal, while a country with 90% independence imports only 10% of its fuel. This index can exceed 100% if exports are larger than imports. Imports for coal are less significant than for hydrocarbons; only 17% of the world's coal was imported in 2017. However, coal is included in the formula because oil, gas, and coal are partial substitutes for each other. SASOL synthesizes oil from coal, and many electric companies can choose between coal and natural gas power plants. During the bitter 1984–1985 UK coal miner strike, the Thatcher government was able to switch from coal to oil and gas from the recently developed North Sea fields.

Figure 1.25 shows the fossil-fuel independence for several regions. Numerical values are given in Table 1.4. North America includes the North American Free Trade Agreement (NAFTA) countries, Canada, Mexico, and the United States. NAFTA is both a tariff treaty and an energy sharing agreement. Norway is not in the European Union, but it is included with it here because its North Sea oil and gas fields contribute to EU consumption. China and North America have the

Table 1.4 Fossil-fuel independence in 2017, based on the data for Figure 1.25. At one time, Japan had significant coal production, but its resources have been exhausted.

	Independence	10-year change
North America	95%	18%
China	76%	−7%
European Union plus Norway	39%	−6%
Japan	0%	0%

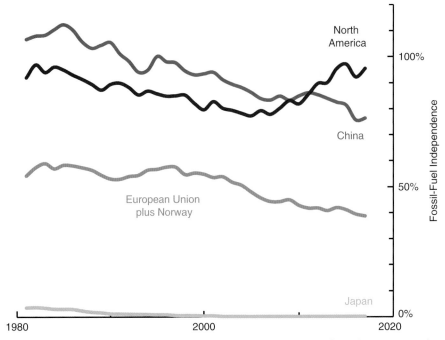

Figure 1.25 Fossil-fuel independence. Production and consumption data from the BP *Statistical Review*. Note that BP oil consumption data includes liquid biofuels and these were subtracted out for the calculations.

highest independence of the four regions, but the trends are different. Chinese independence has dropped 7% over the last ten years from 83% in 2007 to 76% in 2017. China now imports more fossil fuels than any other country in the world. On the other hand, North American independence was 95% in 2017, up 18% in ten years. There are several contributing factors. The US has become a significant coal exporter. Oil production from the tar sands[10] in Alberta has been increasing. Most importantly, US oil and gas production has risen because of fracking technology and the development of horizontal wells.

1.8.1   Europe Turns toward the New Alternatives

In contrast, the European Union's independence has fallen. Oil, gas, and coal production have all been dropping. The EU has shown little interest in the shale gas and tight oil technologies that have revolutionized oil and gas production in the United States. Instead the EU has emphasized wind, solar and biofuels (Figure 1.26). From 2007 to 2017, the share of primary energy for the new alternatives has risen 7.0% for the EU, compared with 3.0% for the world (Table 1.5).

---

[10] It would be more accurate to call these deposits "bitumen sands," but "tar sands" is the common usage. Bitumen is the viscous fraction of oil, while tar refers to the leftovers in coke manufacturing. The Canadian deposits are natural bitumen. Both have historically been combined with gravel to make road surfaces and roofing products.

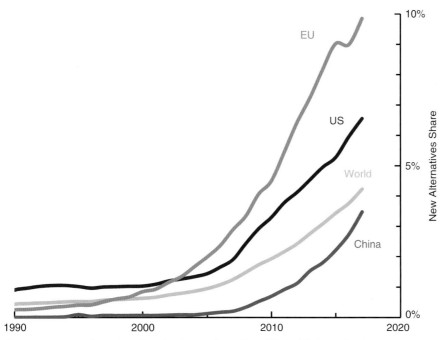

Figure 1.26 New alternatives (wind, solar, geothermal, and biofuels) share of primary energy. Data from the BP *Statistical Review*.

Table 1.5 New alternatives share of primary energy in 2017, based on the data for Figure 1.26.

	Primary energy share	10-year change
EU	9.9%	7.0%
US	6.6%	4.7%
China	3.5%	3.3%
World	4.2%	3.0%

1.8.2   Calculating Annualized Growth Rates

Energy production growth rates fluctuate considerably from year to year. To smooth out the fluctuations, we will ordinarily calculate annualized growth rates over a ten-year period. The *annualized* growth rate g is given by

$$g = \sqrt[t]{p / p_0} - 1 \qquad\qquad 1.15$$

where p is the latest production, p_0 is the production in the initial year, and t is the time period in years. The formula is equivalent to calculating the geometric mean of the $1+g_i$ factors, where g_i is the annual growth rate in year i. Note the distinction between annual and annualized. The annual growth is calculated for a single year from Equation 1.8. The annualized growth is calculated for a number of years. A similar formula is used in finance for calculating annualized stock returns.

As an example we calculate 10-year annualized growth and decline rates for US and EU natural gas production. From the BP *Statistical Review*, US natural gas production in 2017 was 734.5 Gm^3, up from 521.9 Gm^3 in 2007. From Equation 1.15, the annualized growth rate is 3.5%/y. In contrast, EU production in 2017 was 117.8 Gm^3, down from 196.8 Gm^3 in 2007. The annualized decline is 5.0%/y.

1.9   Carbon-Dioxide Emissions

Starting in the 1980s, there have been concerns that rising carbon-dioxide levels could cause harmful climate change. Oil, gas, and coal are made up primarily of carbon and hydrogen and when they burn, the carbon combines with oxygen to form carbon dioxide, and the hydrogen is oxidized to make water. One can calculate the carbon-dioxide emissions from burning fossil fuels from carbon coefficients. The carbon coefficients are conventionally given as the ratio of the mass of carbon in the fuel to the combustion heat expressed in toe (Table 1.6). Formally the carbon coefficients have no units; tonnes in the numerator cancel tonnes in the denominator. However, it makes the expressions clearer if we write the units as tC/toe, where tC is the carbon mass. Natural gas, which is mainly CH_4, has the lowest carbon-dioxide emissions per toe because it produces a higher fraction of its heat from burning hydrogen than oil and coal.

Figure 1.27 shows the world fossil-fuel carbon-dioxide emissions over the years. They have increased steadily. The emissions are calculated from fossil-fuel consumption data using the carbon coefficients. One factor that is not considered in the graph is that a fraction of the carbon in the fossil fuels is sequestered in plastics, resins, and filters. For many of these products, this sequestration is temporary. Often they are degraded over time by microorganisms, releasing carbon dioxide. Also, there are other human activities that are associated with carbon-dioxide emissions that are not included. The most important of these is agriculture. In addition, chemical reactions in the cement making produce carbon dioxide. This process is undone to some extent over time because cement absorbs carbon dioxide as it ages.

Table 1.6 The carbon coefficients used for calculations in this book.

	Carbon coefficient, tC/toe
Oil	0.84
Gas	0.64
Coal	1.08

These are converted from the carbon-dioxide coefficients in the 2015 BP *Statistical Review* by multiplying by the carbon mass fraction of carbon dioxide, 12/44.

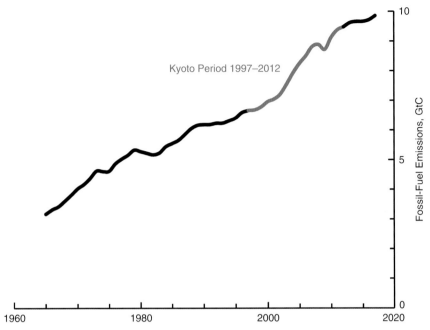

Figure 1.27 World fossil-fuel carbon-dioxide emissions, calculated from the consumption data in the BP *Statistical Review* and the carbon coefficients in Table 1.6.

1.9.1   The Kyoto Protocol and the Paris Agreement

The first international treaty to try to reduce fossil-fuel carbon-dioxide emissions was the 1997 Kyoto Protocol. Under the agreement, higher-income nations set greenhouse gas emissions targets for 2012. The lower-income countries were not obligated to reduce greenhouse-gas emissions, but they could receive grants from the higher-income countries if they did. Ratification of the Kyoto Protocol was almost universal, with the notable exception of the United States. Acting UN Ambassador Peter Burleigh signed the agreement for the US, but President Bill Clinton did not submit the treaty to the Senate for ratification. The State Department offered the explanation, "President Clinton has made it clear that he will not submit the Protocol to the Senate unless there is meaningful participation by key developing countries in addressing climate change."[11] The next President, George W. Bush, pulled the US out of the treaty. Then events took a surprising turn. From 1997 to 2012, US fossil-fuel carbon-dioxide emissions dropped 6%, while the emissions for the rest of the OECD rose 5%. The reason for the reduction in US emissions was shale gas produced by hydraulic fracturing and horizontal wells.

At the world level there was no significant difference between the years covered by the Kyoto Protocol and the earlier years. If anything, emissions accelerated

---

[11] The US Department of State, 1998, "United States signs the Kyoto Protocol." Available at https://1997-2001.state.gov/global/global_issues/climate/fs-us_sign_kyoto_981112.html

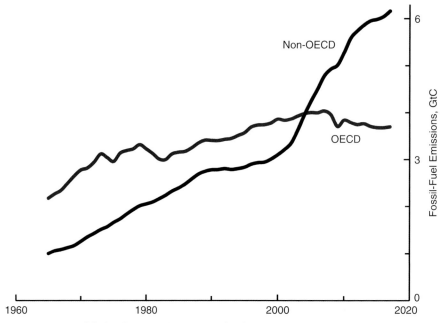

Figure 1.28 Fossil-fuel carbon-dioxide emissions for the OECD and the non-OECD countries, calculated from consumption data in the BP *Statistical Review* and the carbon coefficients in Table 1.6.

during the Kyoto Period. Figure 1.28 shows what happened. OECD emissions, including those from the United States, were flat at 0% change from 1997 to 2012, while non-OECD emissions rose 96%. World CO_2 emissions are now dominated by the lower-income countries.

There was a second agreement in Paris to reduce emissions in 2015. The goal was "to limit the temperature increase to 1.5 °C above pre-industrial levels."[12] Secretary of State John Kerry signed for the United States. President Barack Obama attempted to implement the Paris Agreement through executive orders and regulations instead of submitting it to the Senate for ratification. However, the next President, Donald Trump, had campaigned against the agreement and his administration moved to rescind the executive orders and regulations.

Figure 1.29 shows the atmospheric carbon-dioxide levels measured since 1959 at Mauna Loa, Hawaii. This may be the best-known graph in all of science. This series was started by Charles Keeling at the Scripps Institute of Oceanography and is currently maintained by Pieter Tans at the US National Oceanic and Atmospheric Administration (NOAA) Earth System Research Laboratory (ESRL).

---

[12] United Nations Framework Convention on Climate Change, 2015, "Paris Agreement." Available at https://unfccc.int/files/meetings/paris_nov_2015/application/pdf/paris_agreement_english_.pdf

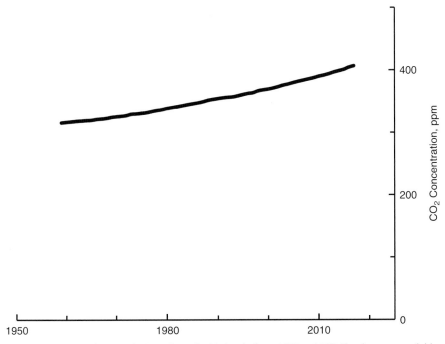

Figure 1.29 Annual atmospheric carbon-dioxide levels from 1959 to 2017. The data are available at ftp://aftp.cmdl.noaa.gov/products/trends/co2/co2_annmean_mlo.txt

The carbon-dioxide level has been rising steadily, and in 2017 the concentration was 407 ppm (parts per million), up 29% since 1959. The rise is seen worldwide.

1.9.2 Fossil-Fuel Production and Carbon-Dioxide Levels

There is an extremely tight relationship between fossil-fuel production and atmospheric CO_2 levels. This is shown in Figure 1.30. On the x-axis is the cumulative fossil-fuel production q_i, defined by

$$q_i = \sum_{j \leq i} p_j \qquad\qquad 1.16$$

where i and j are year indexes, and p_j is the annual fossil-fuel production expressed in terms of the carbon content. This formula is the discrete equivalent of a time integral. The gray dashed trend line shown for comparison is a fitted regression line. The value of r^2 is 0.9992. The y-intercept is 298 ppm. This intercept can be interpreted as an effective pre-industrial CO_2 level. The slope of the line is 0.254 ppm/GtC. In Section 10.5, we will use this slope to estimate the future CO_2 level from a projection of fossil-fuel burning.

There is another way we can interpret the slope. The atmospheric carbon mass for 1 ppm of CO_2 is 2.12 GtC.[13] If we multiply this by the slope, we get a ratio of

---

[13] This value has been recommended by Pieter Tans at the US National Oceanic and Atmospheric Administration (NOAA) Earth System Research Laboratory (ESRL). Dr. Tans maintains the CO_2 data in Figure 1.29.

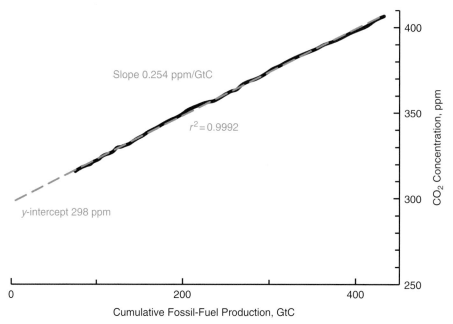

Figure 1.30 Atmospheric CO_2 levels versus cumulative fossil-fuel production from 1959 to 2017. The CO_2 data are those in Figure 1.29.

54%. This is the ratio of the annual gain in atmospheric CO_2 to the emissions. It is called the *airborne fraction*. The remainder, 46%, is called the *sink fraction*. The largest sink component is photosynthesis, where land plants and marine phytoplankton use atmospheric carbon dioxide to synthesize glucose during the growing season. There have been concerns that the sink processes would saturate and that the sink fraction would decline. However, this has not happened, even though fossil-fuel emissions in 2017 were six times those in 1959. This is an indication of increasing photosynthesis. We will return to this in Section 10.2.

The linear relationship between world cumulative fossil-fuel production and CO_2 levels means that it is not the production in a single year or in a single country that matters, but rather world production over an extended time period. If Norway switches to electric cars that run off of hydroelectric power, but redirects the oil it produces to other countries, then the eventual CO_2 levels could be unchanged. If through technology improvements, the United States develops new natural gas resources, current carbon-dioxide emissions would drop when the natural gas substitutes for coal in electricity generation. However, if burning natural gas as a substitute only postpones the coal burning, the long-run carbon-dioxide levels could increase because of the additional natural gas.

Concepts to Review

- Energy supply and demand
- Units and prefixes
- Capacity factor
- Efficiency improvements vs. conservation
- The Jevons Paradox
- Evolution of energy supplies: rise, shock, plateau
- Pseudoscience and junk science
- Electricity generation and income
- Coal in non-OECD electricity generation
- Oil and income
- Yield history for cereals
- Calculating doubling times
- Growth-rate plots for population
- Excel regression line functions: SLOPE, INTERCEPT, RSQ
- Thomas Malthus on population
- Fossil-fuel independence
- Europe turns to alternatives
- Carbon coefficients
- Fossil-fuel burning and atmospheric carbon-dioxide levels
- The Kyoto Protocol and the Paris Agreement

Problems

Problem 1.1   Estimating the population peak for China

Locate the population numbers for China at the UN Population Division website.

a. Make a growth-rate plot using the population data through 2015 and use the plot to make a projection for the peak population.

b. Compare your estimate with the peak in the UN's medium projection for China.

Problem 1.2   Historical US inflation rates

The BP *Statistical* Review gives annual oil prices in nominal dollars and dollars for the most recent year and these series allow one to make inflation adjustments. The United States ended fixed conversions between dollars and gold in 1972, during the Nixon administration. Critics of this action predicted an increase in inflation. To test this idea, calculate the annualized inflation rate from 1900 to 1972 and compare with the inflation from 1972 to the most recent year.

Problem 1.3   Energy prices

Calculate the current prices for oil, natural gas, and coal on an energy basis in $/GJ.

a. For oil use the crude-oil price ($/barrel) in the *Wall Street Journal*'s Futures table at www.wsj.com/market-data/commodities

b. For gas use the natural gas price on the same page. The price is quoted for 1 million Btu.

c. For coal, use the price for Powder River Basin coal (8,800 Btu per pound). The price is quoted for a short ton, 2,000 pounds. These prices are on the cash prices page at www.wsj.com/market-data/commodities/cashprices

Further Reading

- The BP *Statistical Review of World Energy* is published each summer. This publication is an excellent source of information for oil, gas, coal, and electricity production and the alternatives. It is available as an Excel workbook at www.bp.com
- Arnulf Grubler, 2003, *Technology and Global Change*, Cambridge University Press. This book includes plots of the transitions from wood to coal and from horses to cars.
- Stanley Jevons, 1865, *The Coal Question*, MacMillan and Company. The Jevons Paradox is discussed in Chapter 7.
- Angus Maddison (1926–2010) was professor of economics at the University of Groningen, in Holland. His life's work was in developing a GDP series for the different countries of the world back through the years. So if someone says that the Dutch were the richest people in the world in the 1600s, you can check this on Angus Maddison's web page at www.ggdc.net/maddison/maddison-project/orihome.htm
- Thomas Malthus, 1798, *An Essay on the Principle of Population*, available as a pdf at www.esp.org/books/malthus/population/malthus.pdf
- Brian Mitchell's *International Historical Statistics*, published by Palgrave-MacMillan, Basingstoke, UK, is a massive compilation, revised frequently, with superb coverage for economic quantities at the national level, and with excellent coverage back in time. It is available through interlibrary loan from research libraries.
- Paul Samuelson and William Nordhaus, 2010, *Economics*, 19th edition, McGraw-Hill. A detailed discussion of GDP calculations and supply and demand curves.
- The United Nations Food and Agricultural Organization (FAO) database FAOSTAT is the best source of information for agriculture, forests, and fisheries. It is available at http://faostat.fao.org
- The United Nations Population Division publishes "World Population Prospects," available in Excel format at http://esa.un.org/unpd/wpp/Documentation/publications.htm
- The World Nuclear Association keeps an online database with information about every nuclear power plant that has ever been connected to the grid at http://world-nuclear.org/NuclearDatabase/Default.aspx?id=27232

2 Horses, Whales, and Wood

*The horse Indians lived beyond the forests
in an endless, trackless, and mostly waterless
expanse of undulating grass that was itself
terrifying to white men.*

S. C. Gwynne, *Empire of the Summer Moon*

The resources of oil, gas, and coal are finite. The time will come when their production will fall, never to rise again. Alternatives are our once and future energy supply. So what did people do before fossil fuels? Many of their needs were similar to ours. People required light at night. They wanted to travel and to ship their goods. Farmers needed help plowing their fields. At high latitudes, people must have heat to survive. We will discuss three of their energy sources: horses, whale oil, and wood. Horses carried riders and they pulled plows, like tractors today. Whale oil provided light and wood produced heat. We conclude the chapter with a look at a society with a limited energy supply – the Norse Greenland colony.

2.1 Horses

People and horses (*Equus ferus caballus*) go back a long way. One way we can tell this is from old teeth. A riding bit grinds teeth down with a distinctive pattern. In Central Asia, horse teeth with this wear pattern have been found that are 5,000 years old. There are no truly wild horses left today. Rather there are feral horses that run free in some countries that descend from escaped domesticated horses.[1] For example, in the American West, there are tens of thousands of mustangs that descend from horses that escaped from ranches hundreds of years ago. Horses easily range fifty kilometers a day, so they can live where water is scarce (Figure 2.1).

---

[1] There is another subspecies of horse, Przewalski's Horse, *Equus ferus przewalskii*. There are several hundred in Mongolian parks and a thousand more in zoos. Until recently, it was thought that Przewalski's Horse was wild, meaning that they were not descended from escaped domesticated horses. However, in 2018 a group led by Charleen Gaunitz of the Natural History Museum of Denmark, published a study of horse DNA that showed that the Przewalski's Horse descended from horses domesticated by the Botai people of Central Asia.

Table 2.1 Words in European languages that mean gentleman.

Language	Word	Origins
Anglo-Saxon	ridere	
English	cavalier	from Middle French, horseman
French	chevalier	from cheval, "horse"
German	ritter	from which "rider"
Greek	ippotis	from which, hippopotamus, "river horse"
Icelandic	riddari	
Latin	eques	from which, "equestrian"
Romanian	cavaler	
Spanish	caballero	from caballo, "horse"
Swedish	riddare	

Figure 2.1 Mustangs in the state of Utah. Credit: Tom Tietz/Shutterstock.

They do eat grass, and American government biologists have decided that there should be fewer mustangs and more grass. However, people like horses, and they do not want them hunted. So the mustangs are moved to ranches to live out their days, but with mares and stallions on different ranches.

Horses are social animals, and with training they readily work in teams with people and with other horses. They utterly outclass us in speed and strength. A champion quarter horse gallops a quarter mile in 20 seconds, with a rider. The fastest human runners take more than twice as long. Two-horse teams pull regularly five-ton sleds in contests. For comparison, a practice sled for two beefy American football players weighs a few hundred kilograms.

Historically, owning horses was associated with high status. The vestigial traces are in the languages where the word for a gentleman also means "rider" or "horseman." Table 2.1 gives a list for European languages.

PONY EXPRESS !

CHANGE OF TIME!   REDUCED RATES!

10 Days to San Francisco!

LETTERS

WILL BE RECEIVED AT THE

OFFICE, 84 BROADWAY,

NEW YORK,

Up to **4** P. M. every TUESDAY.

AND

Up to **2½** P. M. every SATURDAY,

Which will be forwarded to connect with the PONY EXPRESS leaving ST. JOSEPH, Missouri,

Every WEDNESDAY and SATURDAY at 11 P. M.

TELEGRAMS

Sent to Fort Kearney on the mornings of MONDAY and FRIDAY, will connect with **PONY** leaving St. Joseph, WEDNESDAYS and SATURDAYS.

EXPRESS CHARGES.

LETTERS weighing half ounce or under............$1 00
For every additional half ounce or fraction of an ounce 1 00
In all cases to be enclosed in 10 cent Government Stamped Envelopes,

And all **Express CHARGES** Pre-paid.

☞ **PONY EXPRESS ENVELOPES** For Sale at our Office.

WELLS, FARGO & CO., Ag'ts.

New York, Ju'y 1. 1861.

SLOTE & JANES, STATIONERS AND PRINTERS, 96 FULTON STREET, NEW YORK

Figure 2.2 An 1861 advertisement for the Pony Express. The agent, Wells Fargo, in recent years has sometimes been the largest bank by market capitalization in the world. Courtesy of the Smithsonian Institute/public domain.

In the Internet era one may forget that in earlier times information traveled at the speed of a horse and rider. The Mongol Empire in the thirteenth century covered a fifth of the earth's land area. It extended from the Yellow Sea to the Black Sea, a distance of 7,000 km. The empire was managed by messengers riding horses in relays. For a short time, the United States had relays like this. Figure 2.2 is a poster from July, 1861. The Pony Express delivered mail between New York and San Francisco in ten days – 5,000 km. The price for a half-ounce letter was a dollar. At that time a factory worker might have earned a dollar for a day's work. The Pony Express was expensive, but it was useful for business and legal communications. However, only a few months later, in October, the transcontinental telegraph began operation, and the Pony Express shut down.

Horses had advantages that made people reluctant to give them up for cars. They give riders freedom because they go off road. Children can manage horses. Figure 2.3 shows a young boy directing a team of four horses. The machine is a

Figure 2.3 A boy operating a harrow in the state of Pennsylvania in the early 1900s. Courtesy of the Pennsylvania State Archives, MG-219.2, Philadelphia Commercial Museum Collection.

harrow that makes grooves in the ground. The seeds will be planted later. Horses have a wonderful spatial memory. This helped with routine deliveries like milk because they learned where to stop. Horses can get home by themselves. Maybe cars will be able to do this one day. Horses also have disadvantages. Like people, they sometimes just do not feel like working. They catch colds. They give other horses colds. Horses produce a lot of manure, 25 kg a day. It was important in cities to have efficient cleanup crews, because flies lay their eggs in manure.

2.1.1 Horses and Mules at Work

In addition to horses, mules were used as work animals on farms. On American farms, 15% to 30% of the work animals have been mules. For people who are not familiar with horses and mules, the simplest way to tell a mule from a horse is that its ears are bigger (Figure 2.4). President George Washington, who was fascinated by new farming ideas, is given credit for introducing mules to the United States. A mule is a hybrid animal. The mother is a mare and the father is a jack, or male

Figure 2.4 A Mongol woman with her mule in 1871. Credit: John Thomson/Wellcome Collection CC BY 4.0, https://creativecommons.org/licenses/by/4.0/

donkey. As with many hybrids, mules are sterile. But they are stronger than donkeys and they live longer than horses. They tolerate heat better than horses. The people who own mules say that they are smarter than horses. Mules make good pack animals because each mule has its own sense of what is safe. It will not simply go forward because the animal in front did.

Today we associate horses with rural areas, but they were also important in cities. The horses were kept in stables and people could go to a stable to pick up a horse and a carriage, like a rental car today. Figure 2.5 shows a horse-drawn tram in Los Angeles in the early 1890s. The horse car is actually running on metal rails, like a train. Running on rails limits where a tram can go, but metal wheels on metal rails have low friction, lower than rubber tires on asphalt. It means that freight trains can have much better fuel efficiency than trucks.

It is interesting to compare horses, people, and cars for their mechanical work output and food and fuel inputs (Table 2.2). It is appropriate here to use units of horsepower, taking a horse to have an output of one horsepower. A person can sustain a tenth of a horsepower. A vehicle engine, on the other hand, can generate hundreds of horsepower. For horses and people, the work per year in the table is based on seasonal farm work. Horses do ten times as much work as people, but they also eat ten times as much. The conversion efficiency for food to mechanical work for horses and people is around 15%. For vehicles the conversion efficiency

Figure 2.5 A horse car in Los Angeles in 1892. This is at the intersection of Main Street and Winston Street. The building in the background is the post office. Credit: Charles Pierce/The California Historical Society Collection in the University of Southern California Digital Library.

Table 2.2 Mechanical output for horses, people, and cars.

	Power, hp	Mechanical work, GJ/y	Food, Cal/d
Horse	1	2	25,000
Person	0.1	0.2	2,500
Vehicle	230	20	100,000

Calories for working horses are from Martin Adams, 2006, "Introduction to Equine Nutrition." Estimates for people are derived from Vaclav Smil, 2007, *Energy in Nature and Society*, MIT Press. Vehicle horsepower from the EPA, 2013, *Light-Duty Automotive Technology*. The power for vehicles is a weighted average for cars, buses, and trucks. Fuel consumption is from the US Bureau of Transportation Statistics, 2015, *National Transportation Statistics*. The mechanical work is calculated for the average fuel consumption of 663 gallons of gasoline per vehicle per year, assuming a 25% conversion efficiency.

from fuel to mechanical work is typically 25%. For the purpose of this comparison, we assume that the vehicles run on ethanol derived from corn where the energy in the ethanol is half that of the corn feedstock.[2] The corn-to-wheels efficiency would be 12.5%, similar to the 15% value for people and horses. From the table, averaged over a year, vehicles are more efficient because horses and people need to eat in the off season. Nevertheless, the average vehicle consumes much more energy than a horse and there are now more vehicles than there ever were horses.

Before farmers had tractors, they had to decide how much land to plant for hay and oats for horses, and how much land to plant for corn and wheat for people. The US Department of Agriculture tracked the amount of cropland that was needed for the horses. It was a lot, one quarter of American cropland. When vehicles were introduced, it opened up this land for food production. However, in recent years the pressure has returned. Large areas are planted for corn to produce ethanol for cars in the United States and for corn and rapeseed to produce methane for electricity generation in Germany. With biofuels, we are making the same kind of judgments our ancestors did. How much land goes for food? How much goes for fuel?

2.1.2   The Transition from Horses and Mules to Tractors and Cars

Over time tractors replaced horses and mules and people on American farms. Figure 2.6 shows the population of horses and mules, and people, along with the number of tractors. From 1925 to 1956, the population of horses and mules dropped from 90% of the peak value to 10%. We will calculate *rise and fall times* as the time it takes to go between 10% and 90% of the way between the initial value and the final value. In this case, the fall time is 31 years, a little longer than the lifetime of a horse. The shift was essentially complete by the beginning of the

---

[2] To produce ethanol from corn additional heat will be needed for distillation which we are not
    including here. We will follow up on this in Section 6.7.2.

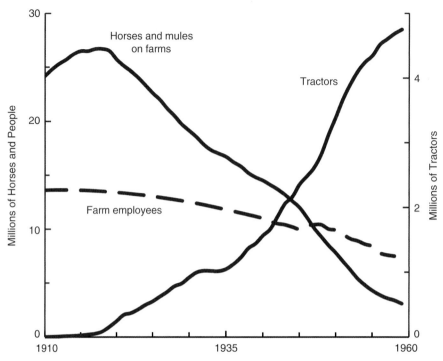

Figure 2.6 Horses, employees, and tractors on American farms. The data come from the US Department of Agriculture, 1960, *Statistical Bulletin 233*, Table 3 and Table 13. Note that the scale is different for tractors.

Green Revolution. The graph also suggests that tractors substitute for employees, but the transition has been slower.

Away from farms, there has also been a transition from horses and mules to cars. Figure 2.7 compares the entire population of horses and mules with the number of vehicles. The data are plotted on a per-person basis to make the comparisons easier. There is also a curve for railroads, with the length of track as the index. Railroads took off after the Civil War ended in 1865. The 10% to 90% rise time was 49 years. The most interesting part of the graph is what happened to the horse and mule population during this rise. One might expect it to fall as the railroads grew. However, it did not. It turns out that railroads and horses and mules were complements rather than competitors. Then and now, trains have a limited number of stations. A pick up is still needed to get to the station and a drop off is still needed to get to the final destination. This gap is called the *last-mile* problem. Horses and mules solved the last-mile problem for railroads. The number of horses and mules per-person actually rose somewhat during this time. At the peak in 1900, there was a horse or mule for every three people. However, this changed with the coming of the automobile.

The United States caught the car bug early and hard. During the 1920s, US car production grew at an annualized rate of 13% per year. Vehicles passed horses and mules

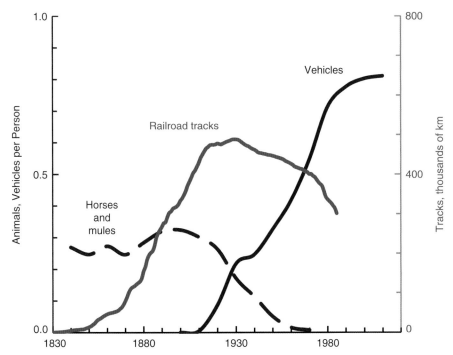

Figure 2.7 Horses and mules, railroads, and cars in the United States. The data come from Dewhurst and Associates, 1955, *America's Needs and Resources*, the US Department of Commerce, Census Bureau, 1975, *Historical Statistics of the United States*, the Bureau of Transportation Statistics, 2012, *National Transportation Statistics*, and Arnulf Grubler, 2003, *Technology and Global Change*, Cambridge University Press.

in 1930. After 1980, the number of vehicles leveled off at eight vehicles for every ten people, essentially one car for every person old enough to drive. This greatly improves our mobility compared to the horse and mule era. We can live farther away from work and we have greater choices for shopping. However, we have lost the connection between horse and rider that people have loved for thousands of years.

2.2   Whales

How did people light their homes before electric lights were invented in the late 1800s? There were several alternatives. In some cities, *coal gas* was available for lamps in homes and offices. This was a mixture of carbon monoxide and hydrogen produced by coal gasification. Outside of the cities with coal gas, people used candles and oil lamps. Some of the best candles were *sperm candles*, made from spermaceti, which comes from the head of sperm whales. The oil lamps had wicks that fed the fuel to the flame by capillary action. They could burn a plant oil like olive oil, or an animal oil like the oil produced by boiling whale blubber. In addition, whale oil is a good lubricant. The *baleens*, which are the bony filters in the whale's mouth that let it feed on krill, were also used in corsets.

2.2.1   Yankee Whalers

Whaling was a big business during the 1800s, and the leading whaling country at that time was the United States. The most important whaling port was New Bedford, Massachusetts. Whalers ranged from the Arctic to the Antarctic, all over the world. Figure 2.8 shows their prey. The sperm whale was the most important, by far. Almost half the whales killed were sperm whales. The blue whale, likely the largest animal that ever lived, was not hunted by American whalers, as it was too fast for rowed whaleboats. Whaling was a high-risk enterprise – 5% of the ships did not return. The financial risk was split between the ship owners and the crew. From an economic perspective, this was an extremely interesting labor market. It is discussed in the reference *In Pursuit of the Leviathan* at the end of the chapter. There was always the possibility that the ship would sink several years into the voyage and the crew would be left with nothing. The contracts gave each sailor a share of the proceeds of the voyage, with different shares for different jobs.

Figure 2.9 shows the production and price history for whale oil. Whale oil was expensive, around $5 per liter in 2017 dollars. Production was flat during the 1840s, and declined sharply during the 1850s and 60s. The production decline continued until 1900. Looking at the curve, one might suppose that the world ran out of whales. It is true that some regional populations were greatly reduced, as were the populations of some whales that have a limited range, like the bowhead and gray.

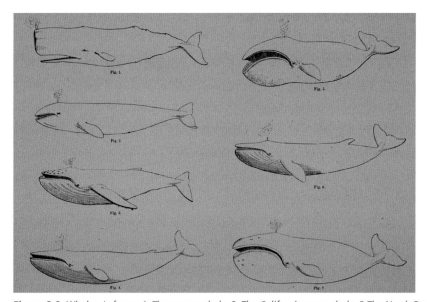

Figure 2.8 Whales. Left top: 1. The sperm whale, 2. The California gray whale, 3 The North Pacific humpback whale, 4. The sulphur-bottom whale. Right top: 5. The finback or Oregon finner, 6. The Pacific right whale, 7. The bowhead whale. The sulphur-bottom whale, now called the blue whale, and the finback whale, now the fin whale, are faster swimmers than the others, and were not captured in large numbers until the twentieth century. Credit: NOAA National Marine Fisheries Service.

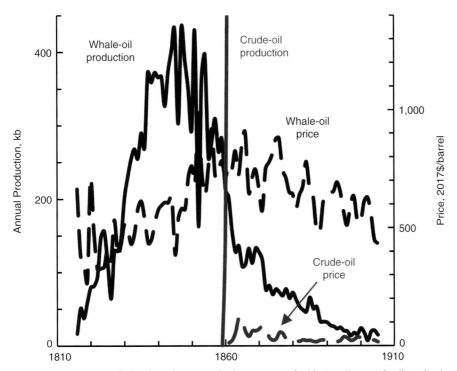

Figure 2.9 American whale oil production and prices compared with American crude oil production and prices. The crude-oil production is easy to miss. It is the almost vertical brown line around 1860. The whale data come from Lance Davis, Robert Gallman and Karin Gleiter, 1997, *In Pursuit of the Leviathan: Technology, Institutions, Productivity, and Profits in American Whaling, 1816–1906*, University of Chicago Press. The crude-oil production data come from the EIA, prices from the BP *Statistical Review*. The whale-oil barrel was around 35 gallons, while the crude-oil barrel was and is 42 gallons. No adjustment was made for this difference in the graph.

However, the primary cause for the decline was different, the loss of the lighting market to kerosene. Later on steel stays replaced the whalebone in corsets. Whale oil in transmission fluid held out until the second half of the twentieth century. Another factor in American whaling's demise was that the wages on land improved to the point that it was difficult to get workers to accept shares in such a risky enterprise when regular wages would allow a worker to buy a house and start a family. Whale populations would come under much greater pressure during the twentieth century from Norwegian whalers with explosive harpoons and powered whaleboats.

2.2.2   Whale Oil and Kerosene

Kerosene is primarily a mixture of alkanes and cycloalkanes with 10 to 15 carbon atoms. Several processes for producing kerosene were developed by different people. It can be distilled from a variety of sources, including cannel coal (a waxy coal formed primarily from pollen and seeds), kerogen shale (often called *oil shale*), bitumen, and petroleum. Abraham Gesner, the provincial geologist of New Brunswick,

Figure 2.10 Abraham Gesner (1797–1864), inventor of kerosene. Photographer unknown, public domain.

Canada, is usually given priority of invention, based on a public demonstration of the distillation from bitumen in 1846 (Figure 2.10). Gesner also came up with the name kerosene. Today it is usually called kerosene in North America, although in earlier times it was sometimes called *coal oil* because some of it was produced from coal. In the UK it is called *paraffin*. Kerosene is still a major petrochemical product. Jet fuel is kerosene. In Japan, kerosene heaters are common in homes. Kerosene lanterns are used in rural areas. The modern kerosene lanterns have metal oxide mantles. These mantles incandesce strongly at visible wavelengths and weakly at infrared wavelengths. This makes them much brighter than the earlier lamps.

In the 1850s, petroleum was produced by skimming it off of ponds at surface seeps. Production was quite limited, and its main use was in patent medicines. This changed in 1859, when Edwin Drake drilled a well for oil in western Pennsylvania. As Figure 2.9 shows, American oil production exploded and the price was dramatically lower than for whale oil. Lighting was the most important market for petroleum until the arrival of the automobile.

2.2.3 A Logistic Model for Whale Oil Production

Figure 2.11 shows the cumulative whale-oil production, calculated from Equation 1.16. Switching to cumulative production changes a bell-shaped curve into an *s*-shaped curve. The cumulative curve also looks smoother. We define the *ultimate production U* as

$$U = \sum p_i \qquad\qquad 2.1$$

where p_i is the annual production, and the sum is over all time. In this case $U = 14$ Mb. This number allows us to estimate that the total number of whales killed was close to one million, reckoned at 15 barrels of oil per whale. We define the 10% time t_{10} when the cumulative production q hits 10% of the ultimate production by the formula

$$q\left(t_{10}\right) = U / 10. \qquad\qquad 2.2$$

The 90% time t_{90} and the halfway point t_{50} are defined in an analogous way. These times are indicated on Figure 2.12. We will write a 10% to 90% lifetime t_l as

$$t_l = t_{90} - t_{10} = 1872 - 1830 = 42 \text{ years.} \qquad\qquad 2.3$$

This definition for lifetime gives one way to answer the question, how long did American whaling last? The definition avoids having to make judgments about the significance of very low production at the beginning and end of the production cycle. One way to interpret t_l is that it took 42 years to come up with an inexpensive substitute for whale oil in the lighting market. The whalers created the market for high-quality lighting, but whale oil was expensive, and this encouraged the

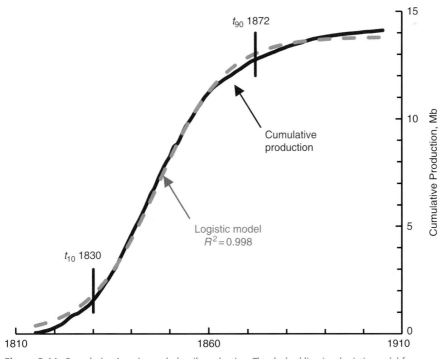

Figure 2.11 Cumulative American whale-oil production. The dashed line is a logistic model for comparison.

kerosene competition. Kerosene allowed a much higher fraction of the population to enjoy good lighting and it relieved the hunting pressure on the whales.

The gray dashed line in the figure is a logistic model for the cumulative production. The logistic function has an exponential growth phase and a matching exponential decline. We will discuss the logistic function in more detail in Section 3.3. We will write the logistic function in several ways. In terms of t_{10} and t_{90} it can be written as

$$q(t) = \frac{U}{1 + 9^{\left(\frac{t_{10} + t_{90} - 2t}{t_{90} - t_{10}}\right)}}. \qquad 2.4$$

When one is first starting to work with the logistic function, it is a good idea to substitute t_{90} into the formula and verify that $q(t_{90}) = 9U/10$ to make sure that the formula is interpreted correctly. In the model, the parameters U, t_{10}, and t_{90} have been adjusted to give the curve the best least-squares fit to the data so they are not exactly the same as the values calculated directly from the data. The value of R^2 is 0.998. Later we will turn this process around and use the logistic function to estimate U, t_{10}, and t_{90} when the production cycle is not complete.

In 1982, the International Whaling Commission voted for a moratorium on killing whales. Part of the motivation was the concern that some whale populations had been greatly reduced, and part was the growing feeling that some animals, like mustangs, simply should not be hunted. Whaling is now limited to modest catches by aboriginal tribes and by countries like Japan, Norway, and the Faroe Islands where whale meat and blubber are a traditional part of the diet. Whale populations are recovering. These enormous creatures can now be seen on whale-watching day cruises from ports around the world.

2.2.4 The Sperm Candle Lighting Efficacy Standard

The catch for whale oil by American whalers during the 1800s and the even larger catch for whale meat in the 1900s are fading from our consciousness. One legacy of the whaling era is our lighting unit, the lumen, abbreviated lm and the lux, abbreviated lx. The lumen is the unit of luminous flux and the lux is the unit of flux density, equal to $1\,lm/m^2$. The lux is the measure of how much light one sees, taking the response of the eye into account. Light meters read in lux. The lux was defined in terms of a sperm candle burning 7.8 grams per hour. The flux density at $1\,m$ from this candle is $1\,lx$, by definition.[3] If the flux density is constant in all directions, the luminous flux would be $4\pi\,lm$. Here 4π is a geometric factor, the area in square meters of a sphere with radius $1\,m$. Lamps are rated in terms of their *efficacy E*, which is the ratio of the total luminous flux in lumens to the input power in watts.

---

[3] As one might guess, standards laboratories do not use sperm candles for calibrating light meters today. The modern definition is in terms of the power in the light itself rather than the power supplied to the source. We will discuss this in Section 8.2.2.

For incandescent bulbs and LEDs, it is easy to measure the input power because it is an electrical measurement that can be done with an inexpensive meter. For the sperm candle, we will take the energy density of the spermaceti to be the same as the ton-of-oil equivalent, 42 GJ/t, or 42 kJ/g. For the standard candle burning 7.8 g/h, the power is 328 kJ/h, or 91 W. The efficacy of the sperm candle is given by

$$E = 4\pi \,/\, 91 = 0.14 \,\text{lm/W}. \qquad\qquad 2.5$$

For comparison a 60-W incandescent lamp has an output of 800 lumens, and an efficacy of 13 lm/W. LEDs reach 200 lm/W, 1,400 times the efficacy of the sperm candle. These LEDs are only a factor of two below the physical limit.

2.3   Wood

Wood is the most significant source of bioenergy by far. At the world level, wood fuel production is three times as large as liquid biofuels. In some lower-income countries, wood is even more important than fossil fuels. The IEA has estimated that in sub-Saharan Africa, wood and agricultural waste account for 70% of primary energy consumption. Most of this is for cooking. Heating food kills parasites and boiling water kills bacteria. In a lower-income country, one advantage of wood fuel is that all family members can gather branches and twigs. A disadvantage is that burning wood in an open fireplace creates smoke that is irritating to the eyes and hazardous to the lungs. It should also be appreciated that commercial logging is a dangerous job. The logging fatality rate in 2017 in the United States was 84 per 100,000 workers.[4] This was the second most dangerous job in the US that year, after fishers at 100 per 100,000 workers. The rate for all American workers was 3 per 100,000.

2.3.1   Wood Fuel and Industrial Roundwood Production

The database of the UN Food and Agriculture organization, FAOSTAT, tracks commercial wood production by country. In forestry statistics, wood production is divided into *wood fuel* and *industrial roundwood*, which includes poles for fences, props in mines, framing wood for houses, and pulp for paper. Wood is an excellent material for making furniture and for building homes. Wood-framed houses have particular advantages in earthquake country, because the studs and rafters can sway without damage. Figure 2.12 shows world production on a per-person basis. Both wood fuel and industrial roundwood are declining slowly, wood fuel a bit faster than industrial roundwood. The ten-year annualized decline rates from 2007 to 2017 were 0.8%/y for wood fuel and 0.5%/y for industrial roundwood. In 1961,

---

[4] Bureau of Labor Statistics 2018 "National Census of Fatal Occupational Injuries 2017." A summary is available at www.bls.gov/news.release/pdf/cfoi.pdf

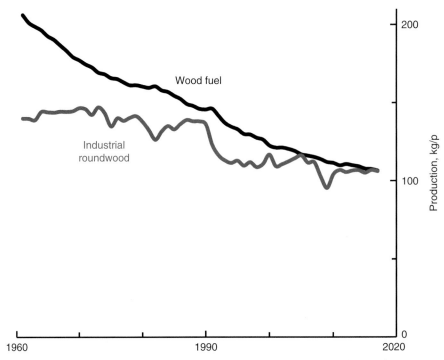

Figure 2.12 World per-person commercial wood production. The wood production data are from FAOSTAT. The population data come from United Nations Populations Division. The FAOSTAT data are converted from volume units to mass units at 425 kg/m³.

wood fuel production was 47% larger than industrial roundwood, but as of 2017 they were equal, at 106 kg/p each.

Figure 2.13 shows the wood production history of the United States and India, the world's two largest producers. In 2017, 86% of Indian production was wood fuel, while 85% of American production was industrial roundwood.

The heat from a wood fire depends on whether the wood has been seasoned to from liquid water. For comparisons, we will use the value 0.35 toe/t for the energy density of commercial wood fuel. This compares with 0.5 toe/t as a representative value for mined coal and 0.71 toe/t for charcoal. On this basis, world commercial wood fuel production in 2017 was 281 Mtoe. This was 2% of world primary energy that year.

A new wood fuel market is emerging, wood pellets. Wood pellets are produced by compressing sawdust and scraps of wood. These may come from waste in logging, sawmills, and furniture factories, but also directly from timber. A wood pellet fire is shown in Figure 2.14. The pellets can be loaded into automatic feeders for wood stoves and for steam boilers in electricity generation. The US EIA tracks wood pellet production. The first full year for data is 2017. The feedstock was 80% residual sawdust and scraps, and 20% timber. US production in 2017 was 6.4 Mt,

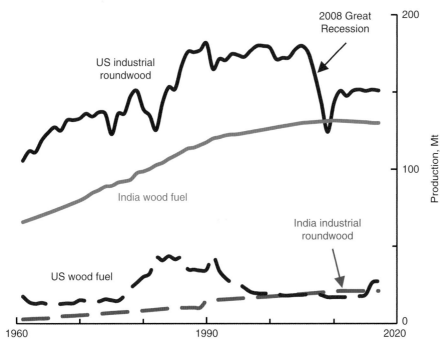

Figure 2.13 Historical wood production in the United States and India. The wood production data are from FAOSTAT. The FAOSTAT production data are converted from cubic meters to mass units at 425 kg/m³.

Figure 2.14 Wood pellets burning. Credit: tchara/iStock by Getty Images.

with 80% exported, primarily for electricity generation. The average energy content of the pellets is 0.43 toe/t, so production in energy terms was 2.7 Mtoe.

2.3.2 Charcoal and Wood Gas

Wood may also be converted into two secondary fuels, charcoal and wood gas. Charcoal is formed by heating wood in limited oxygen over a period of several days. Figure 2.15 shows kilns for making charcoal in Wildrose Canyon, Death Valley, California. From a chemical perspective, wood is mainly cellulose, a carbohydrate. The heat in the kiln drives out the hydrogen and the oxygen in the cellulose, as well as sap and volatile organic compounds (VOCs). This gives a black solid that is 80% carbon, similar to a high-rank coal. In fact, the word "coal" originally meant charcoal. What we now call coal was earlier called sea coal or pit coal. A person who made charcoal was called a collier. If your family name is Collier, you know what an ancestor did for a living. The meaning of the word "collier" has drifted since then. It now means a ship that carries coal.

Charcoal making is not efficient. More than half of the energy in the wood is lost in the process. However, charcoal is an excellent fuel. The energy density is 0.71 toe/t, twice that of wood. This increased energy density makes it much easier to transport the fuel (Figure 2.16). The FAO tracks commercial charcoal production. In 2017 it was 36 Mtoe, 13% of wood fuel production in energy terms. The largest

Figure 2.15 Kilns for making charcoal for smelting lead and silver for the Modoc Mining Company. The kilns are 8 m high. They were built in 1877. Photograph by Dale Yee.

producers were Brazil, Nigeria, and Ethiopia. Charcoal burns with less smoke than wood, which makes it attractive for an indoor fire. Charcoal also burns hot. This has made it the preferred fuel for blacksmiths for thousands of years. Charcoal was also a critical component in extracting metals from ore. The charcoal undergoes a partial combustion to form carbon monoxide to provide heat for the process, and the carbon monoxide acts as a reducing agent to remove oxygen from the ore. Charcoal has other applications. In gardening it is used as a soil conditioner to help roots get better access to nutrients. *Activated charcoal* is a form with high porosity that is often used as a filter to remove impurities from solutions.

Wood gas is made by heating charcoal, coal, or wood with steam. This produces carbon monoxide and molecular hydrogen according to the reaction

$$C + H_2O \implies CO + H_2. \qquad 2.6$$

Wood gas has many other names: producer gas, syngas, coal gas, illuminating gas, and town gas. It was used for street lighting, residential lighting, and heating in

Figure 2.16 Moving charcoal by bicycle in Uganda in 2014. Credit: Rod Waddington/CC BY-SA 2.0, https://creativecommons.org/licenses/by-sa/2.0/

the 1800s and 1900s. However, wood gas leaks are dangerous indoors, because the carbon monoxide is poisonous. When electric lights and natural gas heating became available towns switched away from wood gas. Some people ran their cars on wood gas during World War II, when gasoline supplies for civilians were extremely limited in many countries. Figure 2.17 shows a modern car that runs on wood gas. The developer, Dutch John, reports,

> After 2,000 kilometers on wood gas everything worked properly. Top speed is 120 km/h. Cruising speed 100–110 km/h. Fuel consumption approximately 30 kilos of wood for one hundred kilometers, which is also one effective filling of the fuel bin. The back seat loaded with sacks of wood make the total range approximately 400 kilometers.

2.3.3   The Transition from Wood to Coal

The UK was the first country to make the transition from wood to coal.[5] One factor that encouraged this was that the relative prices of wood and coal shifted in favor of coal. Figure 2.18 shows price indexes for wood and coal for a college in central London

Figure 2.17 A 2008 wood gasifier that provides fuel for a Volvo in Holland. Credit: Dutchjohn1st at http://woodgas.nl/GB/project.html

---

[5] The Song Dynasty in China and the Dutch Golden Age could also be considered the first to move away from wood to fossil fuels. There will be more discussion of this in Section 4.5.

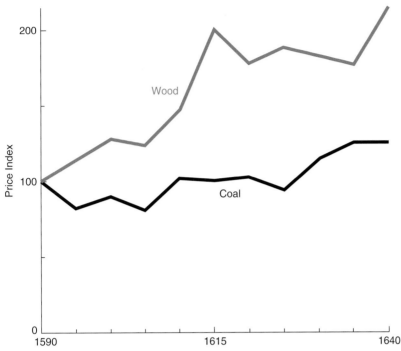

Figure 2.18 Price indexes for fuels purchased by Westminster College, in the London area, from John Hatcher, 1993, *The History of the British Coal Industry*. The absolute price differences on an energy basis are more speculative. Hatcher suggests that in 1640 wood might have been 50% more expensive than coal on an energy basis.

from 1590 to 1640. It is remarkable that these records were preserved for four hundred years. The indexes are set to 100 in 1590. By 1640, wood was at 220, while coal was at 120. Coal mines have a much smaller land footprint, then and now, than the forest needed to supply wood fuel and charcoal. Using coal instead of wood allowed woodlands to be available for other purposes. Some were maintained as hunting preserves, while others were plowed to plant crops or converted to grazing lands.

The transition to coal did not happen everywhere in the UK at the same time. One reason for this is that land transportation before railroads was extraordinarily expensive compared to water transportation. The Scottish political economist Adam Smith wrote in his 1776 book *Wealth of Nations*,

> A broad-wheeled waggon, attended by two men, and drawn by eight horses, in about six weeks' time carries and brings back between London and Edinburgh near four ton weight of goods. In about the same time a ship navigated by six or eight men, and sailing between the ports of London and Leith, frequently carries and brings back two hundred ton weight of goods.

Smith's observation helps us understand the map in Figure 2.19. Coal use was favored over wood near the coal fields and near the coast and rivers.

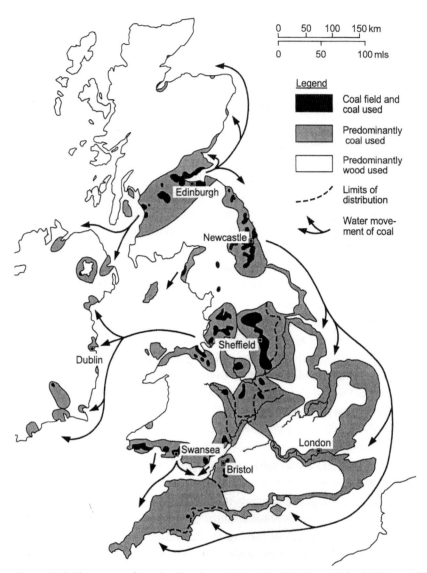

Figure 2.19 The pattern of wood and coal use in Britain in 1680. From Michael Williams, 2003, *Deforesting the Earth*, University of Chicago Press, © 2003 by the University of Chicago. The figure is adapted from a more detailed figure by John Nef, *The Rise of the British Coal Industry*, 1932, George Routledge and Sons.

Trees are often cut down to create arable land and pastures, and this has a major effect on forests. Figure 2.20 and Figure 2.21 give a sense of the enormous scale of the deforestation in the United States. The first figure shows the forest in 1620, when the first successful English colony was founded, and the second shows 1920, when American farming had been fully established. One should be aware that deforestation did not start with the European immigrants. The earlier inhabitants have burned forests to make fields to plant crops and to make hunting areas for

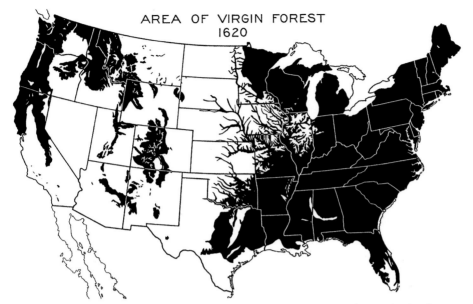

Figure 2.20 The forest when the Pilgrims landed in in Massachusetts in 1620. The map is taken from W. B. Greeley, 1926, "The Relations of Geography to Timber Supply," *Economic Geography*, vol. 1, pp. 1–14.

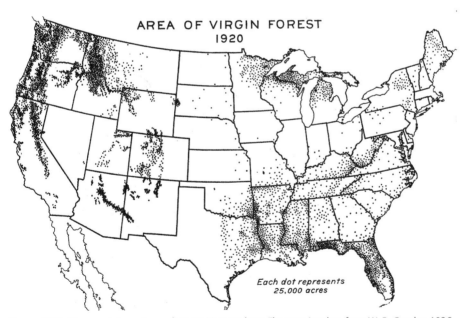

Figure 2.21 The remaining primary forest 300 years later. The map is taken from W. B. Greeley, 1926, "The Relations of Geography to Timber Supply," *Economic Geography*, vol. 1, pp. 1–14.

thousands of years. One should also not get the idea that there are no forests now in the United States. In many areas, farms have been abandoned and the forest has returned. In other areas where there is excellent rainfall, like the Southeast, pine forests are harvested and replanted on a regular cycle.

2.3.4   Timber Density

Figure 2.22 shows the more recent history of timberland area and timber density in the United States. The timberland designation applies to forests capable of producing more than 140 m³ of wood per square kilometer per year. Alaska is included, although only a small part of its forest meets this standard. The timber density is the volume of wood divided by the land area. The resulting unit is length. It can be visualized by imagining that the wood is spread out in a uniform layer over the area. The timber density is the thickness of that layer. It is plotted here in mm. The timberland area has been relatively constant over the last sixty years at 2,000,000 km², 20% of the US land area. In contrast, the timber density has increased dramatically, with a 48% increase from 1953 to 2007. Tree growth also appears to have accelerated in Europe. The article by Hans Pretzsch et al. in the references at the end of the chapter has a discussion. Like the cereal yield increase in Figure 1.21, the increase in timber density was not predicted. On the contrary, in the 1970s there were concerns that acid rain caused by power-plant emissions would reduce tree growth.

What is happening in American forests is different from what is happening in the Green Revolution. It is not based on artificial fertilizers or plant breeding.

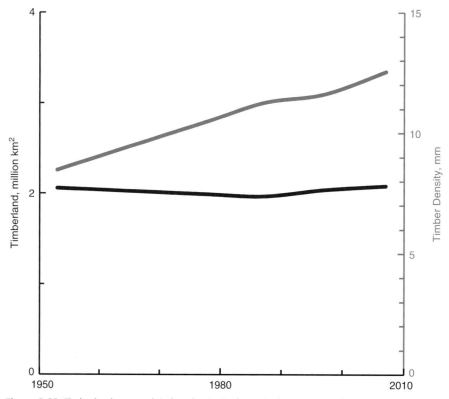

Figure 2.22 Timberland area and timber density in the United States. From the US Department of Agriculture, 2012, *Forest Resources of the United States.*

In some areas the age of the trees is likely a partial explanation for the rise, as in the Northeast, where the timber density increase has been extremely large, 121%. Factors that may have contributed to timber density are longer growing seasons, reactive nitrogen oxides emitted by power plants that are entrained in rain, and increasing CO_2 levels in the atmosphere.

2.4 The Norse Greenland Colony

Thomas Malthus' *Essay on a Principle of Population* (Section 1.6) kicked off a stream of apocalyptic energy literature that has persisted to the present day. In this section, we will study a society with marginal energy supplies. The Viking Greenland colony was founded by Erik the Red in the year 985 or thereabouts. He had been living in western Iceland, but was banished for fighting. Following up on sailor reports of seeing skerries to the west, Erik sailed to Greenland.[6] He established a colony that became known as the Eastern Settlement (Figure 2.23). Figure 2.24 shows a recon- struction of a Viking sod house at the site of Erik the Red's farm. Another colony, the

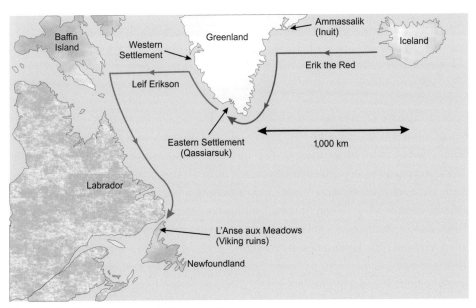

Figure 2.23 The Viking voyages to North America.

---

[6] A skerry is a rocky islet. The word "skerry" comes from the Norse. The skerries themselves do not exist. Moreover, Greenland is 300 km from Iceland, and under ordinary atmospheric conditions, it is not possible to see that far because of the curvature of the earth. However, atmospheric inversions sometimes allow light to travel great distances in the Arctic. Some mirages distort the view of a coastline so that it appears as a group of islands. The Norse would be familiar with these mirages and would likely associate them with land.

Figure 2.24 Reconstruction of a Viking sod longhouse at the site of Brattahlid, Erik the Red's farm in the Eastern Settlement. This site is in the modern village of Qassiarsuk, Greenland. Credit: University of Massachusetts at Amherst.

Western Settlement, was started further north. Eventually thousands of people were living in the two settlements. One advantage the Vikings had in traveling between these islands was an innovation in ship building. Viking ships had keels. Other sailing ships had round bottoms, which made it difficult to sail at a large angle from the wind. The keel gave an advantage in maintaining a course on their voyages. Erik's son Leif explored the Canadian coast. From the Icelandic sagas, it appears that he visited Baffin Island and Labrador, and that he established a settlement at a place called Vinland. The etymology of the name Vinland is not well established. Earlier historians had interpreted it as "wine land," and this led to unsuccessful searches in New England. The Norwegian scientists Ann and Helge Ingstad argued for "meadows" and searched for the remains of the settlement further north. In 1960, at L'Anse aux Meadows on the northern tip of Newfoundland, they found the remains of sod houses in the traditional Norse design. Radioactive carbon dating put the age of the ruins at a thousand years, which is consistent with the sagas.

The wood supply was extremely limited in Greenland. This meant that grass had to serve many purposes, sod for the longhouse walls, fuel for heat, grazing for sheep and goats in the summer, and hay for the animals to eat in the winter. The energy supply was marginal, but the Norse survived in Greenland for more than

400 years. Why did the colony eventually die out? This is one of the great history puzzles. There was no boat out with the last survivors to tell the story.

2.4.1 The Greenland Colony and Climate

There is a regional temperature record for this period that was derived from oxygen isotope analysis of ice cores from central Greenland (Figure 2.25). The temperature on the ice cap was much lower than what the Norse experienced on the coast, so the graph can only be used to infer the timing of changes in the temperature, not absolute levels. The year-to-year details are not preserved, but we can see variations on fifty-year time scales and longer. There were warm periods during Roman times. One should be cautious about making much of the temperature difference in the plot between Roman times and 1850 because there are uncertainties caused by changes in the elevation of the ice surface over time. The Greenland colony started near the end of a period of rising temperatures, 1.7 °C over 240 years. However, after the peak, the temperature fell over the next seven hundred years. The period from 1300 through 1850 is called the Little Ice Age, with the end date conventionally marked by retreating ice in Europe. For more information about

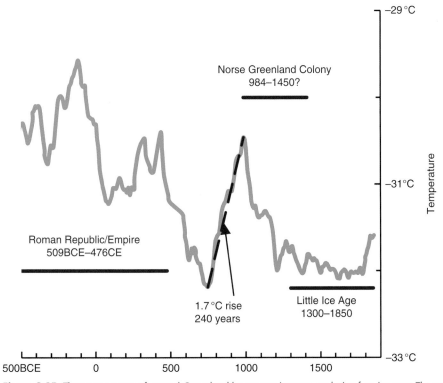

Figure 2.25 The temperature of central Greenland by oxygen isotope analysis of an ice core. The source of the data is R. B. Alley, 2000, "The Younger Dryas cold interval as viewed from central Greenland," *Quaternary Science Reviews*, vol. 19, pp. 213–226.

glaciers and the Little Ice Age, see the book by Cambridge geographer Jean Grove listed in the references at the end of the chapter.

Did the Little Ice Age kill off the Norse in Greenland? Probably. The Norse were utterly dependent on their hay crop. Hay was their heating fuel, their fodder, and their insulation. Hay production is extremely sensitive to temperature and to the length of the growing season. If this was the problem, was there something the Norse could have done to save the colony? Quite possibly. One of the mysteries for anthropologists is that they have not found evidence that the Greenland Norse ate significant amounts of fish. The UCLA anthropologist Jared Diamond suggested that the Greenland Norse may have had a taboo against eating fish. It is also possible that the fish remains were used in some way that they would not be detected, say as fertilizer. Today the Greenland rivers and fjords are teeming with tourists fishing.

2.4.2   The Inuit in Greenland

At the time the Norse settlements disappeared, there was another group of people living in Greenland, the Inuit. They mostly lived in northern Greenland. The Inuit made it through the Little Ice Age and they are still there. The Inuit did not have grazing animals. They hunted seals, whales, and birds, and they fished. The seal and whale fat was burned for heat and light. They constructed snow houses during the winter. The Inuit developed a fast one person hunting boat, the *kayak* (Figure 2.26), that is imitated around the world, albeit in fiber glass rather than seal skin.

Figure 2.26 An Inuit hunter in 1915 near Ammassalik, Greenland (Figure 2.23). At his back is a float that is attached to the harpoon line so that his prey does not sink. Credit: William Thalbitzer/The Danish Arctic Institute.

Are there messages for us today as we contemplate a shift away from fossil fuels? We have more energy choices than the Greenland Norse, and in the Internet era, information about these choices is available everywhere. One message would be to be flexible in adopting technology. Another might be to be cautious before accepting a taboo against a particular form of energy. The danger is that people would be forced to use more expensive energy sources that may make it difficult for them to heat and cool their homes and to get to work.

Concepts to Review

- Horses in history
- Horses and railroads
- Transition from horses to vehicles
- Why did American whaling decline?
- The invention of kerosene
- The discovery of oil in Pennsylvania
- Defining the lumen
- Industrial roundwood and wood fuel
- Charcoal
- Wood gas
- The Norse Greenland colony
- Inuit technology

Problems

Problem 2.1   Growth-rate plot for American whaling
Before you start, download the Excel file for the whaling production history at energybk.caltech.edu. Take the cumulative production through 1815 to be 50 kb.
a. Make a growth-rate plot for the cumulative whaling production q.
b. Make an estimate for the ultimate production U using only the growth rates for the years from 1827 to 1840 and compare with the actual value. Note that 1840 is before kerosene entered the lighting market.

Problem 2.2   Heating with wood
In rural areas, homes are often heated with wood stoves like the one shown in Figure 1.1. Wood stoves put less smoke into the room than the traditional open fireplace, and they are more efficient. In some areas, there are sufficient dead trees available to supply the firewood. Firewood typically is dried for a few months to

reduce the water content before it is burned. For this problem, we will analyze the heating requirements for the cabin shown in Figure 1.3. In the United States, wood fuel is measured in *cords*. A cord is a volume measure equal to 128 cubic feet, or 3.6 m³. It is nominally a stack of firewood 8 feet wide, 4 feet high, and 4 feet deep. The dominant tree in the area around the cabin is the Engelmann spruce. Search online to find the heat content and density for Engelmann spruce.

a. Calculate the heat content in GJ/t.
b. Over a period of 40 days, a stack of firewood 3 feet wide, 3 feet high, and 2 feet deep was burned in the stove. This raised the average temperature of the cabin by 4 K. Calculate the thermal resistance R_t of the cabin in K/kW using Equation 1.7. You may neglect the heat loss up the stove chimney and other sources of heat in the cabin, like lighting and electronics.
c. Without heating the average temperature in the cabin from October 1 to June 30 would be 3 °C. Calculate the number of cords of wood that would be required to raise the average indoor temperature over this period to 20 °C.

Problem 2.3   Fuel and food for the Norse Greenland colony

Read Jared Diamond, 2012, *Norse Greenland*. This is available inexpensively in Kindle format from Amazon, and it can be read with the free Kindle PC reader. Write an essay, no more than one printed page, on energy in the Greenland colonies. You should consider energy in a broad sense, including both fuel and food. How were fuel and food linked? How were fuel and shelter linked? From the latitude difference between the two colonies, what can you say about the effect of latitude on the energy supply, again considering both fuel and food? How did the energy supply evolve over time? Do you have any criticisms of Diamond's analysis? You are free to use other references. If you do, please cite them properly, and print a reference list on a second page.

Further Reading

- Lance Davis, Robert Gallman and Karin Gleiter, 1997, *In Pursuit of the Leviathan: Technology, Institutions, Productivity, and Profits in American Whaling, 1816–1906*, University of Chicago Press. This is the most important whaling reference for this chapter. The discussion is particularly deep on the economics of whaling and its labor market.
- Jared Diamond, 2006, *Collapse: How Societies Choose to Fail or Succeed*, Penguin Books. Diamond argues that two of these societies, the Anasazi in the American Southwest, and the Norse in Greenland were affected by climate changes, drought for the Anasazi and cold for the Norse. The book *Norse Greenland* referenced in Problem 2.2 consists of the chapters on Greenland from this book.

- EIA, 2016, "New EIA survey collects data on production and sales of wood pellets" at www.eia.gov/todayinenergy/detail.php?id=29152
- Michael Faraday, 1861, *The Chemical History of the Candle*. This is a chemistry classic. Michael Faraday gave this set of six public lectures many times at the Royal Institution in London. He emphasized experiments one could do at home. It is easy for us to lose sight of the fact that at one time the candle was a considerable improvement over a burning pine brand.
- Jean Grove, 1988, *The Little Ice Age*, Routledge. One of the challenges of understanding sea level rise is to sort out which parts come from the fact that the Little Age is over and which parts one could associate with fossil fuels. Jean Grove was a Cambridge University professor and her book is the classic reference on glaciers during the Little Ice Age.
- Ann and Helge Ingstad, 2001, *The Viking Discovery of America*, Checkmark Books. The Ingstads found the Viking ruins in Newfoundland. This is one of the greatest archeological discoveries of all time.
- Paul Lucier, 2010, *Scientists and Swindlers: Consulting on Coal and Oil in America, 1820–1890*, Johns Hopkins University Press. An excellent history on Abraham Gesner's career and his work on kerosene.
- Magnus Magnusson and Herman Palsson (translators), 1965, *The Vinland Sagas*. This includes the Greenland Saga and Erik's Saga. The Norse sagas are an important early literature form and they contain information about the Norse explorations that is independent of the archeological evidence.
- Hans Pretzsch, Peter Biber, Gerhard Schutze, Enno Uhl, and Thomas Rotzer, 2014, "Forest stand growth dynamics in Central Europe have accelerated since 1870," *Nature Communications*, available at www.nature.com/ncomms/ 192014/140912/ncomms5967/ full/ncomms5967.html
- Vaclav Smil, 2008, *Energy in Nature and Society*, MIT Press. The book includes a detailed discussion of the energetics of people and work animals, and much else.
- UN Food and Agriculture Organization, 2015, *Global Forest Assessment*. An excellent overall collection of forestry statistics, including wood production.
- US Department of Commerce, Census Bureau, 1975, *Historical Statistics of the United States*. The best single source of early US statistics. Available in various formats at https://archive.org/details/HistoricalStatisticsOfTheUnitedStatesColonialTimesTo1970
- Michael Williams, 2002, *Deforesting the Earth: From Prehistory to Global Crisis*, University of Chicago Press. This book covers much more than the title suggests. It is the definitive history of how people have used wood at all times and at all places.

3 Models

In desperation I asked Fermi whether he was not impressed by the agreement between our calculated numbers and his measured numbers. He replied, "How many arbitrary parameters did you use for your calculations?" I thought for a moment about our cut-off procedures and said, "Four." He said, "I remember my friend Johnny von Neumann used to say, with four parameters I can fit an elephant, and with five I can make him wiggle his trunk."

Freeman Dyson,
"A meeting with Enrico Fermi"

3.1 Introduction

Models can help us understand what has happened in the past and help us make statements about what might happen in the future. As examples, the growth-rate plot for world population in Section 1.5.3 gave us a projection for the world population peak, and the broken-line electricity model in Section 1.6.2 established a strong relationship between electricity generation and income. Our emphasis will be on linearized models. These models start with a transformation of the data. If the model is a good one, the plot will look like a straight line. These models take advantage of the fact that one's eyes can judge well whether data are linear. In addition, a shock like the Iranian Revolution can often be characterized in a simple way as a change in slope. For linearized models, the model parameters can be determined by Excel's SLOPE and INTERCEPT functions. The index that we will use for the quality of the model is R^2, calculated by the Excel function RSQ. We will introduce two different production models: the declining exponential and the logistic model for cumulative production, which we will compare with reserves. We close the chapter with a discussion of an influential economic model for unemployment and a list of statistical formulas for reference.

3.1.1 Types of Statements about the Future
We distinguish between three types of statements about the future.

1. The most restrictive is a *prediction*, where there is no qualification. Paul Erhlich's declaration of imminent deaths from starvation (Section 1.7.1) is an example of a prediction.

2. Next is a *projection*, where there is a qualification. The results from our linearized models are projections. These are based on historical data, so they are subject to social and technical shocks. For example, the Iranian Revolution put world hydrocarbon production on a slower path. In addition, the development of fracking technology has likely increased the ultimate hydrocarbon production in the long run.

3. Finally there is the *scenario*, which is a plausible view of the future. The value of a scenario is that it can help us think about what the future might look like. The standard for political statements is also plausibility, so it is common for governments to use scenarios to justify policies. Economists have a concept called *economic rent* or more commonly just *rent*, which is a payment for a product or service that is larger than would be needed to induce a seller to provide it. *Rent seeking* is trying to persuade the government to adopt rules that enable sellers to gain rent. An historically significant example is the hard coal miners in Germany who for many years were paid several times the prevailing European import prices for their coal. The intergovernmental structure that was established in 1951 to make this subsidy possible was called the European Coal and Steel Community. Its descendant is the European Union.

3.1.2   The R/P Ratio

The R/P ratio (spoken as "R P ratio"), is equal to $R \div p$, where R is the reserves and p is the current annual production. The units of the R/P ratio are years. We will have much more to say about reserves in Section 3.4, but for now we will take reserves to be oil, gas, and coal that are thought likely to be produced in the future. The units are the standard production ones: barrels for oil, cubic meters for gas, and metric tons for coal. For example, from the BP *Statistical Review*, the world natural gas reserves are $193.5\,Tm^3$ and the 2017 production was $3.68\,Tm^3$. The R/P ratio is 53 years. Informally people express this by saying, "We have 53 years of natural gas left."

The R/P ratio is convenient to calculate. However, there are limitations. In practice, production is only flat in unusual situations. For example, the production in an oil field may be set by the capacity of its output pipeline, or the production in a coal mine might be set by the speed of its shearer. One should be cautious in assuming that the reserves are a good estimate of future production. In many cases, oil fields have yielded more than their original reserves, while coal producing countries that have completed the production cycle have historically produced only a small fraction of their original reserves. Even when reserves are a good match to future production, production changes can lead to dramatically different lifetimes than implied by the R/P ratio.

3.2   The Decaying Exponential Model

In this model, annual production declines exponentially with time. We take the production to be of the form

$$p_i = a\left(U - q_i\right) \qquad\qquad 3.1$$

where p_i is the production in year i, a is a constant, U is a projection for ultimate production, and q_i is the cumulative production in year i calculated from Equation 1.16. To see how Equation 3.1 gives exponential decline, we rewrite Equation 1.16 as

$$q_i = q_{i-1} + p_i \qquad\qquad 3.2$$

and substitute this formula into Equation 3.1 to get

$$p_i = a\left(U - q_{i-1}\right) / \left(1 + a\right) = p_{i-1} / \left(1 + a\right). \qquad\qquad 3.3$$

The production falls by a factor of $1 + a$ each year, which is the pattern for a decaying exponential.

In Equation 3.1, if we take p as the y-coordinate and q as the x-coordinate, this is the equation of a line, where U is the x-intercept and a is the negative of the slope. Jean Laherrere of the Total oil company popularized this type of plot in fossil-fuel production modeling, so we will call it the *Laherrere linearization*.

3.2.1   The Prudhoe Bay Oil Field

As an example let us consider the Prudhoe Bay Oil Field on Alaska's North Slope. This field is, by far, America's largest oil field. It was an important addition to the world oil supply after the shock of the 1979 Iranian Revolution. Oil was discovered there in 1968. The North Slope is icebound much of the year, so it is not possible to ship the oil out from Prudhoe Bay directly by tanker. A pipeline was planned to go 1,300 km to the port of Valdez on the south coast (Figure 3.1). A major obstacle to building the pipeline was that the Minerals Leasing Act of 1920 limited the right of way on each side of the pipeline to 25 feet. After years of contentious legal and political battles, the US Congress narrowly approved a revision to loosen the right-of-way limits in the summer of 1973. Then in October 1973 the Yom Kippur War began, followed by the Arab Oil Embargo. That swept away the opposition to the pipeline, and in November Congress overwhelmingly approved legislation that removed all of the remaining obstacles. Construction began the following year and finished in 1977.

The construction of the pipeline, shown in Figure 3.2, was an outstanding engineering achievement. The inside diameter was 1.2 m. The oil is heated to 44 °C at the beginning of the pipeline. Much of the pipeline was elevated above ground to avoid melting the permafrost. The pipeline service company reported in 2007 that it took 12 days for oil to move from the oil field to Valdez. There has been

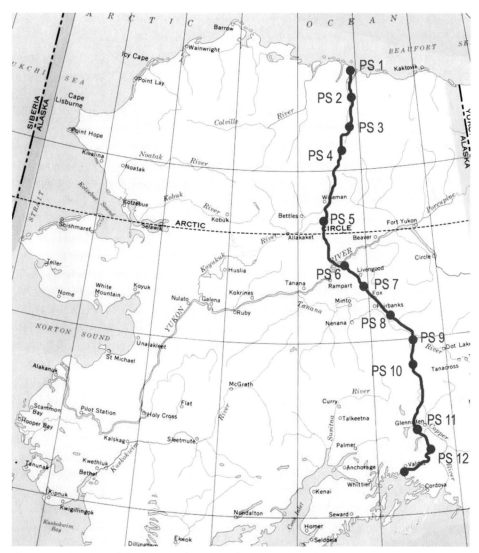

Figure 3.1 The route of the Trans-Alaska Pipeline (TAPS). PS stands for pumping station. Pump stations 1, 3, 4, and 9 were operating in 2018. The pump station 11 site is currently being used as an emergency staging area. Credit: Flominator/CC BY-SA_3.0 https://creativecommons.org/licenses/by-sa/3.0/

one major environmental disaster associated with Alaskan oil. In 1989, the tanker Exxon Valdez ran into Bligh Reef in Prince William Sound after leaving the port of Valdez and released 35 kt of oil into the sound. The spill killed wildlife and ruined fishing in the sound for years. The cause of the wreck was a navigational error. The captain had given an order to steer out of the shipping lane to avoid icebergs and then went to his cabin to work on paperwork. The third mate did not steer back into the shipping lane in time to avoid the reef. As a result of the spill, tankers have been required to have double hulls. However, the grounding was so severe

Figure 3.2 The Trans-Alaska Pipeline at Prudhoe Bay, Alaska, the largest oil field in the United States. Presumably the bears find it easier to walk on the pipeline than across the tundra. Credit: Alaska Dept. of Natural Resources, http://dnr.alaska.gov/

that it is possible that the spill would have occurred even if the Exxon Valdez had had a double hull. On the other hand, navigation has been revolutionized by GPS since that time. Small recreational boats today have better navigational equipment than the Exxon Valdez did.

Figure 3.3 shows the production history. Production was already significant by the time of the Iranian Revolution in 1979. Until 1988, production was limited by the capacity of the pipeline. Since then, the production has declined greatly, and by 2017, it was down 85% from the peak. The cumulative production was 12.47 Gb through 2017. This is half of the original oil in place.

3.2.2   The Laherrere Linearization for Prudhoe Bay

The Laherrere linearization for Prudhoe Bay is shown in Figure 3.4. There is a linear trend downward that began in 1988. Before this time, the production was limited by the size of the pipeline, so the data are not useful for projections. The x-intercept of the regression line gives us the projection for ultimate production U. Following Equation 1.12, the formula for the x-intercept is given by

$$U = -\frac{\text{INTERCEPT}(\mathbf{p}, \mathbf{q})}{\text{SLOPE}(\mathbf{p}, \mathbf{q})} \qquad\qquad 3.4$$

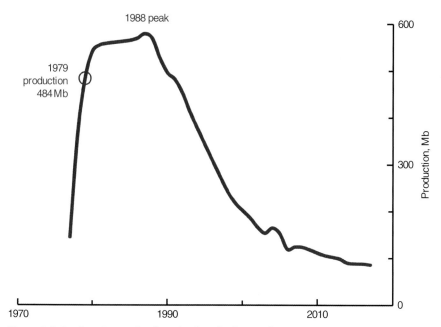

Figure 3.3 Prudhoe Bay crude-oil production. The first production was in 1976. Production through 2003 from the Alaska Department of Revenue, Tax Division. Later production from the Alaska Oil and Gas Commission.

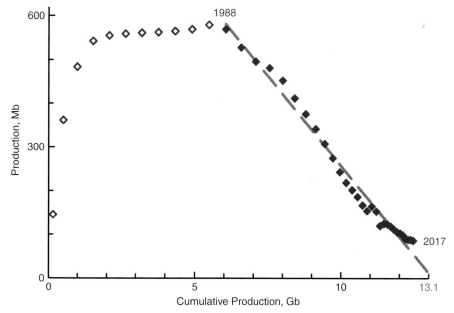

Figure 3.4 The Laherrere linearization for the Prudhoe Bay oil field. The regression line is fitted for the data starting in 1988.

where **p** is the set of cells for the annual production and **q** is the set of cells for the cumulative production. The constant a is given by

$$a = -\text{SLOPE}(\mathbf{p}, \mathbf{q}) \qquad\qquad 3.5$$

The calculations give $a = 0.08$, and $U = 13.1\,\text{Gb}$. We define an exhaustion factor e given by

$$e \equiv q/U. \qquad\qquad 3.6$$

The exhaustion factor shows how close we are to the ultimate production. For the most recent data year, 2017, we have $e = 95\%$. We define the projection for the total future production f as

$$f \equiv U - q. \qquad\qquad 3.7$$

For 2017, we have $f = 680\,\text{Mb}$. One of the advantages of the Laherrere linearization is that a projection for future production can often be made even if early production data are missing, because the missing data reduce both terms on the right side of the equation by the same amount.

Figure 3.5 shows the cumulative production q over time. In addition, the plot shows how the projection for U has evolved starting in 1989. The range of U is

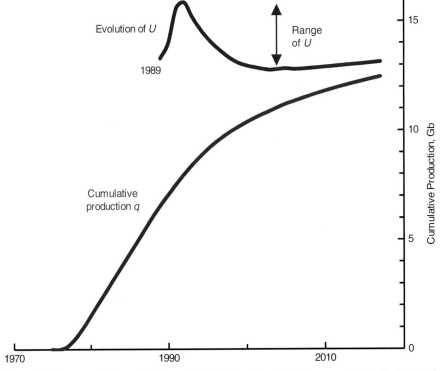

Figure 3.5 The cumulative production q of the Prudhoe Bay oil field, compared with the historical evolution of the projections for ultimate production U.

given by the maximum and minimum values, 12.8 to 15.9 Gb. Time will tell if this range will capture the actual ultimate production. Expressed as a percentage by Equation 1.13, the range is ±11%.

The Laherrere linearization gives an estimate of the ultimate production U and the decay parameter a. An advantage of the linearization is that it gives a good visual check on whether the model is appropriate. However, the decaying exponential model can use data only after the peak in production. The next model, the logistic, removes this limitation.

3.3 The Logistic Model

The logistic function was invented by Pierre Verhulst in 1838. He applied the logistic function to model the populations of France, Belgium, and Russia. We used it to analyze whale-oil production in Figure 2.11. Verhulst defined the logistic function by the differential equation

$$q'(t) = a\,q\,(t)\,(1 - q\,(t)\,/\,U)$$ 3.8

where the prime denotes a time derivative. If the time units are taken to be years, we can write the corresponding discrete equation

$$p_i = a\,q_i\,(1 - q_i\,/\,U).$$ 3.9

The factor of q_i gives exponential growth in the early part of the production cycle. As q_i approaches U, this reduces to the formula for the decaying exponential, Equation 3.1.

3.3.1 The Hubbert Linearization
We linearize Equation 3.9 by dividing by q_i to get

$$p_i\,/\,q_i = a\,(1 - q_i\,/\,U).$$ 3.10

We can think of $p\,/\,q$ as the growth rate for the cumulative production. If we plot $p\,/\,q$ on the y-axis and q on the x-axis, the x-intercept is U. This kind of plot for fossil fuels is called a Hubbert linearization, after King Hubbert of the Shell Oil Company, who first applied it to fossil-fuel production modeling.

As an example, we consider nineteenth-century coal production in the United Kingdom. It is useful to make projections from old data, because we know how things turned out. Figure 3.6 shows the production from 1854 to 1900. During this time, the UK was the world's largest producer of coal every year except for 1893, 1899, and 1900, when it ran second to the United States. There was a rise during the entire period, except for a dip in 1893, when there was a strike by miners. We would not be able to use a declining exponential model in this situation because

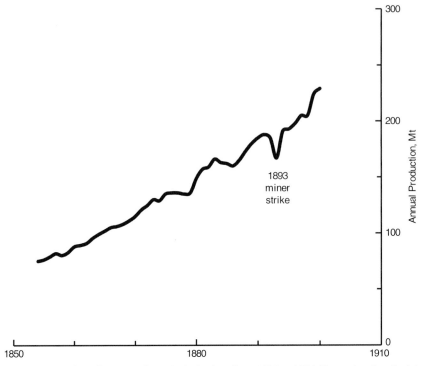

Figure 3.6 Coal production in the United Kingdom from 1854 to 1900. The national coal mining data collection began in 1854. The data come from Brian Mitchell, 2007, *International Historical Statistics*, Palgrave Macmillan.

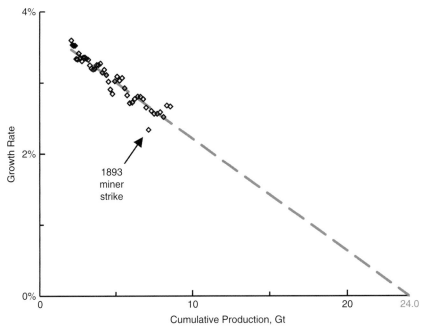

Figure 3.7 Hubbert linearization for British coal using the production data from 1854 through 1900. The estimate for cumulative production before 1854 comes from Sidney Pollard, 1980, "A New Estimate of British Coal Production 1750–1850," *Economic History Review*, vol. 33, pp. 212–235. The later production data come from Brian Mitchell, *International Historical Statistics,* Palgrave Macmillan.

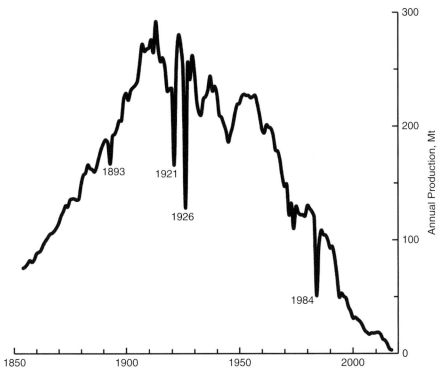

Figure 3.8 British coal production from 1854 to 2017. The data through 1980 come from Brian Mitchell, 2007, *International Historical Statistics*, Palgrave Macmillan, and afterward from the BP *Statistical Review*. The large drops in 1893, 1921, 1926, and 1984 are from strikes.

the production is not declining. Figure 3.7 shows the Hubbert linearization. The x-intercept gives U as 24.0 Gt.

Now fast forward 117 years to 2017. Figure 3.8 shows the full production history. Production peaked at 292 Mt in 1913, on the eve of World War I. At that time, there were 3,024 producing mines and 1.1 million miners. However, at the beginning of the war, 40% of the miners of military age volunteered for service, and the 1913 production was never exceeded. There has been a steady decline since the peak, with the last deep mine closing in 2015. This leaves only a few small surface mines and drift mines producing. The actual cumulative production through 2017 was 27.5 Gt, so our projection of 24.0 Gt based on the production from 1854 to 1900 was 13% low.

Figure 3.9 shows the full Hubbert linearization. It is remarkable that the linear relationship is maintained throughout the entire period for which we have data.

The advantage of the logistic model over the decaying exponential model is that it can make use of data before the peak in production. The Hubbert linearization gives an excellent way to check visually whether the logistic model is appropriate. However, the Hubbert linearization has limitations. One is that the annual production p appears in the formula. The production data points for the strikes in 1893, 1921, 1926, and 1984 are far away from their neighbors. This is a problem, because

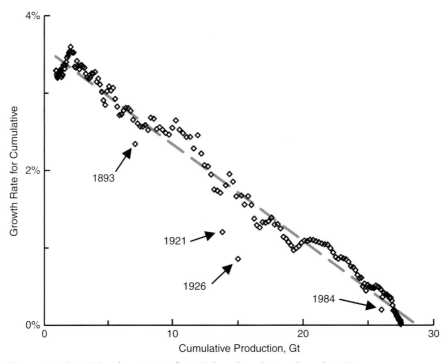

Figure 3.9 The Hubbert linearization for British coal production through 2017.

a regression line minimizes the squares of the distances from the line. Minimizing the squares of the distances rather than the distances themselves is convenient mathematically. Equation solvers converge reliably and many useful statistical measures can be applied. However, it does give the strike years a much larger influence than the other points. It would be possible to do the calculation without these points, but then one would need to decide on criteria for discarding a year. For example, 1974 was also a significant strike year. The drop in production was not large, but the government of Prime Minister Edward Heath was brought down. More fundamentally, the Hubbert linearization does not give us a natural way to make projections for t_{10} and t_{90}.

3.3.2   The Logit Transform Linearization

The logit transform is an alternative way to linearize the logistic function that overcomes many of the limitations of the Hubbert linearization. To develop the logit transform, we rewrite the Verhulst differential equation, Equation 3.8, in the form

$$\frac{q'(t)}{q(t)\big(1-q(t)/U\big)} = a. \qquad\qquad 3.11$$

This equation can be integrated and the result written as

$$-\ln\big(U/q(t)-1\big) = a\big(t-t_{50}\big). \qquad\qquad 3.12$$

It is a good exercise to differentiate Equation 3.12 as a check. The parameter t_{50} is a constant of integration, defined by

$$q(t_{50}) = U / 2. \tag{3.13}$$

The *logit transform* $L(x)$ is defined as

$$L(x) \equiv -\ln(1 / x - 1). \tag{3.14}$$

We can rewrite Equation 3.12 in terms of the logit transform as

$$L(q / U) = a(t - t_{50}). \tag{3.15}$$

If we plot L on the y-axis and t on the x-axis, we have the equation of a line with slope a and x-intercept t_{50}. The parameters a and t_{50} are alternatives to t_{10} and t_{90} in writing the logistic function in Equation 2.4. It is useful to derive equations for converting back and forth between the two sets of parameters. We can write two equations

$$L(0.1) = a(t_{10} - t_{50}) \tag{3.16}$$

$$L(0.9) = a(t_{90} - t_{50}) \tag{3.17}$$

which give us

$$-\ln(9) = a(t_{10} - t_{50}) \tag{3.18}$$

$$\ln(9) = a(t_{90} - t_{50}) \tag{3.19}$$

that we can solve for t_{10} and t_{90} in terms of the x-intercept t_{50} and the slope a as

$$t_{10} = t_{50} - \ln(9) / a \tag{3.20}$$

$$t_{90} = t_{50} + \ln(9) / a. \tag{3.21}$$

The 10% to 90% lifetime t_l is given by

$$t_l = t_{90} - t_{10} = \ln(81) / a. \tag{3.22}$$

Now we go the other way. The formulas for t_{50} and a in terms of t_{10} and t_{90} are given by

$$t_{50} = \text{average}(t_{10}, t_{90}) \tag{3.23}$$

$$a = \frac{\ln(81)}{t_{90} - t_{10}} = \frac{\ln(81)}{t_l}. \tag{3.24}$$

Finally, we can rewrite the logistic function, Equation 2.4, in terms of a and t_{50} as

$$q(t) = \frac{U}{1 + \exp\left(a\left(t_{50} - t\right)\right)}. \tag{3.25}$$

To make the projection for U, we vary U to maximize R^2 for L and t. Mathematically, varying U changes the curvature of the plot for L. When U is too small, the plot curves upward. When it is too large, L bends down. The Excel function call is $RSQ(L_U, t)$, where t is the set of cells with time values and L_U is the set of cells with logit transform values evaluated at a particular value of U. It is convenient to find the projection for U using Excel's Solver program. The values of R^2 will be quite close to 1, so the convergence in Solver should be set to a small value like 10^{-13}.

As an example, we return to nineteenth-century British coal production. Figure 3.10 shows the logit transform from 1854 to 1900 for $U = 22.7$ Gt, the value that gives the maximum value of R^2, 0.999987. Maximizing R^2 brings the curvature close to zero. The actual cumulative production in 2017 was 27.5 Gt. The x-intercept of the regression line, t_{50}, is 1913 and the slope a is 0.039/y. From Equation 3.21, we get the historical projection for t_{90}, 1970. The actual year that cumulative production reached 90% of the 2017 cumulative production was three years later, in 1973.

We associate t_{90} with falling production that encourages switching to different energy sources. The production in 1973 had dropped to 45% of the 1913 peak.

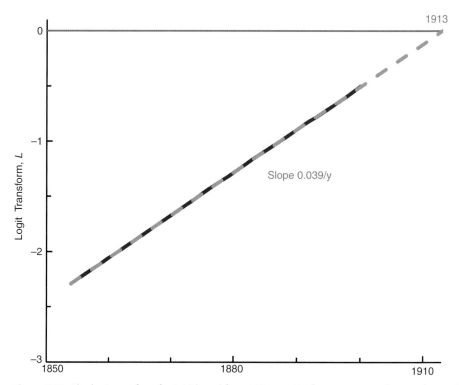

Figure 3.10 The logit transform for British coal from 1854 to 1900 for $U = 22.7$ Gt. The transform values are overlaid with the dashed gray trend line. R^2 is exceedingly close to 1, so the transform values fall close to a straight line.

It also can be a time of social stress because of lost mining jobs. There were strikes by Arthur Scargill's National Union of Mineworkers against the Conservative governments of Edward Heath in 1972 and 1974 and against Margaret Thatcher's government in 1984. The miners' defeat left a legacy of bitterness that lingers today. Moreover, the strikes encouraged the electrical utilities to switch to North Sea gas. Electricity providers place a high importance on having a secure supply because their customers expect the grid to work continuously.

In Figure 3.11 the projection calculation is repeated with the ending year varied. When we plot U versus the ending year, it shows the historical evolution of the projection for ultimate production. This can be helpful in giving us a range for projections. It is convenient to automate this kind of repetitive calculation within Excel using VBA.[1] The plot is unstable in the early years, starting at very low values. This section is shown with a dashed line. Eventually the plot stabilizes as a decaying oscillation. This is plotted with a solid line, starting in 1871 and continuing to 2017. The year 1871 was the year that a Royal Commission on Coal Supplies published a comprehensive reserves study. In Section 4.5, we will compare the

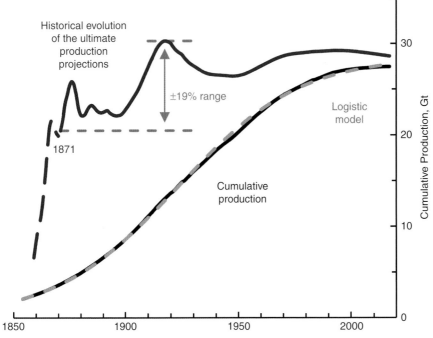

Figure 3.11 Cumulative production compared with the logistic model and with the historical evolution of the ultimate production projections calculated from the logit transform. The logistic model is calculated with the logit transform projections for production data through 2017.

---

[1] VBA (Visual Basic for Applications) is a programming language within Excel. Programming in VBA is extremely productive compared to traditional programming languages because the Excel cells and functions that users are already familiar with are available as function calls. It also takes advantage of the fact that the data are often initially in Excel format.

evolution of the projections with reserves. The range for U is 20.5 Gt to 30.3 Gt, ±19% in percentage terms. We will see in Section 4.6.3 that this range is typical for the mature coal regions. Given the closure of the British coal mines, it is clear that the range will capture the actual ultimate production.

The logit transform allows the projection for ultimate production U to be calculated as a single-parameter optimization. The time parameters t_{10} and t_{90} are found from the slope and intercept of the transform's regression line. A different approach to characterizing future production, complementary to these models, is based on geological measurements of fossil-fuel deposits.

3.4   Reserves, Resources, and Occurrences

Sometimes there is evidence of a fossil-fuel deposit at the surface. The tar pits of Los Angeles hint at the oil fields below. The Pittsburgh coal seam outcrops on the cliffs above the Monongahela River. However, planning production from individual wells and mines requires geological measurements by drilling wells and taking core

Figure 3.12 Photograph of a natural-gas shale core from the Barnett Shale. Credit: Professor Peter Fleming at the University of Texas at Austin.

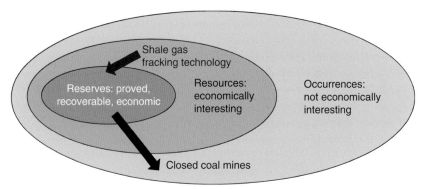

Figure 3.13 Reserves, resources, and occurrences.

samples. Figure 3.12 shows a core from the Barnett shale in north Texas, ground zero for the fracking revolution. The black color is a sign of organic material.

The language that companies and government agencies have used to characterize hydrocarbon and coal deposits has changed over time. The general term is *resources*. We distinguish between three specific classes: reserves, resources, and occurrences. The word "resources" here is both a general term and a specific class. One needs the context to figure out which is meant. The relationship between the three classes is shown in schematic form in Figure 3.13.

The most restrictive category is *reserves*. There are three requirements to qualify as reserves. The deposits must be *proved, recoverable,* and *economic.*

1. Proved. There must be a measurement within a specified distance. For example, for American coal reserves the distance to the measurement point must be 1/4 mile or less to qualify.
2. Recoverable. It must be legal to produce the coal or hydrocarbons. For example, there are restrictions in national parks. In addition, reserves are quoted on a recoverable basis, allowing for coal and hydrocarbons that are left behind when the production is done. In most oil fields more than half of the original oil in place remains in small drops adhering by surface tension to the rock. In a coal mine, pillars of coal may be used to support the roof and these are often left when the mine closes.
3. Economic. One can make money producing the deposits. This depends on geological factors like the thickness and depth of a coal seam, and the permeability and porosity of oil reservoir rock. It also depends on non-geological factors like the coal and oil prices, the cost of transportation, government taxes and subsidies, the skill and the price of labor, and the cost of capital to develop the infrastructure. In practice, many of these factors may not actually be considered in calculating reserves. In particular, national coal reserves have traditionally used simple fixed criteria for seam thickness and depth for decades.

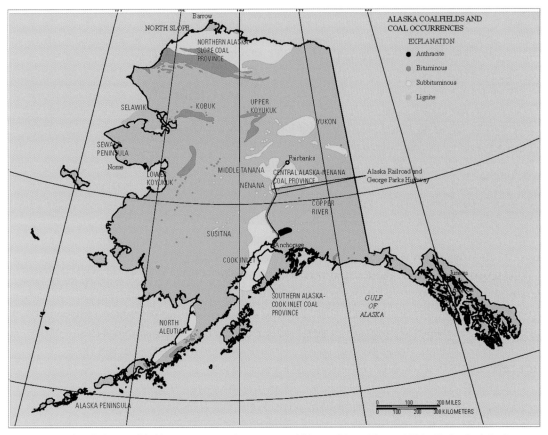

Figure 3.14 The Alaskan coal fields. From Romeo Flores, Gary Strickler, and Scott Kinney, 2004, *Alaska Coal Geology, Resources, and Coalbed Methane Potential*, US Geological Survey DDS 77.

For *resources* we only require that the deposits be of economic interest. In practice, resources may be quite speculative, based on measurements hundreds of kilometers away. Resources can be converted to reserves. Prices may rise, there may be new measurements close by, or new technology may become available. For example, the development of hydrofracturing technology has caused some natural gas resources to become reserves. Another example is oil from the Canadian tar sands. These deposits used to be considered resources, but new production technology based on steam injection led to a reassessment. Much of the tar sands are now considered reserves. One warning is that journalists and politicians often do not distinguish between reserves and resources.

Occurrences refer to deposits that are not economically interesting. An example of an occurrence would be coal on the Alaska North Slope that is so remote that it is unlikely to be mined (Figure 3.14). To the south, the region is hemmed in by national parks. The nearby ocean is not navigable most of the year because of pack ice. It would be exceedingly difficult to develop trains to move the coal because of permafrost. It is common for reserves to be converted to occurrences. When underground coal mines shut down, it is appropriate to classify the coal that is left behind as an occurrence. One should be aware that in much of the professional literature no distinction is made between resources and occurrences.

Not all technical improvements lead to an increase in reserves. As an example, horizontal drilling can increase the contact length of a well with an oil reservoir by a large factor. This gives the ability to produce the oil faster, but it may not change the ultimate production of the field. The longwall mining system for coal greatly increases safety and productivity and for these reasons it dominates underground coal production. However, many seams are not regular enough for longwall mining and as a result may never be produced.

Several organizations publish tables of oil and gas reserves. These do not differ significantly, and we will use the ones in the BP *Statistical Review*. Coal reserves have had a more organized survey process. The first detailed survey of world coal reserves was published in 1913 for the 12th International Geological Congress in Toronto. The reserves were updated for the World Power Conference in 1924 and then again in 1929. Beginning in 1936, this was put on a more regular basis with a series of *Statistical Yearbooks*. These surveys used the same criteria for seam thickness and depth as the British Royal Commission of 1871. The work continued under the World Energy Council (WEC), published as *Surveys of Energy Resources*. In recent years each country has set its own standards for reserves. The last survey was published in 2013, 100 years after the Toronto meeting.[2] The WEC surveys are the most important primary reference for the evolution of world coal reserves, and the results have often been simply copied by other agencies.

3.4.1   Original Reserves

We will often make comparisons between projections for the ultimate production U and reserves, or more precisely, original reserves. *Original reserves* refer to the oil, gas, and coal in the ground before production began. No deduction is taken for the production that follows, although adjustments are made to reflect improved geological surveys. In the modern resources literature, reserves are adjusted for the current cumulative production. We can relate the two by the formula

$$R_o = R + q \qquad\qquad 3.26$$

where R_o are the original reserves, R are the reserves, and q is the cumulative production. In applying this formula we need to consider that both R and q evolve

---

[2] From 1998 to 2010 the World Energy Council *Surveys of Energy Resources* were done at three year intervals by Alan Clark and Judy Trinnaman. Clark and Trinnaman set a high standard and they retired in 2010. The coal reserves were updated for the 2013 World Energy Council report. However, no new survey was done for the 2016 report. The world coal reserves of 891,531 Mt from the 2013 report were simply repeated in the 2016 report.

over time and we need to evaluate them at the same time. For example, if we are quoting year-end 2014 reserves, we also need to calculate the cumulative production through the end of 2014. In the early reserves literature, it was conventional to use original reserves. In principle, original reserves are not affected by the ongoing production. However, they do change with new discoveries and revisions from new surveys. Both original reserves and reserves are useful concepts. The advantage of the modern convention for reserves is that we can make comparisons with projections for future production. We write

$$f \Leftrightarrow R \qquad\qquad 3.27$$

where f is the projection for future production defined by Equation 3.7 and the symbol $\Leftrightarrow$ indicates a comparison. We will compare original reserves with the projections for ultimate production U, writing

$$U \Leftrightarrow R_o. \qquad\qquad 3.28$$

The quantity on the left is derived from the production history, while the quantity on the right is based on geological measurements. In practice U does not have to be close to R_o. For example, the British original reserves established by the 1871 Royal Commission on Coal Supplies amounted to 148.8 Gt, immensely larger than our 1900-based projection for U, 22.7 Gt and the actual 2017 cumulative production,

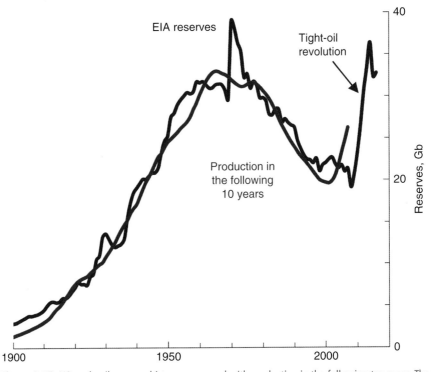

Figure 3.15 US crude-oil reserves history, compared with production in the following ten years. The data come from the EIA.

27.5 Gt. We will see that this is typical for coal. It reflects the fact that coal is easy to find, but production needs significant capital for infrastructure. In addition, the society must accept the safety risks in mining and the environmental impacts.

3.4.2   Oil Reserves History

For oil and gas, it is the opposite. Typically U is greater than R_o. By comparison with coal, oil and gas are hard to find, but easier to produce. Figure 3.15 shows the history of US crude-oil reserves. For comparison a curve is drawn that shows the crude-oil production in the following ten years. The close correspondence between the two curves means that we should think of US crude-oil reserves as a kind of warehouse with a 10-year supply. As oil is produced, there is ongoing development to convert resources to reserves.

Governments determine reserves at the national level. This introduces a political component to reserves. It is important to appreciate that there is no penalty when a government overstates reserves. This is in stark contrast to the situation for a private company. In 2004, the Shell Oil Company admitted that it had overstated reserves by 25%. In the aftermath, the managing director was fired, the stock price fell, and the company made large payments to unhappy investors. Initially the oil fields in OPEC (Organization of Petroleum Exporting Countries) were developed as concessions by private foreign oil companies and their reserves were calculated according to the rules of the national regulators in the home countries of

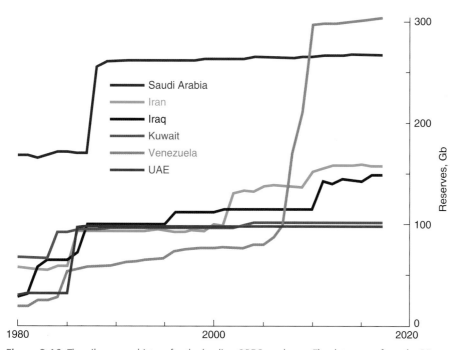

Figure 3.16 The oil reserves history for the leading OPEC producers. The data come from the BP *Statistical Review.*

the oil companies. However, in the 1970s, the OPEC countries nationalized their wells. The result was enormous unjustified increases in reserves. The history is shown in Figure 3.16. In addition, reserves should be adjusted downward for production, but OPEC reserves do not go down. One can see jockeying for position between Iran and Iraq. Venezuela is particularly problematic. The country claims the largest oil reserves in the world in 2017, and the R/P ratio is stupendous, 400 years. However, production has been falling at a 10-year annualized rate of 4%/y. Knowing Venezuela's reserves tells us nothing about future production.

The OPEC reserves increases are extremely significant, accounting for 69% of the world increase from 1980 to 2017. In spite of these problems, the practice has been for groups that tabulate reserves like the BP *Statistical Review* to use the numbers provided by the national governments. The problem is that the underlying data are considered state secrets, and there is simply no way an outside group can make credible independent reserves estimates.

3.4.3 Coal Resources History

The criteria for the resources category are often loosely applied, and historically resources have been subject to arbitrary changes. Figure 3.17 shows a history of coal resources developed by the German resources agency BGR. In the figure,

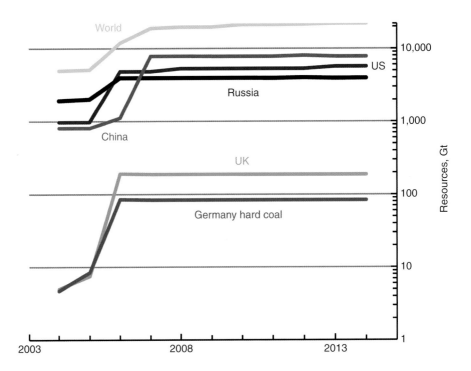

Figure 3.17 The assessment of coal resources by the BGR (Bundesanstalt für Geowissenschaften und Rohstoffe), the resources agency of Germany. The scale is logarithmic. The data come from various editions of the BGR *Energy Resources Report*.

British coal resources rose from 8 Gt to 191 Gt. For perspective, 191 Gt is half the cumulative production of the world. But the UK has no deep underground mines left. This coal should be classified as an occurrence rather than a resource. US coal resources rose from 800 Gt to 8,000 Gt. About half of this amount, 3,600 Gt, is coal north of the Brooks Range in Alaska. It is unlikely that a significant amount of this coal will be produced.

3.5 An Economic Model

Next we consider an economic model. As the Great Recession was unfolding in 2009, economists Christina Romer and Jared Bernstein analyzed the effect of a plan to increase American federal spending by 775 billion dollars. Romer was the incoming chair of the Council of Economic Advisors. The model results are shown in Figure 3.18. With the proposed spending under the plan, the maximum unemployment rate would be limited to 7.6%. Without the additional spending, unemployment would rise to 8.9%. Congress did actually appropriate 800 billion dollars for the plan. By any measure this was an extraordinary amount of money,

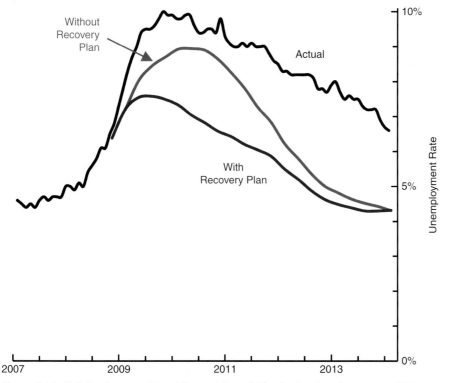

Figure 3.18 Christina Romer and Jared Bernstein's model for the American Recovery and Reinvestment Plan. From Christina Romer and Jared Bernstein, 2009, "The job Impact of the American Recovery and Reinvestment Plan." The actual unemployment data come from the St. Louis Fed.

$2,600 for every man, woman, and child in America. It was the largest program of its kind in American history. It may also be the best opportunity to compare the output of an economic model with what actually happened afterward.

The actual unemployment data are plotted on the same graph for comparison. The actual unemployment peaked at 10.0%. It was consistently higher than the modeled unemployment without any recovery plan. Over the five-year period from 2009 to 2013, the actual unemployment was 2% higher than the model. The labor force at this time was 150 million people so this represented an additional 3 million people out of work. One can easily come up with hypothetical situations where government spending increases unemployment rather than reduces it. For example, a typical expenditure under the American Recovery and Reinvestment Plan would be to resurface a street. The initial stage would be to rip up the street. This is a mechanized process that needs few workers. After this there would often be a delay to wait for the next funding increment. Ripping up a street in a commercial area makes it difficult for customers to get to the stores. Any construction delays cause sales to drop and the stores to lay off employees, possibly more than the number that were hired to rip up the street. As another example, the additional spending included many research grants to universities. Under a research grant, a university might hire a skilled engineer from industry as a research assistant for a fundamental research project. If the engineer had stayed with the company instead, that same engineer might have designed a new product that would have led to greater sales for the company and the company would have hired factory workers to manufacture the product. The fundamental research project might also lead to new jobs, but many years in the future rather than within the time frame of the American Recovery and Reinvestment Plan.

If this were a science-based model, like a computer program that predicts the position of a planet, or an engineering model like a circuit simulator that predicts the transmitter power for a cell phone, this large disagreement between theory and experiment would lead the modelers to acknowledge that the models are wrong. Physicist Richard Feynman expressed forcefully his views about this in a 1964 lecture series at Cornell University, "The Character of Physical Law."

> In general we look for a new law by the following process. First we guess it. Then we compute the consequences of the guess to see what would be implied if this law that we guessed is right. Then we compare the result of the computation to nature, with experiment or experience, compare it directly with observation, to see if it works. If it disagrees with experiment it is wrong. In that simple statement is the key to science. It does not make any difference how beautiful your guess is. It does not make any difference how smart you are, who made the guess, or what his name is – if it disagrees with experiment it is wrong. That is all there is to it.

When an economic model fails, there may be no public acknowledgement. On the contrary, supporters of the plan may argue that result would have been even worse without the plan. From a political perspective the model result and the excuse for its failure only need to be plausible. In fact, politicians might consider the model to be a success because it allowed them to distribute money in visible projects to people who might be encouraged to vote for them in the future. On the other hand, from our modeling perspective this means that it may be appropriate to treat the outputs of economic models as scenarios rather than projections.

3.6 Statistical Formulas

In this section, we list for reference the formulas underlying the statistical quantities that are used in this book and the corresponding Excel functions. They are not derived here, but the derivations can be found in William Navidi's excellent statistics book that is listed as a reference at the end of the chapter. We consider two series x_i and y_i, where i is an index. The average (or mean) value is written as $\bar{x}$, read as "x bar," and defined as

$$\bar{x} \equiv \left(\sum_i x_i \right) / n. \tag{3.29}$$

The Excel function call is AVERAGE (**X**), where **X** is the set of cells for x. The variance is written as σ_x^2, and defined as

$$\sigma_x^2 \equiv \left(\sum_i \left(x_i - \bar{x} \right)^2 \right) / n. \tag{3.30}$$

The variance is a measure of the spread in the data. It is non-negative. In Excel, this is calculated from the population variance function VAR.P(**X**). We write the regression model value $\widehat{y}_i$, read as "y hat," for the observed value y_i as

$$\widehat{y}_i = \bar{y} + s \left(x_i - \bar{x} \right) \tag{3.31}$$

where s is the slope. The *residual* d_i is defined as the difference between the observed value y_i and the model value $\widehat{y}_i$, given by

$$d_i \equiv y_i - \widehat{y}_i. \tag{3.32}$$

The regression calculation minimizes the sum of the squares of d_i. The slope s is given by

$$s = \frac{\sum_i \left(x_i - \bar{x} \right) \left(y_i - \bar{y} \right)}{\sum_i \left(x_i - \bar{x} \right)^2}. \tag{3.33}$$

In the numerator the terms where x_i is far away from the mean $\bar{x}$ are weighted more heavily than those that are close to the mean. The same goes for the y_i. The Excel function is SLOPE(**Y**, **X**). Note that the order of x and y in the formula matters in the formula and this reflects the fact that we are using x to make an estimate of y, not the other way around.

The parameter r^2 is defined as

$$r^2 \equiv \frac{\left(\sum_i (x_i - \bar{x})(y_i - \bar{y}) \right)^2}{\sum_i (x_i - \bar{x})^2 \sum_i (y_i - \bar{y})^2}.$$ 3.34

This is the formula for the square of the correlation coefficient, but usually people just say "r-squared." The Excel function call is RSQ(**Y, X**). Note that this formula is symmetric in x and y, so we can interchange the order of the two variables without changing r^2. In addition r^2 is non-negative. It will be left an exercise for the reader to show that we can write r^2 in terms of the variance of the residuals as

$$r^2 = 1 - \frac{\sigma_d^2}{\sigma_y^2}.$$ 3.35

The interpretation is that r^2 is the proportion of the variance of y that is accounted for by the model. The better the model, the closer r^2 is to 1. Conventionally the notation r^2 is used for a single linear regression, while R^2 is used for a more complex model. As an example of a more complex model, we will consider in Section 8.6 an electricity demand model where there are two variables, price and degree cooling days to estimate demand. In this situation, Equation 3.34, which has only the single input variable x, does not work. However, the more general Equation 3.35 is still appropriate as a measure of the quality of the model.

Concepts to Review

- Predictions, projections, and scenarios
- The R/P ratio
- The declining exponential model and the Laherrere linearization
- The logistic model and the Hubbert linearization
- Making projections with the logit transform
- Reserves vs. resources vs. occurrences
- Testing the economic model behind the American Recovery and Reinvestment Plan

Problems

Problem 3.1   Deriving the form of the logistic function
Derive Equation 3.25 from Equation 2.4.

Problem 3.2   Laherrere linearization for the Kern River Oil Field

The Kern River Oil Field is California's third largest oil field, with a cumulative production of more than 2 Gb since production began in 1900. The oil in this field is heavy oil, meaning that it is viscous and does not flow freely. There have been two production peaks. The first was in 1902 at 15 Mb, shortly after production began. In 1962, operators began to inject steam into the wells to improve the recovery. This was successful, and there was a second peak in 1985 at 52 Mb. Production has been declining since then. Before you start, download the Excel file with the production data through 2009 from energybk.caltech.edu.

a. From the production data make the Laherrere plot. Make a projection for U in barrels.

b. Calculate the exhaustion factor e for 2009.

c. Make the projection for future production f for 2009. This could be compared with the California Department of Conservation reserves at year-end 2009 as 569 Mb.

Further Reading

- BGR (Bundesanstalt für Geowissenschaften und Rohstoff). The German resources agency. Their energy studies with reserves and resources for oil, gas, and coal are available at www.bgr.bund.de/EN/Themen/Energie/energie_node_en.html
- Richard Feynman, the late Caltech professor of physics, made vivid the distinction between science and pseudoscience. The transcript of his 1981 BBC interview discussing pseudoscience is available in Richard Feynman, 2005, *The Pleasure of Finding Things Out.*
- Romeo Flores, Gary Strickler, and Scott Kinney, 2005, "Alaska Coal Resources and Coalbed Methane Potential," *US Geological Survey, Bulletin 2198.*
- William Navidi, 2011, *Statistics for Engineers and Scientists*, 3rd Edition, McGraw-Hill. Excellent coverage of line fitting, models, and correlation coefficients.
- Christina Romer and Jared Bernstein, 2009, "The Job Impact of the American Recovery and Reinvestment Plan." We have discussed the predictions for unemployment, but the paper also discusses effects on GDP, individual industries, and gender impacts. This white paper has become scarce online. Write to the author at rutledge@caltech.edu for a copy.
- David Rutledge, 2011, "Estimating long-term world coal production with logit and probit transforms," *International Journal of Coal Geology*, pp. 23–33. More information about using the logit transform to make projections for U and t_{90}. The paper also includes a discussion of the choice of the starting year for the calculation. This paper is available at the journal web site as a free download.
- Pierre Verhulst, 1838, "Notice sur la loi que la population suit dans son accroissement," *Mathematique et Physique*, pp. 113–121. The mathematics is accessible even if one does not read French. This paper is available at energybk.caltech.edu.

4 Coal

Traced from a map, the Coal Line has the raceme structure of a bluebell or a lily of the valley, as dainty an image as nature can provide for a stem whose flowers are coal mines. Black Thunder Junction, 5:45 P.M., nineteen degrees, dark, and snowing.

John McPhee "Coal Train"

4.1 Introduction

Coal is a black rock that burns (Figure 4.1). The surface of coal can be shiny, and in the 1800s it was popular to polish it to make a gemstone called *jet*. Queen Victoria had a jet mourning necklace. Figure 4.2 shows a representative portion of the chemical structure of coal. It is formed of irregular sheets of linked carbon rings. When coal burns the carbon atoms are oxidized to form carbon dioxide and the hydrogen atoms produce water vapor. There is also sulfur, which produces sulfur dioxide, a major pollutant. Coal reigned as the most important energy source from 1888 through 1956 (Figure 1.4). During this time coal provided double the energy of wood and water (Figure 1.6). Coal-powered locomotives and ships enabled fast long-distance transportation over land and sea for the first time. Coal stoves allowed people to heat their homes comfortably. Steam engines powered by coal made factory mass production possible. For the first time the average family in the coal-producing countries could buy factory-made clothing, dishes, and furniture.

4.1.1 Coal consumption

Coal has been superseded by oil for transportation and by natural gas for heating. Its main use today is for electricity generation. World coal consumption per person was steady from 1913 to 2000 at 0.8 t/p (Figure 4.3). Since 2000, the consumption has risen to around 1.0 t/p. This increased consumption is associated with electrification in lower-income countries, particularly China and India, which were the largest consumers of coal in the world in 2017. The advantages for

Figure 4.1 A piece of anthracite, hewed by hand by Andrea Schiappa. Anthracite is black, but a fracture surface can be smooth enough to act like a mirror. The first significant coal production in the United States was from the anthracite fields of eastern Pennsylvania. Photograph by the author.

Figure 4.2 A representative portion of the molecular structure of coal. The black balls represent carbon atoms, the white ones are hydrogen, the red are oxygen, the blue nitrogen, and the yellow sulfur. Photograph by the author.

lower-income countries are that coal is the cheapest fossil fuel and it requires the least infrastructure.

Figure 4.4 shows a comparison of per-person coal consumption in the OECD countries with non-OECD consumption. The curves are converging. The consumption ratio was 3.7:1 in 2000 and 1.5:1 in 2017. The OECD per-person consumption has dropped sharply, from 1.0 toe/p in 2000 to 0.7 toe/p in 2017. This is associated

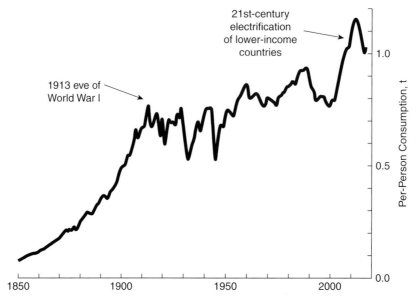

Figure 4.3 World per-person coal consumption. This is the same data as in Figure 1.6, but plotted in production units rather than energy units.

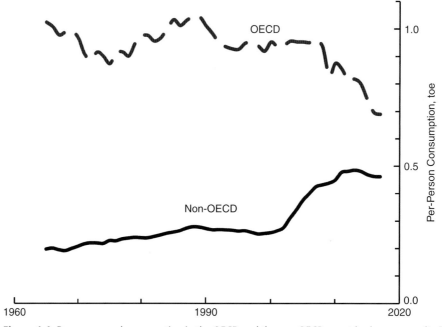

Figure 4.4 Per-person coal consumption in the OECD and the non-OECD countries in energy units. The consumption data come from the BP *Statistical Review* and the population data come from the United Nations Population Division.

with switching to natural gas and the new alternatives for electricity production. At the same time, consumption in non-OECD countries has risen from 0.26 toe/p

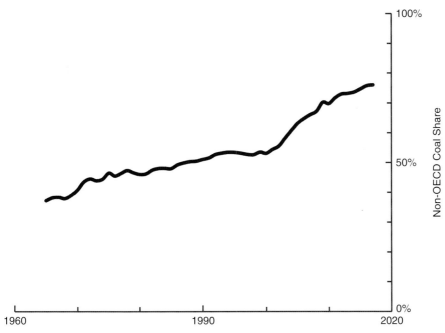

Figure 4.5 Non-OECD share of world coal consumption. The data come from the BP *Statistical Review.*

to 0.46 toe/p, and because the non-OECD population is larger, world coal consumption overall has increased rather than fallen.

Collectively the non-OECD countries completely dominate world coal consumption, with a 76% share in 2017, up from 53% in 2000. This is shown in Figure 4.5. This dominance limits the effect of policies aimed at reducing carbon-dioxide emission by reducing coal consumption in OECD countries. It should also be appreciated that even with the recent drop in coal consumption in OECD countries, consumption is still larger on a per-person basis than in the non-OECD countries.

4.1.2 Coal in Electricity Generation

Coal has been the largest source of energy for generating electricity for many years. Figure 4.6 shows the shares of world electricity generation from 1985 to 2017, characterized as coal, hydrocarbons, and alternatives. It is remarkable how little the shares have changed during the time of this series. Coal was 38% in 1985 and it was 38% in 2017. Similarly, hydrocarbons were 26% in 1985 and 27% in 2017, while alternatives were 36% in 1985 and 35% in 2017. A major advantage of coal for generating electricity is that it can be stored at the plant. Figure 4.7 shows the Navajo Generating Station in Page, Arizona. The plant supplies electricity to the states of Nevada, Arizona, and California. In the background is Lake Powell, which provides cooling water. The coal comes from the nearby Black

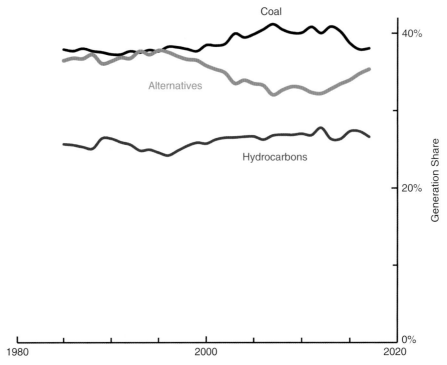

Figure 4.6 World electricity generation shares from the BP *Statistical Review*.

Figure 4.7 The Navajo Generating Station in Page, Arizona in 2009. This is the largest coal plant west of the Rocky Mountains, with a capacity of 2.3 GW. The operating utility has judged that it is too expensive to install the pollution control equipment required to meet the Environmental Protection Agency (EPA) rules for reducing haze at the Grand Canyon, and the plant is scheduled for shutdown in 2019. Photograph by the author.

Figure 4.8 Coke for steel making. Photograph by the author.

Mesa Mine on the Navajo Reservation. The coal stocks are in the foreground. Typically, utilities store a month's supply of coal. In contrast, natural gas storage requires an extensive infrastructure of caverns, wells and connecting pipelines. For alternatives, it is possible to store energy electrically in batteries, but they are expensive.

Coal is also a critical industrial chemical. Iron ores are primarily iron oxide, and the challenge in producing steel is to reduce the iron ore without adding impurities like sulfur that ruin the steel. By far the most important reducing agent is *coke*, a stiff black porous material made by heating coal to drive off volatile organic material and impurities (Figure 4.8). The carbon in the coke is both the reducing agent and the fuel for heating to drive the reaction. Another interesting early application of coke was roasting malt for beer. The beer acquired a wonderful honey brown color with no taste of smoke. This is the celebrated English bitter. The residue from making coke from coal is a mixture of organic compounds called *coal tar*, which provided the raw material for the early European dye industry.

Coal mining has played an important part in our social and legal history. The leading industrial powers, like the United Kingdom, Germany, the United States, and Japan, all started their development with coal. China and India are following this path today. Compared with the hydrocarbons, coal is easy to find. It is often visible as a gray stripe outcropping on cliffs. However, there is nothing easy about going down into an underground mine and hewing coal. The work is difficult and dangerous, but it has also been respected and well paid. Sons followed fathers and grandfathers down into the mines. Coal towns provided strong social support.

The experience of underground coal mining promotes intense loyalty among miners, and they were the first to organize labor unions in the United Kingdom and in the United States. Burning coal produces smoke, sulfur oxides, and nitrogen oxides. The first environmental laws to control air pollution were for coal. Finally, coal is not forever. For various reasons, sometimes geological exhaustion, sometimes environmental regulations, sometimes fading social support, societies move on from coal. Coal has had no second acts comparable to the fracking revolution in hydrocarbons.

The first section in this chapter is on the origins of coal. Then we discuss mining technology. This will be followed by a look at environmental impacts of mining and combustion. There has been a long-standing assumption that there is an enormous supply of coal, sufficient for several centuries even with increasing production. Because of this assumption, coal dominates future carbon-dioxide emissions in climate scenarios. We will see that the history contradicts this assumption. Historically, countries complete the full cycle of coal production while only producing a small fraction of their early original reserves. To understand this behavior, we will study the history of coal in Britain, where the cycle is complete. Logistic models have worked well for coal production. The goal of the final section of the chapter is to make projections for world coal production and to make comparisons with reserves.

4.2   Origins

4.2.1   Formation of Coal

The formation of coal is outlined in Figure 4.9. Coal begins on land with plants (Stage 1 in the figure). This is in contrast to oil, which starts with plankton in the sea. When plants die, ordinarily they undergo aerobic decomposition by microorganisms. In this process, the carbon atoms in the plant are oxidized to form carbon dioxide. Aerobic decomposition requires oxygen. However, in some environments, particularly bogs and swamps, not enough oxygen is available to decompose the plants, and the organic material accumulates. This accumulation is called *peat*, the second stage in Figure 4.9. Peat has the potential to become coal. The next step is for the peat to be buried by sediment (Stage 3). This could happen when a river nearby changes direction and dumps sand or mud over the peat. Peat along a coast can be buried when sea level rises during an ice age deglaciation. Ice age cycles can happen repeatedly, so that there might be ten or more peat layers alternating with sand or mud, thousands of meters below the surface. At depth, the temperature and pressure rise. The layers of sand are transformed to sandstone, the mud layers become shale, and the peat changes to coal (Stage 4).

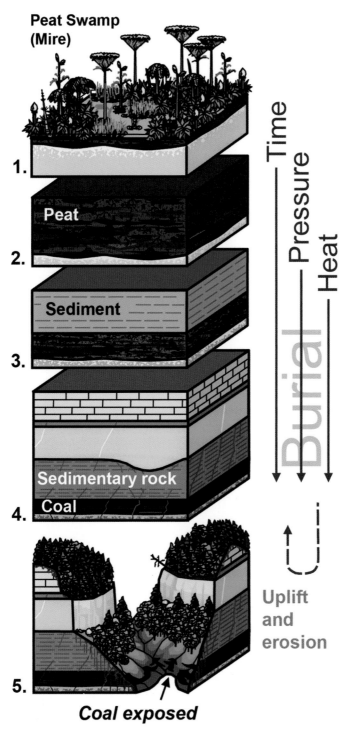

Figure 4.9 The stages of the formation of coal. Credit: Stephen Greb of the Kentucky Geological Survey and the University of Kentucky.

4.2.2 Ranks of Coal

The process of converting peat to coal is called *coalification*. Geologists define ranks that reflect the structural changes in the coal under heat and pressure over time. Figure 4.10 shows the ranks in terms of carbon content, water, volatile organic compounds, and energy density during the process. Ordered from low to high carbon content, the ranks are *lignite, sub-bituminous, bituminous*, and *anthracite*. Anthracite can be 90% carbon. In Europe it is common to give coal production statistics in two categories, *brown coal* and *hard coal*, where brown coal includes lignite and sub-bituminous coal, and hard coal includes bituminous coal and anthracite. As we move from lignite through sub-bituminous to bituminous, the water content drops and the energy density rises. The energy density peak at 36 GJ/t is actually for bituminous coal rather than anthracite, which has higher carbon content. The reason for this is the hydrogen content of the volatile organic compounds in the bituminous coal. Hydrogen forms water when it burns and it has a higher energy density than carbon. If we go past anthracite, we get graphite. Graphite is a useful material, but it is not a good fuel because it has a high ignition temperature. Pencil leads are made of graphite. It also provides a low-friction electrically conducting surface. The contacts in DC motors, called brushes, are made of graphite.

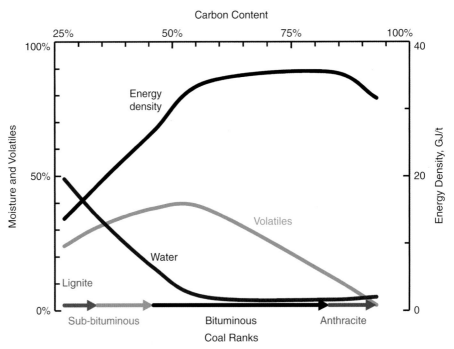

Figure 4.10 The coalification process. The data come from Paul Averitt, 1974, "Coal Resources of the United States," *US Geological Survey Bulletin 1412*.

Lignite is used entirely for electricity generation, primarily in Germany, which has been the largest producer. It has a low energy density and it tends to catch on fire spontaneously. This makes it difficult and expensive to transport. For these reasons lignite power plants are usually built close to the mines. Sub-bituminous coal, like lignite, is used almost exclusively for electricity generation, most importantly in the US. There are extensive low-sulfur seams more than 20 meters thick in the Powder River Basin in Wyoming (Figure 4.11). These were laid down 60 million years ago during the Paleocene epoch. The seams are shallow and the coal can be extracted inexpensively and safely by surface mining. The coal travels by unit trains directly from the mine to power plants all over the eastern United States. Bituminous coal production dominates in the rest of the world. The majority of the bituminous reserves were formed during the Carboniferous Period 300 million years ago. Bituminous coal has a higher energy density than lignite and sub-bituminous coal and this reduces the transportation cost. It is used for electricity, and in lower-income countries for heating and cooking. Coke is made from bituminous coal. Anthracite, the highest rank of coal, burns almost without smoke. A common use for anthracite today is in heating homes. In the era of steam ships, navies often chose anthracite as a fuel so that the exhaust would not be seen by their enemies.

Figure 4.11 A wall of sub-bituminous coal at the North Antelope Rochelle Mine in Wyoming, in the western United States. The people and the car at the lower left give the scale. This mine is the most productive in the United States, producing 102 Mt in 2017, 13% of the production for the entire country that year. Credit: US Bureau of Land Management (BLM).

4.3   Mining Coal

Figure 4.12 is a photograph taken in an underground anthracite mine in 2006 in eastern Pennsylvania. In this region, anthracite is used for heating homes. The coal is sold in pieces 5 to 10 cm in diameter. Coal is softer than most rocks, and coal miners ordinarily work within a coal seam. This was a small mine where the coal was hewed by hand. The miner with the pick cuts a slot, or *kerf*, at the bottom of the seam. The driller in the background then makes a shot hole for explosives. The kerf is required for safety. It gives room for expansion during the blast and it allows the weight of the coal to help in knocking it loose. This is an uncomfortable, dangerous way to spend your life. You never get to stand up and without the lamps, it is utter darkness. The mine was operated by the S&M Coal Company, and it produced 3,000 t per year with three miners. It closed in 2014.

4.3.1   Longwall Mining

Today most coal is produced in highly mechanized underground mines. Figure 4.13 shows a radio-controlled shearer in operation. On the left, a rotating drum fitted with picks breaks up the coal and the coal falls onto a conveyor belt. The operator can move the drum up and down across the seam face and along the conveyor belt. In this mine, the height of the cut is 5 m. The exposed coal face is called the *longwall*, and this system of mining is called *longwall mining*.

Above the operator's head are a line of roof supports called *shields*. The shields pull themselves forward hydraulically in interleaved waves by steel loops attached to the conveyor belt. When the shearer reaches the end of the belt, hydraulic pistons push the shearer and the belt forward to set up the next pass. As the shields advance the roof collapses behind them. The collapsed material is called the *gob*. One problem with longwall mining is that the surface above the gob will drop. This *subsidence* is more severe than for hand hewing where supporting coal pillars are left in place. The subsidence causes cracks in buildings and roads.

Figure 4.14 shows the layout of a longwall mine. First a rectangular block of coal is selected for mining. The block is called a *panel*. The panels in the colliery in Figure 4.13 were 300 m wide by 2,500 m long and they included 3 Mt of minable coal. Two access tunnels are dug along the long sides of the panels. There must be two tunnels rather than one to enable ventilation along the face and to allow the miners to get out if there is a collapse in one of the tunnels. Ventilation is critical in a coal mine to prevent methane explosions. The ventilation requirement is the ultimate limit on how far miners can go from the shaft entrance, and it is this distance that determines the reserves of a mine. Ten kilometers is typical. A conveyor belt is set up in one of the tunnels to transport the coal to the shaft. The shearer starts at the end of the tunnels and works backwards toward the beginning of the panel. The shields greatly improve the safety of the mine, compared with

Figure 4.12 Bob Hubler (left) and Justin Koperna (right) in the Buck Mountain Slope anthracite mine, Pennsylvania in 2006. Credit: Christian Abraham.

Figure 4.13 A coal shearer 800 m below the surface in the Daw Mill Colliery, Warwickshire, England. In its heyday, this was one of the most productive mines in the UK, with a maximum production of 2.5 Mt/y. It closed in 2013. Credit: UK Coal.

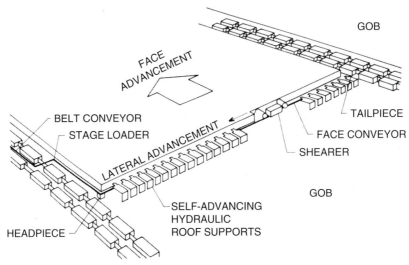

Figure 4.14 Layout of a longwall mine. From Robert Stefanko, *Coal Mining Technology*, the Society of Mining Engineers, 1983.

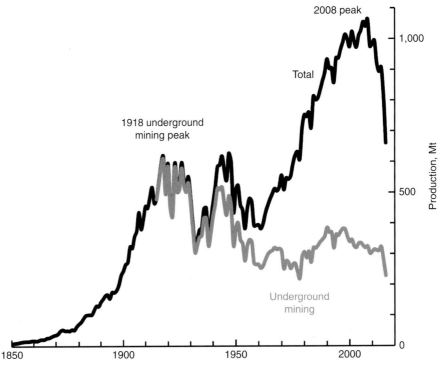

Figure 4.15 American coal mining production from the US Department of Commerce and the EIA.

the anthracite mine in Figure 4.12, where the roof is supported by pillars of coal. In addition, the longwall system is relatively automated, so few people are at risk. One disadvantage of longwall mining is that it is not flexible. The panels need to be relatively flat and regular. However, because of the improvement in productivity and safety, when countries can afford it, they shift to the longwall mining system. As a consequence, however, the ultimate production drops, because much coal that could have been mined by hand is not suitable for longwall production.

4.3.2 Surface Mining

Since World War II, surface mining of coal has become more important, particularly in the United States, where it was 65% of total production in 2016. Underground mining in the US actually peaked a century ago in 1918 (Figure 4.15). The increase in production since then has come entirely from surface mining. Surface mining has major advantages over underground mining. The equipment can be made much larger, and larger explosive charges can be used. For these reasons surface mining is much more productive. The US average for underground mining is 3 t/h, while the surface mining average is 10 t/h. Surface miners are not exposed to roof falls, and they work inside steel cabs. This makes surface mining much safer than underground mining, and also safer than working on oil and gas rigs. However, in surface mining, the rock above the coal, called the *overburden*, is removed temporarily, and this gives it a larger environmental impact than underground mining. The ratio of the thickness of the overburden to the thickness of the coal seam is

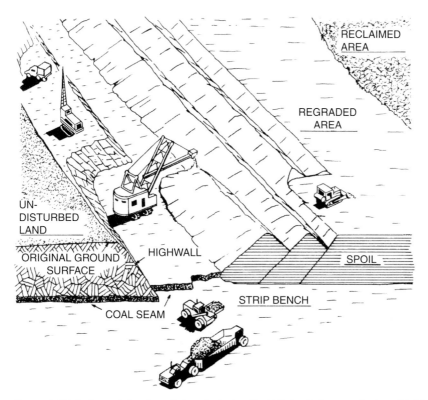

Figure 4.16 Surface mining. From Robert Stefanko, *Coal Mining Technology,* Society of Mining Engineers, 1983.

called the *stripping ratio*. If the stripping ratio is above ten, it is difficult for the coal miners to make money. This means that most of the world's coal is not accessible by surface mining. When countries shift their emphasis to surface mining, the ultimate production drops.

Figure 4.16 shows the layout of a surface coal mine. The first step is to remove and store the topsoil. Then a drilling rig, second from the left in the figure, makes holes in the overburden for explosives to break up the rocks above the coal. A common explosive is a mixture of ammonium nitrate and diesel fuel (ANFO). After the blasting, a shovel, third from the left in the figure, moves the overburden to the right. At the bottom of the figure, a loader puts the coal in a hauler. It is efficient if the overburden is removed directly to an area that was previously mined, so that it becomes the first step in the restoration of the site. The bulldozer on the right makes the surface flat again. Finally the topsoil is returned, and trees and grass are planted.

Restoration is required by government authorities, and it is a significant part of the expense of running a mine. The effectiveness of restoration varies. In mountainous areas, the overburden often ends up in stream valleys, and this alters the drainage pattern. Sometimes it is possible to change the purpose of the land so that it can be used for a new development, like houses or recreational activities. Areas that were previously ranch pastures can ordinarily be restored to run cattle on them again. Figure 4.17 shows reclamation following surface mining in Wyoming. There

Figure 4.17 Reclamation for a surface coal mine in Wyoming. Credit: Ken Takahashi, from the US Geological Survey, 2005, *Coal Assessment Northern Rocky Mountains*, US Geological Survey Professional Paper 1625-A.

is little difference between the reclaimed land and the undisturbed land. This is high plains grassland, sparsely populated, and unbearably cold in winter. Before coal mining, it was poor ranchland. After coal mining, it will be poor ranchland again.

There is a pattern in moving from mining by hand to longwall mining to surface mining. At each step the mining becomes safer and more productive. At the same time, however, the environmental impacts increase, and the ultimate production declines. Surface mining also differs from underground mining as a social experience. People work in separate machines, and their skills in operating loaders, hydraulic shovels, drag lines, dump trucks, and bulldozers are also valuable at gravel pits and in civil engineering projects. Surface mines are often short term, and the miners understand from the beginning that they will need to find another job when the mine closes. This is not the situation for underground mines, where a miner may work at one mine for decades with the same group of people. Underground miners' jobs are highly specialized and well paid, but they often have no close outside equivalents. This means that in a region where underground coal mining is declining, the laid-off miners may not find new jobs at anything like the same pay and status.

4.4 Impacts

Coal is the most accessible of the fossil fuels. It can be mined by pick and shovel, moved in a hand cart, and burned in a stove. However, the environmental impacts are more difficult to manage than for hydrocarbons. Burning coal produces smoke and sulfur dioxide. It is not feasible to reduce the pollution from burning coal in homes and shops, so as countries become richer they ban this use in cities. It is possible to greatly reduce coal air pollution from electricity generating plants, but the equipment to do this must be installed and operational. Mining coal generates a great deal of dirt, coal dust, and water. The water may contain metal compounds that are a risk if the water gets into streams and aquifers. In many situations, it is possible to manage the waste so that it does little damage. However, this costs money, which means that companies have incentives to cut corners. Successful waste management generally requires a good framework of government regulations, together with inspectors who have the authority to shut down production and who do not take bribes. However, it would be a mistake to associate environmental impacts exclusively with private companies. Many of history's worst environmental disasters are associated with state enterprises.

4.4.1 The Aberfan Disaster

The piles of spoil themselves can be a risk. The saddest day in the history of mining involved a colliery waste pile, a *tip* in British English. It happened in 1966 in the town of Aberfan in Wales. The spoil from the Merthyr Vale Colliery had been dumped in tips on a slope above the town. However, there were springs underneath, and the water formed a slippery layer. And one day down it all came in a mighty roar. The spoil swept into the Pantglas Junior School (Figure 4.18), killing 116 children and 28 adults. No miners were hurt, but many of their children were in the school. In this case no private companies were involved because the British government had nationalized all underground coal mines in 1947. This was part of a broader

Figure 4.18 The Aberfan disaster, October 21, 1966. Unknown photographer. From www.codex99.com/photography/images/aberfan/aberfan_1_lg.jpg

Mr. D. Roberts,

Area Chief Mechanical Engineer, National Coal Board,

ABERAMAN, Aberdare

20th August 1963

Dear Sir,

Danger from Coal Slurry being tipped at the rear of the Pantglas Schools

In connection with the above my Public Works Superintendent has been in touch with Mr. Wynne, Manager at the Merthyr Vale Colliery in connection with the deposit of slurry on the existing tip at the rear of the Pantglas Schools.

I am very apprehensive about this matter and this apprehension is also in the minds of the local Councillors and the residents in this area as they have previously experienced, during periods of heavy rain, the movement of the slurry to the danger and detriment of people and property adjoining the site of the tips.

I understand that Mr. Wynne has told my Superintendent that the slurry is 40% dewatered before being tipped but he agrees of course, that this would not be a solution to the movement of the slurry in the winter time due to the absorption of storm water.

You are no doubt well aware that the tips at Merthyr Vale tower above the Pantglas Area and if they were to move a very serious position would accrue.

I should like your observations upon this as soon as possible.

Yours faithfully

D.C.W. Jones

Borough and Waterworks Engineer

Figure 4.19 One of the letters from D. C. W. Jones calling attention to the waste tip danger. This letter and others are available online in a collection for the Aberfan disaster maintained by Professor Iain McLean at Oxford University. The link is www.nuffield.ox.ac.uk/politics/aberfan/home2.htm

nationalization program by the Labour Government of Clement Attlee. The government also took over the steel industry, electricity generation, airlines, railroads, and trucking. In the nationalization, the coal mines were taken away from their owners, who were compensated. The mines were run by the National Coal Board.

The risk of the spoil tips had been recognized and acknowledged by the Coal Board. In 1963 and 1964, Mr. D. C. W. Jones, a civil engineer working for the town, wrote letters to different authorities to alert them to the danger. One is reproduced in Figure 4.19. The letter emphasizes the risk that arose from the combination of the slope, the waste, and water. Mr. Jones had suggested that rain would provide the water. In the actual slide, the water came from springs.

D. Roberts replied the following spring to a later letter from Jones, acknowledging the risk.

With regard to disposing of slurries this is, at present, still being disposed of on the tipping site, via the local tramway but it is our intention to discontinue this and dispose of the slurries mixed with Washery Shale at Plymouth Colliery Site until such time as a new tipping site can be found. As you will appreciate, these tailings are very difficult to handle and we are very careful in disposing of this material, so as not to inconvenience any person or persons and, therefore, we would not like to continue beyond the next 6/8 weeks in tipping it on the mountain side where it is likely to be a source of danger to Pantglas School.

Notwithstanding this letter, the dumping continued right up to the day of the disaster. Even though the government Coal Board was completely responsible for the disaster, no one was disciplined, much less prosecuted. The Board's behavior in the aftermath was atrocious. The Coal Board contributed £160,000 to compensate the victims, but then this was almost completely offset when it extorted £150,000 from a private Aberfan disaster charity fund to move the remaining tips above the town. This was not the British government's finest hour.

4.4.2 Air Pollution from Coal

The smoke from burning coal is irritating to breathe. It reduces visibility and blackens walls. The sulfur dioxide reacts with molecular oxygen and water vapor

Figure 4.20 Coal smoke at the corner of Liberty and Fifth Avenues, Pittsburgh, Pennsylvania, in the 1940s. Courtesy of the University of Pittsburgh/Smoke Control Lantern Slides Collection.

to form sulfuric acid, which causes lung damage and etches stone buildings. Coal smoke is a severe problem in China today, but it was also a problem earlier in the US and the UK. Figure 4.20 is a photograph that was taken in Pittsburgh, Pennsylvania, in the 1940s. At the time Pennsylvania produced more coal than any other state. The photograph was taken at 10:35 in the morning, but the smoke makes it look like night. For healthy people, it is unpleasant to breathe in this environment. For people with poor lung function, it is dangerous.

In London, bituminous coal burning was associated with a mixture of smoke and fog that people called *smog*. Smog was a particular problem when there was a temperature inversion. In a temperature inversion, temperature rises with altitude, rather than the normal decrease. This shuts down vertical convection, and smoke and sulfur dioxide build up over time. The most notable of these accumulations was the Great Smog in December, 1952. Visibility was essentially zero and transportation was shut down. Figure 4.21 shows the evolution of the smoke and sulfur-dioxide concentrations. Thousands of people died. The deaths tracked the rise and fall of the pollution levels.

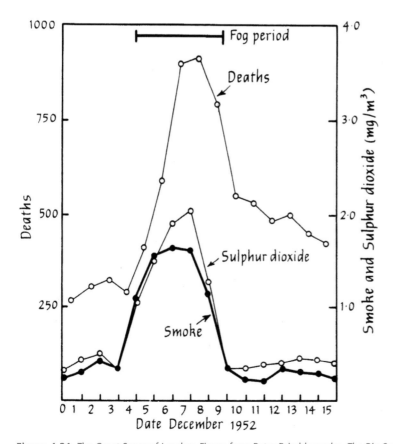

Figure 4.21 The Great Smog of London. Figure from Peter Brimblecombe, *The Big Smoke*, 1987, Methuen and Company.

The sulfur content of coal can be reduced somewhat by washing it with water and drying it before burning. That is because much of the sulfur is in the form of pyrites precipitated in fractures in the coal. However, this only removes part of the sulfur, and it does not stop the smoke. The British government passed the Clean Air Act in 1956. This law allowed local governments to establish zones where smokeless fuels like anthracite were required. In 1972, the London Act allowed the local governments to limit the sulfur content of fuels to 1%. Over time the homes and businesses have shifted from coal to electricity and natural gas for heat. These policies have been successful, and London is smog free today, as is Pittsburgh.

Today most coal is burned in power plants. These can be sited away from cities and fitted with pollution control equipment to reduce sulfur dioxide and nitrogen oxide emissions. Figure 4.22 shows the power plant emissions in the United States over time. Both SO_2 and nitrogen oxides (NO_x) are down dramatically. For SO_2 the reduction from 1990 to 2017 is 89% and for NO_x it is 77%. A part of the sulfur reduction comes from switching from coal to natural gas.

Today China and India have severe air pollution problems and coal is a major contributor. The examples of London and Pittsburgh, as well as the history of sulfur dioxide and nitrogen oxide emissions from American power plants point to the solution. History gives us every expectation that as these countries become richer, they will improve their air quality.

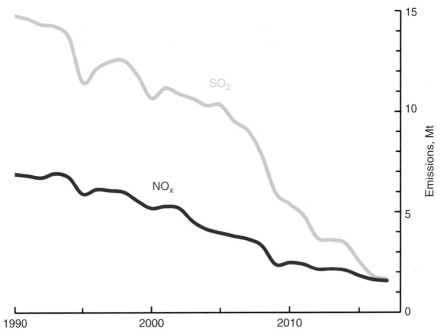

Figure 4.22 Sulfur dioxide and nitrogen oxide emissions from American power plants. The data come from various editions of the EIA, "Electric Power Annual."

4.5   British Coal

The UK was the largest producer of coal in the 1800s. However, it was not the first country to make significant use of coal. The first was China, which has extensive bituminous and anthracite resources. The Chinese invented a process for making coke from bituminous coal. During the Song Dynasty from 960 to 1279, coke was used for smelting iron on a large scale. Song China was then the most industrialized country in the world. The Mongols overthrew the Song and Kublai Khan established the Yuan Dynasty. Marco Polo, a merchant from Venice, traveled in China during his reign. He wrote,

> It is a fact that all over the country of Cathay there is a kind of black stones existing in beds in the mountains, which they dig out and burn like firewood. If you supply the fire with them at night, and see that they are well kindled, you will find them still alight in the morning; and they make such capital fuel that no other is used throughout the country. It is true that they have plenty of wood also, but they do not burn it, because those stones burn better and cost less.[1]

Coal was not the only fossil fuel burned before the Industrial Revolution. The Dutch Golden Age in the 1600s was a time of great prosperity and remarkable artists, notably Rembrandt, Vermeer, and Hals. It also produced an early speculative bubble, in tulips of all things. The primary energy source in the Netherlands at that time was peat cut from bogs. The peat was sliced in long bricks that were dried in the sun. These bricks were called *turves*. The singular is "turf." The word "turf" today is more commonly used for an area of grass. Peat has a lower energy density than coal, but the Dutch developed an efficient canal system for transporting the turves around the country. The peat production was equivalent to about a million tonnes of coal a year, a little less than a tonne per person. There was significant production of peat in the Netherlands up to the time of World War II.

  In energy terms British coal production greatly surpassed Dutch peat, hitting a peak of 292 Mt in 1913. Figure 4.23 shows the evolution of energy shares for coal, wood fuel, and draft animals in England and Wales from 1550 to 1850. This shift to coal took 250 years to move from a 13% share in 1560 to 90% in 1819. A key factor was James Watt's development of the steam engine in the second half of the eighteenth century. The steam engine enabled the Industrial Revolution, and coal provided the fuel. Great Britain under Queen Victoria dominated the 1800s. Half of cumulative world coal production through 1900 was British and the Royal Navy ruled the seas with coal-fired warships. Their empire was the largest in the world, in fact the largest the world has ever known. Its citizens were the wealthiest of any country. The Victorians also dominated science. Charles Darwin and Alfred Russell Wallace revolutionized biology with their theory of evolution. Michael Faraday

---

[1] Marco Polo, *The Travels of Marco Polo*, translated by Henry Yule, 1903.

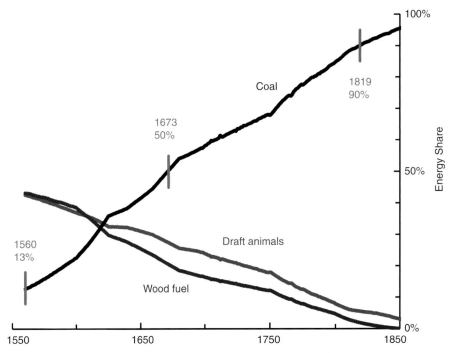

Figure 4.23 Energy shares for England and Wales over time. Wind and water energy are included in calculating the shares, but are small and not shown separately. In the plot the energy for draft animals is calculated as the energy in the feed. The data come from Paul Warde, 2007, *Energy Consumption in England and Wales*, Consiglio Nazionale della Ricerche, Naples.

invented the electrical generator and James Maxwell gave us the equations that describe radio waves and light.

4.5.1   Stanley Jevons and *The Coal Question*

In 1865, the economist Stanley Jevons (Figure 4.24) started an intense debate about how long British coal would last with his book *The Coal Question*. Jevons had a broad background that made him uniquely suited to write the book. He was born in 1835 in Liverpool. His father was an ironmonger.[2] His mother was a poet. He studied botany and chemistry at University College London. In 1853 he became the Assayer to the Australian Royal Mint in Sydney. In 1858 he returned to University College, studying logic, philosophy, and political economy, finishing his Master's degree in 1862. After graduating, he taught at Owens College in Manchester. He returned to University College in 1876 as Professor of Political Economy. Jevons died young, drowning while on vacation near Hastings in 1882.

In *The Coal Question*, Jevons broke new ground in several ways. He was energetic in collecting data and in making tables – coal prices, shipping records, town populations, emigration records, steam engine efficiencies, and much more. To

---

[2] In American English, he sold hardware.

Figure 4.24 William Stanley Jevons (1835–1882) at the age of 22, while he was Assayer to the Australian Royal Mint. Reproduced by courtesy of the University Librarian and Director, The John Rylands University Library, The University of Manchester.

make his points, he drew time-series graphs. These are graphs with time on the x-axis and data on the y-axis. Time-series graphs allow one to visually identify trends and locate bad data points. Jevons was not the first to use time-series graphs, but at this time they were novel.[3]

Before Jevons, people typically calculated a static reserves lifetime based on the annual production. In modern language this is the R/P ratio (Section 3.1.2). For example, Edward Hull, the pre-eminent coal geologist of the day, in his 1861 book, *The Coal Fields of Britain*, posed the question, "How long will our coal fields last?" Hull reckoned the British reserves at 79,843 million long tons.[4] In

---

[3] The Scottish political economist William Playfair's 1786 book, *The Commercial and Political Atlas* showed the first time-series graphs of economic data. Playfair also invented bar graphs, pie charts, and timelines.

[4] The long ton, abbreviated lt, is 2,240 pounds. American production data are often given in short tons, abbreviated st, of 2,000 pounds. One long ton is 1.016 metric tons, so Hull's 79,843 million long tons is 81.1 Gt.

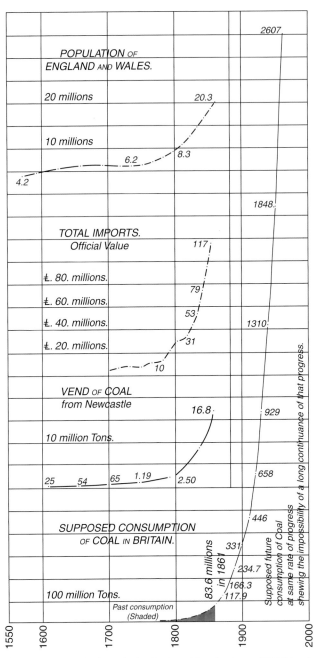

Figure 4.25 Graphs from the beginning of Stanley Jevons' 1865 book, *The Coal Question*, MacMillan and Company. The caption on the far right reads "Supposed future consumption of Coal at same rate of progress shewing the impossibility of a long continuance of that progress."

his assessment, Hull included seams that were two feet thick or more, down to a depth of 4,000 feet. Dividing by the production in 1859, 72 million long tons, Hull arrived at 1,100 years. However, Jevons was intrigued by Thomas Malthus' argument that unchecked human populations grow exponentially (Section 1.7.2). Jevons deduced that the British population, the value of imports, and the shipments of coal from Newcastle had all been growing this way. The graphs for each are shown in Figure 4.25. Furthermore, Jevons calculated that British coal consumption had been rising exponentially at 3.5% per year. He argued that consumption would continue to rise in this fashion, but that this would become unsustainable sometime in the twentieth century when it hit the limits of Hull's reserves. In the right-hand graph of the figure, the exponential growth is extended to 1961, when Edward Hull's reserves would be exhausted. He noted, "I draw the conclusion that I think any one would draw, that we cannot long maintain our present rate of increase of consumption; that we can never advance to the higher amounts of consumption supposed." He added, "But how shortened and darkened will the prospects of the country appear, with mines already deep, fuel dear, and yet a high rate of consumption to keep up if we are not to retrograde."

Today Jevons' model would be called a *demand model*. In a demand model, one selects a growth rate for the consumption and extrapolates it into the future. Jevons assumed that the historical 3.5% annual growth rate would continue. Modern economists are usually reluctant to run demand models so far into the future that resource exhaustion has to be taken into account. However, Jevons was most interested in the question of how long British coal would last, so he could not ignore exhaustion.

Often consumption statistics are not available, and we calculate consumption from production by subtracting exports and adding imports. At the world level energy production and consumption are essentially the same. There are some stocks of fossil fuels, but the changes from year to year are small enough that they are ordinarily neglected in making plots like Figure 1.4 and Figure 1.6, where the y-axis could be labeled as either production or consumption without significant error. However, for individual countries, coal production and consumption can be quite different. Some major coal producers like Australia and Indonesia export most of the coal they produce. Some major consumers like Japan and South Korea import almost all of the coal they consume. In the early years, the UK was the world's leading exporter. However, in recent times it has imported most of the coal it burns.

Figure 4.26 shows the history of British coal production, exports, and imports. Production rose until the beginning of World War I. In 1913, the UK was the world's largest coal exporter, exporting 26% of its production. Production has been falling ever since, the decline halted by neither the 1947 nationalization by Clement Atlee's Labour government nor the 1994 privatization by John Major's Conservative government. Production has fallen faster than consumption, so over time exports evaporated and imports began. In recent years imports have dominated consumption. The import share of consumption was 74% in 2017. It is not

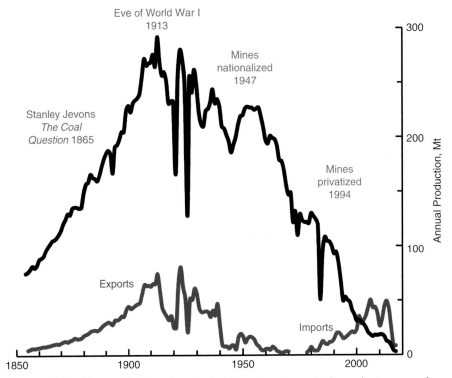

Figure 4.26 The history of British coal production, imports, and exports. The production comes from Figure 3.8. Exports through 1938 from Brian Mitchell, 1988, *British Historical Statistics*, Cambridge University Press. Afterwards the data come from the UK Department of Energy and Climate Change. The import data come from the IEA *Coal Information* series.

easy to overstate the extent of the production collapse. In 1973 there were 803 active longwalls. In December 2015 the last longwall mine in the UK, Kellingley Colliery, shut down.

4.5.2   The Royal Commission on Coal Supplies

In 1866, the year after *The Coal Question* was published, Parliament established a Royal Commission on Coal Supplies. The Commission labored five years and produced an extensive report in three volumes in 1871. Most of the Commission's work was in developing a new reserves estimate. The commissioners were more optimistic about reserves than Edward Hull. They included seams that he had rejected as speculative and they used higher recovery factors. They retained Hull's 4,000-foot depth limit, but reduced his 2-foot minimum seam thickness to 1 foot. As a result, the Commission's estimate of minable coal was higher than Hull's, 148.8 Gt instead of 81.1 Gt. As we saw in Section 3.4.3, this was not the last time a government body was enthusiastic about coal resources. The report was extremely influential. The Commission's reserves stood essentially unchanged for almost one hundred years, and the criteria they chose were widely adopted by other countries.

The United States had nothing of comparable quality until Paul Averitt completed his reserves studies for the US Geological Survey in 1974, a hundred years later.

The commissioners did not accept Jevons' argument that coal consumption was likely to grow at 3.5%/y until production was limited by lack of resources. They presented a new model, developed by Richard Price-Williams, in which coal consumption would rise to 4.7 t/p by 1890 and would stay at this level until reserves were exhausted in 2139. Figure 4.27 compares Jevons' model, the Commission's model, and the actual per-person consumption. We can give credit to Jevons for deducing correctly that consumption had risen and would continue to rise. The Commission predicted correctly that consumption would level off at 4.7 t/p at the beginning of the twentieth century. This consumption is extraordinarily high. Only the US has had higher coal consumption, hitting 6 t/p in 1918. China's consumption appears to be peaking at 3 t/p. The UK began to substitute oil for coal

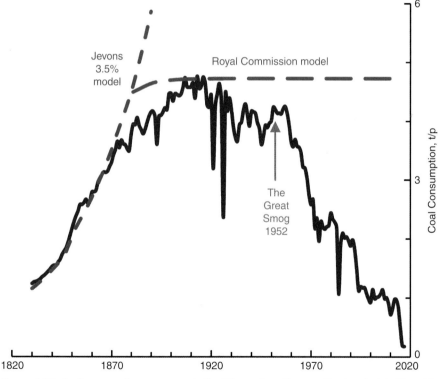

Figure 4.27 Coal consumption per person in the UK over time. Consumption is calculated by adding imports and subtracting exports from production. The Royal Commission model comes from the Royal Commission on Coal Supplies, 1871, *Report of the Commissioners appointed to inquire into the several matters relating to Coal*, printed by George Edward Eyre and Spottiswoode, London. British population through 2008 comes from Angus Maddison, afterwards from the United Nations Population Division. Consumption in the Jevons model starts with the 1865 consumption, adjusts it by 3.5% per year, and then divides by the actual population for that year.

after the Great Smog of 1952 and gas for coal after North Sea production began in the 1970s. Consumption has fallen from 4.3 t/p in 1957 to 0.2 t/p in 2017.

Today it is hard to argue with Jevons on his main thesis. British coal production did start its terminal decline in the twentieth century. He is sometimes criticized because he did not consider oil as a possible alternative fuel to coal. Oil seeps were recognized, but the main use of oil had been in medicines. The modern era of oil production as a fuel began with a well drilled in Pennsylvania in 1859. The first kerosene distilled from Pennsylvania crude arrived in London in 1861. It was sold for lighting. However, it should be appreciated that Jevons was a university trained chemist, and at that time the leading scientific explanation for the origin of petroleum was that it was distilled in coal seams, a natural analog to the process people used for distilling kerosene. This would have indicated to him that petroleum resources would be limited compared to those for coal.

Figure 4.28 is a comparison between the historical evolution of the ultimate production projections for British coal from Figure 3.11 and the original reserves from the international surveys. The original reserves in the international surveys are somewhat larger than the Royal Commission reserves. This is mainly an accounting artifact. The Royal Commission applied recovery factors that allowed for coal that would be left behind when the mines closed down. The early international

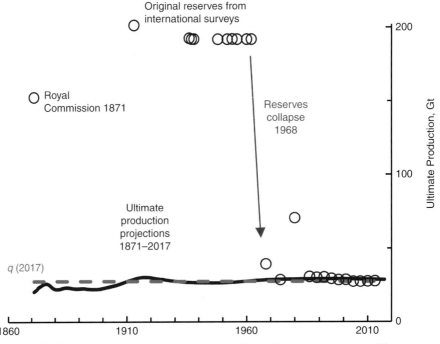

Figure 4.28 Comparing the ultimate production projections with original reserves and the current cumulative production. Reserves from the 12th International Geological Congress, the World Power Conference *Statistical Yearbooks,* and the World Energy Council *Surveys of Energy Resources.*

surveys used an in-place accounting basis without a recovery factor. Notice that there is a collapse in the original reserves. In 1962, R_o = 191.9 Gt, while in 1968 it is 39.4 Gt. By 1986 it is at 30 Gt. Other mature coal regions show similar behavior. It reflects a realization that a country has limited prospects for new coal mines. Countries track two sets of reserves. The normal one includes all of the coal in the country that meets the reserves criteria. The other is the reserves that are accessible from the producing mines. Reserves are intended to include coal that is economic, that is to say, one could make money mining it. Without prospects for new mines, the reserves accounting basis shifts from all the coal in the country to just the coal at the working mines. The current cumulative production is also shown as a dashed line in the figure.

The Royal Commission's original reserves were not a good estimate of ultimate production. Only 18% of their original reserves have been produced. We will also see this behavior in other coal regions. In addition, the original reserves stay too high until very late in the production cycle. When the reserves finally come down in 1968, 88% of the coal had already been produced. Why did this happen? One problem is that the reserves criteria were too optimistic. It is not clear how a 1-foot seam at 4,000 feet could ever be mined.

We can get further perspective by considering the seams of one mine in detail. Table 4.1 lists the coal seams at Welbeck Colliery, in Nottinghamshire, England. Welbeck was a successful mine with a production of 1 Mt per year. It was in operation from 1915 to 2011. There are ten coal seams ranging in thickness from 0.5 m to 2.3 m, for a total of 14 m. All of these seams would have met the Royal Commission's criteria for seam thickness and depth. However, only three of the

Table 4.1 Coal seams at the Welbeck Colliery, in Nottinghamshire, England.

Seam	Depth, m	Thickness, m
Coal	189	1.4
Wales	376	0.9
Swinton Pottery	441	1.8
Clowne	466	0.7
Main Bright	487	1.1
Two Foot	500	0.5
High Hazles	557	1.1
Top Hard	635	1.8
Deep Soft	808	2.0
Parkgate	842	2.3

The seams that were worked are shown in red. The data come from International Mining Consultants (IMC), 2003, "A Review of the Remaining Reserves at Deep Mines for the DTI (Department of Trade and Industry)."

thickest seams were mined, Top Hard, Deep Soft, and Parkgate, 6.1 m together. The thinnest of these, Top Hard, was 1.8 m thick, seven times the Commission's minimum, 0.3 m. After the mine closed, the access shaft was filled in to stabilize the site. The network of roadways will collapse without maintenance and there will be no possibility of reopening the mine to work the remaining 7.5 m of seams.

4.6   The Mature Regions

A good way to get a perspective on how coal production evolves is to study mature coal regions. What do we mean by a mature region? We will consider five areas where the current annual production has fallen to less than 1/1,000 of the cumulative production. They are pretty much dead. These regions are the United Kingdom, German hard coal, Pennsylvania anthracite, France and Belgium, and Japan and South Korea. The advantage of starting with the mature regions is that we know what happened, so we can see how good the projections of ultimate production from the early data are. The goal of this section will be to answer three questions. How well did the logistic models project ultimate production? How did the early original reserves compare with the ultimate production? How did the production at t_{90} compare with the peak production? We have already answered these questions for the UK, where the logistic model predicted the ultimate production within a ±19% range, the cumulative production is 18% of the original reserves, and the production at t_{90} was 45% of the peak production. While some of these results may seem surprising, they will turn out to be typical of the mature regions.

4.6.1   German Hard Coal

In the German classification, *steinkohle*, literally "stone coal" but usually rendered as *hard coal*, includes bituminous coal and anthracite. Most of the German hard coal production came from the Ruhr Valley. Figure 4.29 shows the history of German coal production, exports after World War I, and recent imports. There are two large drops in production. After World War I, Germany was forced to pay reparations to the victors, including money, coal, and timber. Germany did not meet its obligations, and as a result, in 1923 the French and Belgian armies occupied the Ruhr. There was a second catastrophic drop in 1945, the final year of World War II. This collapse was associated with horrible suffering in Germany because of the lack of heating fuel that winter. Since 1960, production has dropped steadily and the last two mines closed in 2018. There is still a market for hard coal in Germany, but in recent years it has been supplied mainly by imports.

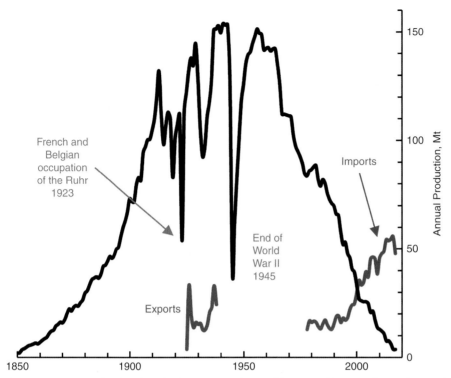

Figure 4.29 German hard coal production, exports after World War I, and imports. The production data come from Statistik der Kohlenwirtschaft e.V. The imports are from the IEA *Coal Information* series. The exports are from Brian Mitchell, *International Historical Statistics*. Silesia was a part of Germany until the end of World War I, when it was transferred to Poland. Its production is included with Poland later.

Figure 4.30 shows the cumulative production, compared with the historical evolution of the ultimate production projections calculated by the logit transform. The range of projections from 1946 on is 8.4 Gt to 12.5 Gt (±20%). This range captures the ultimate production of 12.1 Gt.

Figure 4.31 shows a comparison between the ultimate production projections and the original reserves. As with the UK, the early reserves are much higher than the production that followed. Only 5% of the 1913 original reserves have actually been produced. As with the UK there is a collapse in the reserves. For Germany, it was extremely late in the production cycle, after 99% of the coal had been produced, and it was the most complete collapse in the reserves literature, from reserves of 23 Gt in the 2001 World Energy Council Survey to 183 Mt in the 2004 survey, a 99% write-off. The German representative gave this remarkable explanation.

> Earlier assessments of German coal reserves (e.g. end-1996 and end-1999) contained large amounts of speculative resources which are no longer taken into account.

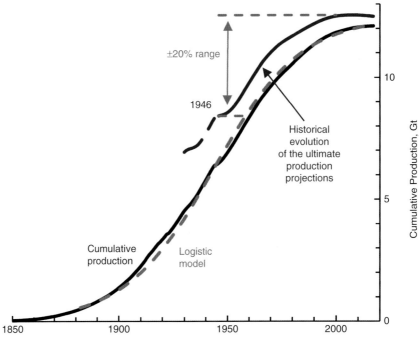

±20% range

1946

Historical
evolution
of the ultimate
production
projections

Cumulative
production

Logistic
model

Cumulative Production, Gt

Figure 4.30 Cumulative production of German hard coal compared with the historical evolution of the ultimate production projections calculated from the logit transform beginning in 1873. The logistic model uses the data from 1873 to 2017.

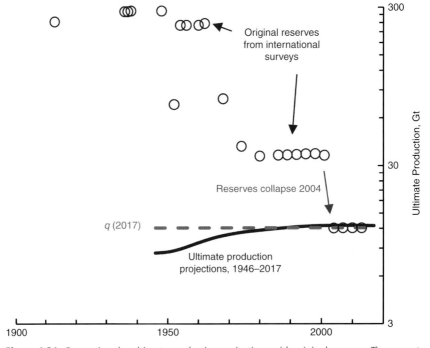

Original reserves
from international
surveys

Reserves collapse 2004

q (2017)

Ultimate production
projections, 1946–2017

Ultimate Production, Gt

Figure 4.31 Comparing the ultimate production projections with original reserves. The current cumulative production is also shown. Reserves from the 12th International Geological Congress, the World Power Conference *Statistical Yearbooks*, and the World Energy Council *Surveys of Energy Resources*.

Left unexplained was why "large amounts of speculative resources" were listed as proved, recoverable reserves in the first place. In complete contradiction to this reserves downgrade, two years later, German hard-coal resources listed by the German resources agency, the BGR, went the other way, jumping from 8 Gt to 84 Gt (Figure 3.17). With absolutely no prospects for new mines in the future, any German hard coal should be classified as an occurrence rather than a resource.

Aside from the resources irregularities, the other notable aspect of German hard coal is the extraordinary subsidy. The effective price for German hard coal has been three times the price of imported coal from the 1980s on. The paper by Karl Storchmann in the references at the end of the chapter gives the details. Several justifications have been given for the subsidy. One is energy security. Germany has poor hydrocarbon resources. Another is that Germany has companies that make excellent mining equipment. It is helpful for them to have German mines to test their products. However, the most important factor may have been that the Germans could see the violent confrontations that occurred in the UK when unprofitable mines were closed. The subsidy allowed these conflicts to be avoided. However, from a social perspective the subsidy is a problem. Coal miners are highly paid industrial workers, so people who made less money than coal miners provided the subsidy.

4.6.2 Pennsylvania Anthracite

The anthracite fields in Pennsylvania were the first coal fields to be developed in the United States. Unlike the British coal mines, the American anthracite mines have never been owned by the government, and unlike the German hard coal mines, they have operated without subsidies. The main demand for anthracite in the United States in recent years has been for residential heating. These fields are the only significant coal fields on the eastern side of the Appalachian Mountains. This meant that the coal could be moved by barge down the Susquehanna River to the Chesapeake Bay and from there to cities on the Atlantic Coast. Production peaked at 90 Mt in 1917, the year the United States entered World War I (Figure 4.32). In the center of the anthracite fields is the city of Scranton, with 138,000 people in 1920. A 1919 article on Pennsylvania in the *National Geographic* magazine noted, "Probably no other city of its class in the world is richer than Scranton." However, since then production has fallen steadily except for a second, lower production peak in 1944 during World War II. In 1959, the Susquehanna River broke into the Knox Mine works and twelve miners were killed. This disaster is often blamed for the demise of the anthracite industry, but the graph makes it clear that there was a substantial decline in production before the tragedy. Production in 2017 was only 2% of the peak production. Today the average per-person income in Scranton is lower than for the rest of Pennsylvania and lower than for the United States. Fossil fuels are not forever.

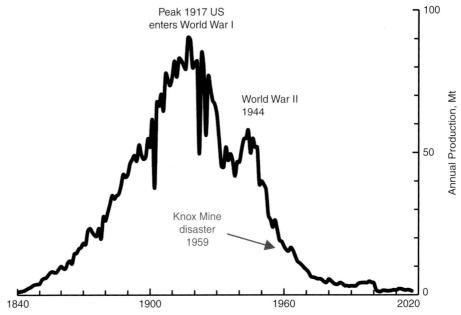

Figure 4.32 Pennsylvania anthracite production. The data from 1949 on come from the EIA Monthly Energy Review. Earlier data come from Robert Milici, US Geological Survey Open-File Report 97-447.

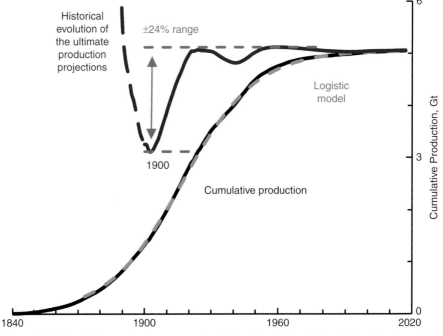

Figure 4.33 Cumulative production compared with the historical evolution of the ultimate production projections calculated from the logit transform, beginning in 1873. The logistic model uses the data from 1873 to 2017.

Figure 4.33 compares the cumulative production with the historical evolution of the ultimate production projections calculated by the logit transform. The range of projections from 1900 on is 3.1 Gt to 5.1 Gt (±24%). As with the UK and German hard coal, it appears likely that this range will capture the ultimate production.

Figure 4.34 shows a comparison between the ultimate production projections and the original reserves. The first reserves were compiled by the Anthracite Coal Waste Commission in 1892. Their original reserves were 7.9 Gt. This was followed by successively higher estimates by Dever Ashmead in 1921, George Ashley in 1944, and Harold Arndt in 1964. Current cumulative production is 42% of the 1921 original reserves. Like the UK and German hard coal, the reserves went through a collapse, falling by 93% from 1964 to 1991.

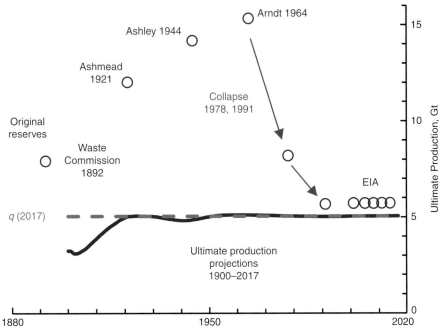

Figure 4.34 Comparing the ultimate production projections with original reserves. The current cumulative production is also shown. Reserves sources: 1892 Anthracite Coal Waste Commission quoted in Marius Campbell and Edward Parker, 1909, *Coal Fields of the United States*, USGS Bulletin 394; 1921 Dever C. Ashmead, "Anthracite Losses and Reserves in Pennsylvania," Pennsylvania Geological Survey, 4th series., Bull. 8; 1926, 1945 G. H. Ashley, "Anthracite Reserves," *Pennsylvania Topog. and Geol. Survey Prog. Report 130*; 1964 reserves: US Geological Survey and the US Bureau of Mines, 1968, "Mineral Resources of the Appalachian Region," *Geological Survey Professional Paper 580*, in the chapter by Harold Arndt, Paul Averitt, James Dowd, and Donald Frendzel, "Coal," p. 133; EIA, "Demonstrated Reserve Base of Coal in the United States," on January 1, 1979, EIA, "U.S. Coal Reserves: An Update by Heat and Sulfur Content February 1993," and various editions of the EIA *Annual Coal Report*.

4.6.3 Summary for the Mature Regions

Table 4.2 shows the results for the five mature regions. All of these regions had good mining engineers, excellent access to capital, and skilled workers. The results are consistent. In every case only a small fraction of the early original reserves has been produced. The median fraction is 20%. In every case the projections for ultimate production appear to have captured the ultimate production. The median projection range is ±21%. And finally, in every case, at t_{90} production is down substantially from the peak. The median t_{90} production is 53% of the peak production.

What have we learned? Coal regions die after producing a small fraction of their original reserves. The logit transform typically provides a ±20% range of projections that captures the ultimate production. If the production history is not long enough to provide this range, we will need to use the original reserves to get a sense of the ultimate production, but they are likely to be too high. The effect of subsidies on the relationship between ultimate production and original reserves appears to be negative. Germany provided a massive subsidy but produced the smallest fraction of the original reserves. The Pennsylvania anthracite fields had no subsidy, but produced the highest fraction of the original reserves. It should be recognized that there is a political component to national reserves. The British original reserves were developed by a Royal Commission that was established because Stanley Jevons challenged the conventional wisdom that there were a thousand years of coal left. In this situation, choosing reserves criteria becomes a political act. At some point the reserves accounting shifts from all the coal in the country to just the coal within reach of the working mines. The timing for this shift is also a political decision, and countries tend to make the change very late in the production cycle. Perhaps surprisingly, one factor in the production shortfall is the development of modern

Table 4.2 Summary of the results for the mature coal regions.

	Ultimate production U, Gt	Ultimate production fraction of original reserves U/R_o (reserves year)	Ultimate production projection range ΔU, Gt (starting year)	% of peak production at t_{90} (t_{90})
United Kingdom	27.5	18% (1871)	20.5–30.3 ±19% (1871)	45% (1973)
German hard coal	12.1	5% (1913)	8.4–12.5 ±20% (1946)	53% (1985)
France and Belgium	7.2	23% (1913)	4.3–8.5 ±33% (1900)	58% (1970)
Pennsylvania anthracite	5.0	42% (1921)	3.1–5.1 ±24% (1900)	41% (1952)
Japan and South Korea	3.6	21% (1936)	2.6–4.0 ±20% (1946)	63% (1986)
Median percentages	na	21%	±20%	53%

The ultimate production values listed here are cumulative production through 2017. A small amount of production, less than a tenth of a percent of the cumulative production, continues in the UK, the Pennsylvania anthracite fields, and in Japan and South Korea. The calculations for this table are available in the Excel workbook fossil fuels at energybk.caltech.edu mining equipment. Mechanized longwall mining dominates world underground coal production. Modern shearers and self-advancing roof supports allow high productivity and greatly improve safety over hewing coal by hand. However, longwall mining is not flexible. It needs large, fault-free, almost-level blocks of thick coal. This rules out much of the coal that would have been considered minable a hundred years ago. However, government resources agencies may not take these factors into account in reserves assessments. In addition, it is common for governments not to make enough money available to gather the data that would be needed to make more realistic assessments – assessments that would likely result in lower reserves. It is a rare government that wants to find out that its reserves numbers are too high.

4.7   The Active Regions

The next goal is to develop an estimate for U and t_{90} for world coal. We start with the plot of annual production in Figure 4.35. The first thing to note is that even though hydrocarbons passed coal in 1956 (Figure 1.4), coal production continued to rise. Currently production is dominated by China. It is possible that Chinese production peaked in 2013, and it is also possible that world production has peaked. Time will tell.

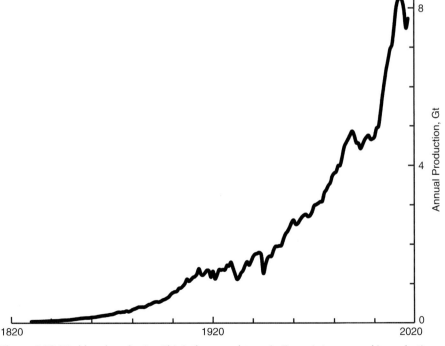

Figure 4.35 World coal production. This is the same data as in Figure 1.4, expressed in production units.

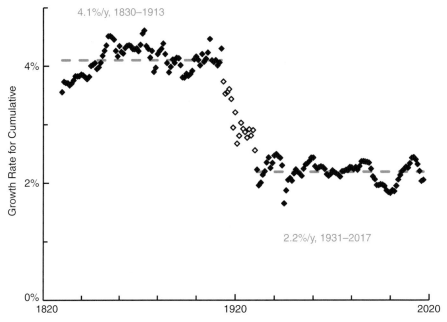

Figure 4.36 The growth rate for the cumulative world coal production over time.

The simplest approach would be to make a logistic model for world production to estimate U. However, this fails. Figure 4.36 shows p/q, the growth rate for the cumulative production, plotted against time. This is different from the ordinary growth-rate plot where the cumulative production is on the x-axis. The growth rate is essentially constant at 4.1%/y from 1830 to 1913. This production was almost entirely in Europe and the United States, which together accounted for 88% of the cumulative production through 1913. We had earlier identified 1913 as a shock for coal production associated with World War I (Figure 1.6). After 1913 the growth rate declines. In 1931, it levels out again at 2.2%/y. The growth rate has been flat ever since. A growth-rate plot uses the x-intercept as the estimate for ultimate production. There will be no intercept if the growth rate is constant. We need a different approach.

Transportation costs for coal are much higher than for hydrocarbons, and as a result only 17% of world coal production was exported in 2017, compared with 56% for hydrocarbons.[5] This makes it appropriate to make projections for U and t_{90} on a regional basis for coal, but on a global basis for hydrocarbons (Section 5.7). Figure 4.37 shows the coal regions in the analysis. In addition to the five mature

---

[5] The BP *Statistical Review* tracks hydrocarbon trade flows and production, and this allows us to calculate an export fraction. A good source for the coal export total is the IEA's annual *Coal Information*, available at www.iea.org/bookshop/

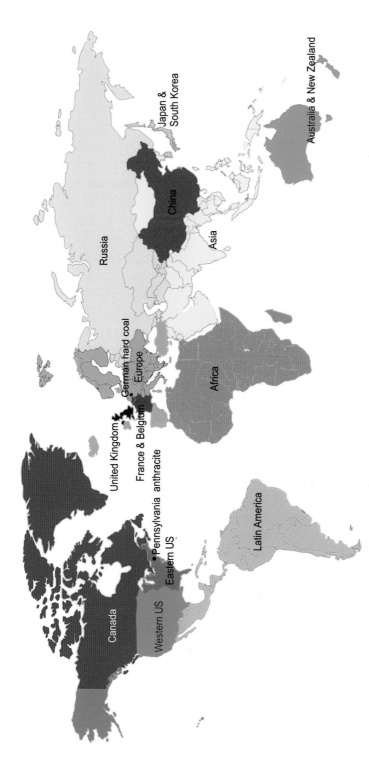

Figure 4.37 The 15 coal regions that are used for the analysis.

regions in Table 4.2, there are ten active regions. We will analyze two active regions in detail, the Western United States and China.

4.7.1 Western US Coal

The United States passed the UK to become the number one coal producer in 1899 and it held this position until 1985, when it was overtaken by China. Figure 4.38 shows a map of the coal fields in the lower 48 states. The American coal fields are enormous, extending over a million square kilometers. Historically the most important region has been the Appalachian Basin, with large production in Pennsylvania, West Virginia, and eastern Kentucky. This is predominantly bituminous coal from underground mines. However, in 1970, during the Nixon Administration, the Clean Air Act Extension encouraged utilities to switch to low-sulfur coal. The low-sulfur resources are limited in the East, but they are abundant in the West, particularly in the Powder River Basin in Wyoming. Most of the Western coal production is surface-mined sub-bituminous coal. Western production increased dramatically. It accounted for 58% of US production in 2017.

Transportation is a critical factor for Western coal. Most US coal power plants are a long way from the Powder River Basin. However, the executives of the Burlington Northern Railroad recognized the opportunity, and made major investments to expand rail access for the heavy coal trains. Figure 4.39 gives a way to visualize the coal rail shipments. The width of a line shows how much coal moved over that line. The movement from the Powder River Basin dominates the figure.

Figure 4.40 shows the production history for the Western United States. This is different from any of our mature regions in that there are two distinct cycles centered on different states. The first cycle peaked in 1918. At that time the western state with the largest production was Colorado. Fuel for locomotives was an important market. In addition, coal was made into coke for the blast furnaces in Pueblo, Colorado. The second, much larger cycle kicks off with the Clean Air Act Extension in 1970. The second cycle peaks in 2008, the last year of the Bush Administration. During the succeeding Obama Administration, coal production dropped by 34% from the peak in 2008 to 2017. Several factors contributed to this drop. The federal Environmental Protection Agency (EPA) proposed CO_2 emissions standards for new power plants that were so strict that based on the chemical composition of coal, the limits could not be met without developing a technology for capturing and burying the emissions. This was a break from previous EPA policies that specified best available technology for emissions reduction, and miners accused the Obama Administration of waging a war on coal. At the same time, natural gas production was increasing dramatically because of fracking. Natural gas is a cleaner burning fuel, and natural gas plants are cheaper to build than coal plants. In addition, the plants are more efficient than coal plants and more flexible in their output. Finally, several states, notably California, have discouraged utilities from buying electricity produced by coal plants.

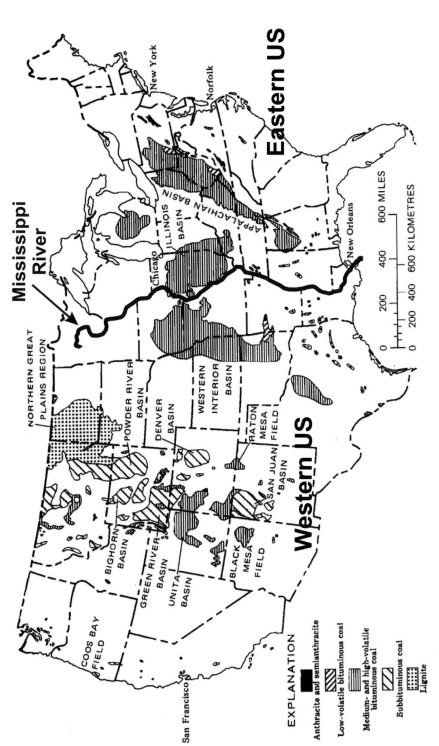

Figure 4.38 Map of US coal fields from Paul Averitt, 1974, "Coal Resources of the United States," *US Geological Survey Bulletin 1412*. The dividing line between eastern and western coal is the Mississippi River.

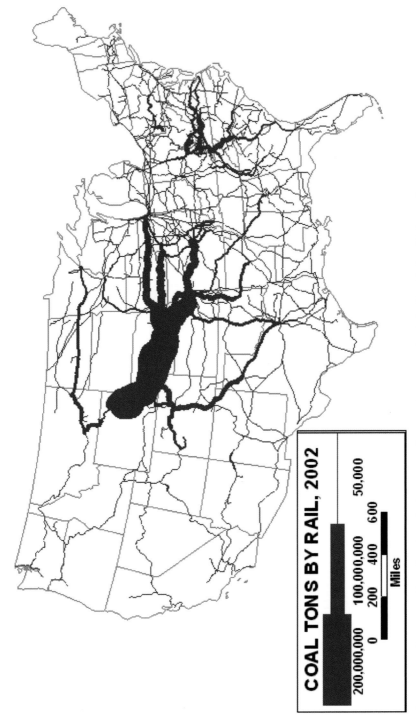

Figure 4.39 Shipping coal by rail in the United States. Credit: Bruce Peterson at the Oak Ridge National Laboratories.

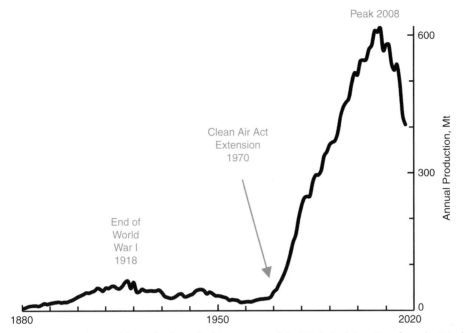

Figure 4.40 Production history for the United States west of the Mississippi river. Data from the EIA, the US Bureau of Mines, from Robert Milici of the US Geological Survey, and from Howard Eavenson, 1942, *The First Century and a Quarter of American Coal Industry*, privately published, Waverly Press, Production Table 20, pp. 426–434.

Figure 4.41 shows the historical evolution of the ultimate production projections for Western US coal calculated with the logit transform. The range from 1994 to 2017 is 37 Gt to 51 Gt, ±17%. The original reserves from the US Geological Survey are 162 Gt, four times the current projection for ultimate production. This is similar to the mature regions. However, Western US coal is the only one of the 15 coal regions that is dominated by surface mining. When underground mines close down, it is usually appropriate to reclassify their resources as occurrences, because it is so difficult to revive an underground mine. In contrast, it is easy to restart surface mines, so their resource classifications should not change.

4.7.2 Chinese Coal

Coal is the foundation of China's industrial power. It provided 60% of China's primary energy in 2017, compared with 15% for the United States and 14% for the European Union. China dominates world coal production; 46% of the world's coal is mined in China. Chinese coal mining is an enormous economic activity, operating on a human scale never seen before. Writing in 2012, political economist Tim Wright made the following estimate, "an output-based calculation suggests around 4 million workers in the 1990s, later rising to over 6 million. Putting together all the evidence, a total level of employment across all enterprises of 6 million in the

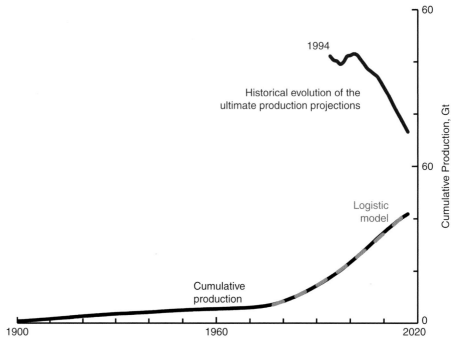

Figure 4.41 Cumulative production for Western US coal compared with the historical evolution of the ultimate production projections calculated from the logit transform, beginning in 1977. The logistic model uses the data from 1977 to 2017.

late 2000s would be a minimum, and the real figure is probably approaching 10 million." No other country has ever had anything like ten million coal miners. For comparison, the peak in the UK was 1.2 million coal miners in 1924. Chinese coal production is divided between state-owned enterprises at the provincial and county level, and township and village mines. Wright gives the 2009 production shares as 62% for the state-owned enterprises and 38% for township and village mines. This large contribution by township and village mines, larger than the production of the United States, is unique to China.

Figure 4.42 shows a Chinese village mine. This type of mine is called a drift mine. *Drift* is a mining term for a horizontal tunnel. In a drift mine, tunnels are driven directly into the side of a cliff. In the photo, several drift entrances are visible at the base of the cliff. Below the entrances, there are two larger openings for the yellow loader. One can also make out a road for trucks leading off to the right. This approach allows production with minimal capital investment and delay. However, only a small fraction of coal resources can be reached by drifts. The mine works also pollute the stream at the bottom edge of the photograph.

By comparison with the state-owned enterprises and with mines in other countries, the Chinese township and village mines are dangerous. Table 4.3 shows fatality rates for the town and village mines, the state-owned enterprises, British mines, and American underground and surface mines. From the perspective of the risk

Table 4.3 Fatalities in coal mines.

	Deaths per 100,000 workers	Deaths/Gt
China, town and village mines (1992–2007)	301	7,050
China, state-owned enterprises (1992–2009)	64	1,090
UK (1963–1979)	33	750
US underground (2006–2015)	38	53
US surface (2006–2015)	18	10

Chinese and British numbers adapted from Tim Wright, 2012, *The Political Economy of the Chinese Coal Industry*. American numbers for production and employment from the EIA *Annual Coal Report*, various years, and fatalities are from the US Mine Safety and Health annual reports. Note that the British numbers are for an earlier period than the other countries, at a time when there was still significant coal production in the UK.

Figure 4.42 A photograph of a village mine in Guizhou province. Courtesy of Edwin Moise.

to miners and to their families, the important number is the number of deaths per 100,000 workers. On this scale, the town and village mines are five times more dangerous than the state-owned enterprises, and eight times more dangerous than the British mines and the American underground mines. The American surface mines are safer than any of the others, but China and the UK have limited resources suitable for surface mining. The table also shows fatalities on an energy basis in terms of the deaths per billion tonnes produced. This is the human cost to society for its energy. On this basis, surface mining is vastly superior to underground mining.

Figure 4.43 shows the Chinese production history. One should be aware that there are major problems with Chinese economic statistics. In her book *The Chinese Coal Industry*, economist Elspeth Thomson writes,

– figures deliberately inflated or deflated by low level bureaucrats trying to get some message across to superiors. For example, in order to get bonuses, they would report that the targets were reached when they weren't, or they would under-report to try to get more funding next year, etc.
– figures for public consumption at home or abroad, deliberately inflated or deflated by authorities, for example, downplaying of accidents.
– discrepancies in figures supplied by Chinese official sources, for example, many typos, and variations found in the statistical yearbooks.
– much fuller information given in Chinese sources, for example, English and Chinese versions of the Coal Industry Yearbook.

The figure shows production tripling during the Great Leap Forward. At that time, bureaucrats felt obligated for their personal safety to provide the high production numbers that their managers wanted to show their superiors. Similarly, there was a

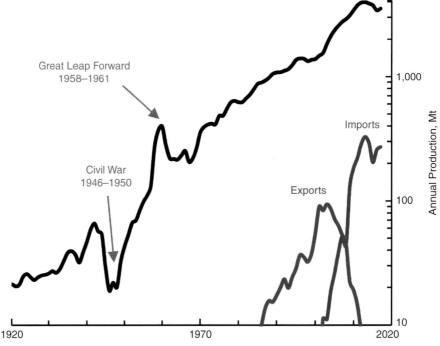

Figure 4.43 Chinese coal production history. The scale is logarithmic. The data from 1920 through 1980 come from Elspeth Thomson, 2003, *The Chinese Coal Industry, An Economic History*, RoutledgeCurzon. The data from 1981 to 2017 come from the BP *Statistical Review*. The import and export data come from the IEA, *Coal Information*.

campaign in the late 90s to close unsafe mines. In many cases the result was that the mines continued to operate, but they stopped reporting production. Statisticians try to correct for these errors by considering other data like electricity production. However, the adjustments may not be correct and they may not be final.

With that advisory, what can we learn from the data? There has been a stupendous increase in production, from 21 Mt in 1920 to a peak of 3,974 Mt in 2013, a gain of almost 200:1. Production was down 11% in 2017 compared with the 2013 peak. Was 2013 the all-time high? We do not know, but there are clues. One is that Chinese coal mines are getting deep. In 2013, Xie Heping, president of Sichuan University and a mining engineer, wrote, "shallow coal resources in the key coal areas have been depleted, leaving an average mining depth of approximately 600 m." The experience with German hard coal was that 1,000 m was about as deep as coal mines go, even with subsidies. The great depth is an indication of substantial exhaustion of Chinese mines. Imports give another clue. There has been a rapid switch from exports to imports. In 2003 China was a major exporter, at 93 Mt, second only to Australia. But by 2011, China had passed Japan to become the world's largest importer of coal. This is a sign that Chinese miners are having trouble meeting demand. The clues suggest an all-time peak. Time will tell.

Making projections for Chinese coal production is not as straightforward as it was for Western US production. The logit transform of the data is shown in Figure 4.44. There is a large dip at the time of the Civil War and the aftermath. For this reason, the period from 1946 to 1960 is not included in the curve fit. It is interesting that the time of the Republic of China and the time of the People's

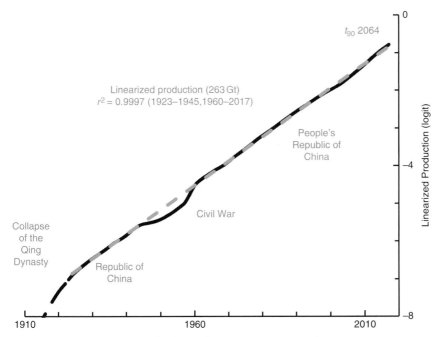

Figure 4.44 Logit linearization for the production in Figure 4.43 with $U = 263$ Gt.

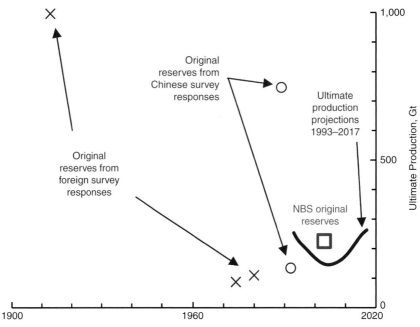

Figure 4.45 Evolution of the ultimate production projections for Chinese coal from 1993 to 2017 versus original reserves. Reserves from the 12th International Geological Congress, the World Power Conference *Statistical Yearbooks*, and the World Energy Council *Surveys of Energy Resources*. NBS stands for the Chinese National Bureau of Statistics.

Republic of China fall on the same line. The economic systems were quite different, but the slopes were the same.

Figure 4.45 shows the evolution of the ultimate production projections from 1993 to 2017, compared with reserves. In contrast to the projections for the Western United States, the projections for China are untamed. Moreover, Chinese reserves are a mess. For many years there were no Chinese sources for the reserves. The estimates were done by foreigners. The problem with this is that China has a hundred significant coal fields. It is just not possible for foreigners to make proper reserves assessments. However, the Chinese themselves have only submitted reserves twice to the World Energy Council, in 1989 and 1992. The two numbers differed by a factor of six. In 2003, the Chinese National Bureau of Statistics suggested reserves of 189 Gt. This corresponds to original reserves of 226 Gt, which is within the range of projections.

4.7.3   Summary for the Active Regions

Table 4.4 is a summary for the active regions. The projection ranges are calculated from 1996 to 2017. For Asia[6] and Latin America, the projections do not converge over this time period. The reason the projections do not converge is exponential

[6] The Asia region does not include countries that are in in other regions, like China, Japan, South Korea, and the former Soviet Union.

Table 4.4 Summary for the active regions.

	Cumulative production q, Gt (2017)	Ultimate production projection U, Gt	Original reserves R_o, Gt	Ultimate production projection range, ΔU, Gt	t_{90} projection
Australia	14	45	88	27–48 ±28%	2072
China	82	263	174	145–263 ±29%	2064
Africa	11	16	41	13–16 ±10%	2065
Europe	78	113	178	82–119 ±18%	2064
Russia	30	48	221	na	2064
Western US	21	37	162	37–51 ±17%	2046
Eastern US	51	72	144	72–78 ±4%	2062
Canada	4	5	10	5–6 ±14%	2036
Asia	27	117	117	81–117 ±18%	2073
Latin America	3	17	18	12–24 ±33%	2082
World	376	787	1,219	658–787 ±9%	2065

The world total includes the mature regions from Table 4.2. The projection ranges are calculated from 1996 to 2017. For consistency with the other regions, the World Energy Council reserves are used for China, rather than the more recent reserves from the Chinese National Bureau of Statistics. The calculations for this table are available in the Excel workbook fossil fuels at energybk.caltech.edu

growth in production. The twenty-year annualized growth rates are 6%/y for Asia, and 4%/y for Latin America. For Asia and Latin America, the original reserves are used for the projection of ultimate production. Note that based on the experience of the mature regions, the original reserves are likely to be higher than the ultimate production. When these are added together with the results from the other regions, they will bias the world projection for ultimate production high. For Russia,[7] the curve fitting for recent production does not converge, and the projection is calculated from 1980 to 1988, before the collapse of the Soviet Union, but with no projection range. This means that when the Russian results are added to those of the other regions, the projection range will be biased narrow. The 2017 projection for the world total is $U = 787$ Gt and $t_{90} = 2065$. The projection range from 1996 to 2017 is 658–787 Gt (±9%).

Figure 4.46 shows a comparison of the evolution of the world ultimate production projections with the original reserves. The most significant feature of the graph is the dramatic reduction in original reserves. The original reserves in 2013 were 1/6 of the original reserves in 1913, one hundred years earlier. In addition, we can see that the projections for ultimate production are lower than original

---

[7] The Russia region includes the countries of the former Soviet Union except for Ukraine, which is counted with Europe.

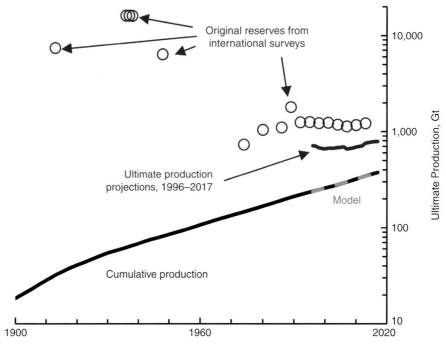

Figure 4.46 Comparing the ultimate production projections for world coal with the original reserves. These are calculated by adding the values for the different regions. The scale is logarithmic.

reserves. We will see that the reverse is true for hydrocarbons. World hydrocarbon original reserves have grown with time. Moreover, the projection for world ultimate hydrocarbon production will turn out to be larger than original reserves.

The ratio of U for world coal to the original reserves is 65%. This is higher than the median 21% for the mature coal regions. The two numbers are not comparable for several reasons. Original reserves were used instead of curve fits for Asia and Latin America. Based on the experience of the mature coal regions, when growth slows and we can get a range for the production curve fits, it will turn out that the original reserves are too high again. Then there is China. For consistency with the other regions, we used the World Energy Council survey for Chinese reserves. However, these reserves were last revised in 1992, and it is not clear what they are based on, because the reserves three years earlier were six times higher. Assessing national reserves for a major producer like China requires decades, not years. Finally, we are using the current values of original reserves rather than the early original reserves. The current original reserves are a small fraction of the early ones.

The time t_{90} is 2065. The remaining large country with plans for growth in production and consumption of coal is India, which is building out its electricity grid. In 2014 India passed Japan to become the world's second largest importer of coal. In 2015, the country passed the US to become the second-largest consumer of coal. In 2016, India passed the United States to become the second largest producer of coal. We should keep in mind that a shock like the collapse of the Soviet Union could change t_{90}. Europe, the US, and China are spending enormous amounts of money on wind turbines, solar panels, and biogas with the goal of displacing coal in electricity production. Moving away from coal reflects society as much as geology.

Concepts to Review

- Coal's role in electricity production
- Coal formation
- Hewing coal by hand
- Longwall mining
- Surface mining
- The Aberfan disaster
- Air pollution from burning coal
- British coal production
- Pennsylvania anthracite production
- German hard coal production
- Mature coal regions
- Western US production
- Chinese coal production
- Active coal regions

Problems

Problem 4.1   Energy production on US federal lands

During the 2016 American presidential election campaign there was a discussion about stopping fossil-fuel production on federal lands. As an example, see Devon Henry, 2016, "Clinton: Banning Fossil Fuels on Public Lands a Done Deal," *The Hill*. One can find a map of "Federal Lands and Indian Reservations" by the Department of the Interior and the US Geological Survey in the online National Atlas. They are mostly in the western United States.

Locate the EIA publication "Sales of Fossil Fuels Produced from Federal and Indian Lands" published in 2015 online. The coal production share for federal lands was steady at 41% in 2004 and in 2014. Calculate the share of US hydrocarbon energy production on Federal lands (not including "Indian lands") in 2004 and 2014.

Problem 4.2   Growth rates and R/P ratios

In 2017, India was the second largest producer, consumer, and importer of coal, after China.

a. Calculate the annualized growth rate for the coal production of India for the latest ten years from the BP *Statistical Review*.

b. If Indian coal production were to continue to grow at this rate, how long would the reserves last? How does this compare to the R/P ratio given in the BP *Statistical Review*? It should be noted that historically there has not been a sudden drop off in production, but rather long periods of decline.

Further Reading

• Peter Brimblecombe, 1987, *The Big Smoke*, Methuen and Company. A history of air pollution in London. A more recent edition was published in 2011 by Routledge, Oxford.

• Gunter Fettweis, 1979, *World Coal Resources*. Comprehensive discussion of world coal reserves through 1979.

• Barbara Freese, 2003, *Coal: A Human History*, Penguin Books. A thoughtful discussion of coal's evolving role in our society, emphasizing the United Kingdom, the United States, and China. Her background is in pollution regulation.

• Phyllis Geller, 2009, *Coal Country*. This is a documentary about surface mining in the Appalachian Mountains in West Virginia. Geller allows the viewers to hear from both sides. There is good coverage of reclamation and the lawsuits over the mines.

• John Maynard Keynes, 1936, "William Stanley Jevons, 1835–1882: A centenary allocation on his life and work as economist and statistician," *Journal of the Royal Statistical Society*, vol. 99, pp. 516–555. This is the best summary available of Jevons' work and enormous impact on the field of economics. Keynes is critical of the *Coal Question*, viewing it as alarmist.

• Royal Commission on Coal Supplies, 1871, *Report of the Commissioners appointed to inquire into the several matters relating to Coal*, printed by George Edward Eyre and Spottiswoode, London.

• Robert Stefanko, 1983, *Coal Mining Technology*, the Society of Mining Engineers. The classic textbook on coal mining engineering.

• Karl Storchmann 2005 "The rise and fall of German hard coal subsidies," *Energy Policy* vol. 33, pp. 1469–1492. A detailed discussion of the German hard coal subsidies.

• World Energy Council (WEC), *Surveys of Energy Resources*. These are the fundamental sources for coal reserves over the years. The BP *Statistical Review* based their reserves on the WEC surveys, while other agencies start with the WEC reserves and then adjust them. Under different titles, these surveys go back more than 100 years to the 12th Geological Congress in Toronto Canada in 1913. The last survey was in 2013. Recent surveys can be found at the World Energy Council website at www.worldenergy.org/publications

• Tim Wright, 2012, *The Political Economy of the Chinese Coal Industry*, Routledge.

• J. W. de Zeeuw, 2006, "Peat and the Dutch Golden Age." This paper gives quantitative estimates for Dutch peat production. It is available on the website of the International Peatland Society at www.peatsociety.org/sites/default/files/Zeeuw.pdf

5 | Hydrocarbons

Oil is a liquid mixture of hydrocarbons, usually found in rock. It is also called *petroleum* (Figure 5.1) The Latin roots of the word "petroleum" mean *rock oil*. In earlier times it was called rock oil to distinguish it from vegetable oils like rapeseed oil and animal oils like whale oil. Many different molecules may be present in petroleum. Examples are heptane and isooctane, which are used to define the octane scale for gasoline (Figure 5.2). Normal heptane is zero on the octane scale and isooctane is 100. Mixtures with higher octane numbers are more resistant to premature detonation in a spark-ignition engine. The isooctane carbon chain has branches that give it a higher octane number. Oil seeps have been known for thousands of years, and on a small scale, oil was skimmed off the surface of ponds and recovered from wells dug by hand. It was mainly used in medicines, particularly for skin ailments. Marco Polo described the oil of the Caspian region, as he did the coal of China (Section 4.5).

> The country [Armenia] is bounded on the south by a kingdom called Mosul, the people of which are Jacobite and Nestorian Christians, of whom I shall have more to tell you presently. On the north it is bounded by the Land of the Georgians, of whom also I shall speak. On the confines towards Georgiania [Georgia] there is a fountain from which oil springs in great abundance, insomuch that a hundred shiploads might be taken from it at one time. This oil is not good to use with food, but 'tis good to burn, and is also used to anoint camels that have the mange. People come from vast distances to fetch it, for in all the countries round about they have no other oil.[1]

5.1 Introduction

In 1853, George Bissell, a Dartmouth College graduate, returned to New Hampshire to visit his old college. He was intrigued by a bottle of oil he saw there from Titusville, a lumber town in northwest Pennsylvania. Bissell persuaded other people to join him in founding a company to produce oil. However, they had trouble

---

[1] Marco Polo, *The Travels of Marco Polo*, translated by Henry Yule, 1903.

Figure 5.1 Petroleum from Daisy Bradford #3, the discovery well of the East Texas Oil Field. From the East Texas Oil Museum, Kilgore, Texas. Photograph by the author.

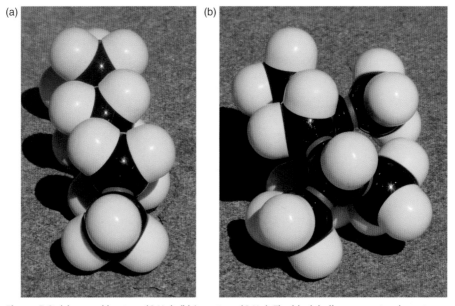

Figure 5.2 (a) normal heptane (C_7H_{16}). (b) isooctane (C_8H_{18}). The black balls represent carbon atoms and the white balls hydrogen atoms. Not all of the balls are visible. Photographs by the author.

Table 5.1 Lamp output and costs.

	Lamp light, lm	lmh/¢
Town gas	95	1.5
Sperm oil lamp	94	0.9
Rapeseed oil	94	1.1
Camphene	138	4.8
Oil distillate	102	?
LED (Walmart 8.5 W)	800	7,300

The data for the early lamps come from Benjamin Silliman's 1855 lighting tests in his *Report on the rock oil, or petroleum, from Venango Co., Pennsylvania: with special reference to its use for illumination and other purposes.* The values are in 2017 cents, with the 2017 to 1861 deflator from the BP *Statistical Review.* For the LED lamp, the price of electricity is taken to be 12.9 ¢/kWh, the US residential average in the 2018 IEA *Electricity Information.*

selling the stock shares. A major problem was that it was not clear what the oil was good for. It burned with a bad smell and a lot of smoke.

The founders decided to hire a chemist as a consultant to analyze the oil. Yale Professor of Chemistry Benjamin Silliman, Jr., studied the oil for six months and in April 1855 he submitted a report. Silliman found that a light distilled fraction of the oil burned well in a lamp designed for camphene, a lighting fuel distilled from pine resin. To compare the light from different lamps, he built a photometer that was calibrated with a sperm candle (Section 2.2.4). The results are given in Table 5.1. The middle column shows the output in modern units. For comparison a modern LED lamp is included. The LED produces much more light than the 1850s lamps. Moreover, the early lamps all produced significant heat, hundreds of watts. This was a disadvantage in the summer, as there was no air conditioning.

Silliman gave the rate of fuel consumption and unit fuel costs in his report. At this stage, there would be no price for the oil distillate. From his numbers, we can calculate the lumen hours that a 2017 penny would buy. These are shown in the right column. Town gas was a mixture of carbon monoxide, hydrogen, and volatile hydrocarbons that was produced by heating coal (Section 2.3.2). It was cheaper than whale oil and it was distributed in urban areas through a pipeline system, like natural gas today. It was used for heating even after the town gas lamps were replaced by electric lights. Camphene produced more light for a penny than town gas, sperm oil, and rapeseed oil, but it is quite volatile, and this made it an explosion hazard. Modern LEDs are thousands of times cheaper to operate than the town gas lamps.

5.1.1   Edwin Drake and His Well

Copies of Silliman's report were printed for potential investors. This enabled Bissel to persuade people to buy the stock. The company hired a retired railroad conductor, Edwin Drake, to drill an oil well. While George Bissel gets credit for his remarkable insight, Edwin Drake gets credit for his remarkable stubbornness. There were

Figure 5.3 The oil boom in 1865, near Titusville, Pennsylvania. Courtesy of the Drake Well Museum, Pennsylvania Historical and Museum Commission.

many obstacles. The locals in Titusville were skeptical that a well would produce oil. After a false start, Drake hired Billy Smith, who had drilled wells for brine. They set up a drill with a steam engine. During the drilling, the walls of the well collapsed. Drake solved this problem by adding a casing for the well. By late 1859 the investors had lost patience and they sent Drake a letter telling him to shut the well down. However, this was in the days before email. On August 27, 1859, before the letter arrived, Drake struck oil at a depth of 69 feet. The news set off a frenzy (Figure 5.3).

One question Silliman could not answer in his report was whether oil would turn out to be cheaper than the other fuels. This was soon resolved. At the time, sperm oil sold for $2.50 per gallon, rapeseed oil for $2.00 per gallon and camphene for $0.68 per gallon. Camphene was produced from pine trees in the South, but the supply was interrupted when the Civil War began in 1861. The BP *Statistical Review* gives the 1861 price of a *barrel* of oil as $0.49, a little more than a penny per gallon. By this time, a lighting market already existed for a chemically similar coal product, known as coal oil or kerosene, that had developed since Professor Silliman wrote his report in 1855. Coal oil was more expensive to produce than the petroleum based lamp fuel. Only a special type of coal called *cannel coal* would work. "Cannel" was an English dialect pronunciation of "candle." Cannel coal has a higher fraction of hydrogen than other coals due to deposits of pollen and spores that have high wax content. The coal was first heated to release liquids and condensable vapors. These products were then distilled to make kerosene. There were additional treatments required of the oil distillate to remove sulfur, and additional refining to make a standardized product. The oil product quickly pushed coal out of the market. However, people still called it coal oil or kerosene.

5.1.2 John D. Rockefeller and Standard Oil

Refining oil was a simple process and many people started refineries. However, the refining industry came to be dominated in the United States by a single company, the Standard Oil Company. Standard Oil was founded by John D. Rockefeller, an outstanding business innovator (Figure 5.4). He developed an efficient management

Figure 5.4 John Davison Rockefeller (1839–1937), taken in 1885. Rockefeller dominated the early American oil industry. Credit: Rockefeller Archive Center.

structure to keep prices low, while maintaining the quality of his products. He was a pioneer in his careful tracking of sales, both his own and those of his competitors. Rockefeller started in refining, but over time he developed a vertically integrated company that owned wells, tank cars, tankers, and gas stations. He appreciated the critical role at that time of the railroads for his business. Rockefeller negotiated rebates from the railroads for his oil shipments, which meant that he had lower shipping rates than his competitors. These rebates could be viewed as volume discounts that simply reflected economic reality and Rockefeller was not the only railroad customer who received them. However, some argued that railroads had a special obligation to have a transparent and consistent pricing structure because they had special privileges from the state in being able to condemn land for railroad tracks. In any event, the rebates were resented by the other oil companies that already had higher refining costs than Standard Oil.

Rockefeller's competitors particularly despised *drawbacks*, where Standard Oil received money from the railroad when *competitors* shipped oil. Rockefeller operated in a completely different business environment from today's and often actions that people complained about were legal then or their legality had not yet been tested.[2] Over time people pushed back against Standard Oil. Other countries and the state of Texas favored local companies. In 1911 the Supreme Court broke the company up into many smaller companies. The descendants of these smaller companies have still been important oil companies. The two largest publicly traded oil companies in the United States by market capitalization, ExxonMobil and Chevron, trace back to the Standard Oil companies created by the breakup. Standard Oil of Indiana, later renamed Amoco, developed the first process for thermal cracking. Thermal cracking breaks large hydrocarbon molecules into smaller molecules, increasing the yield of gasoline.

Rockefeller was a devout Baptist, and his influence on American philanthropy was as important as his influence on American business. His donations established the University of Chicago and Spelman College. Spelman was his wife's family name. Spelman College is an historical black women's college that has been ranked #1 in the United States among historical black colleges and universities by *US News and World Report.* The Rockefeller Foundation made the grant to Norman Borlaug that started the Green Revolution.

---

[2] Whether they were legal or not, drawbacks were utterly unfair to Rockefeller's competitors by the standards of either our century or his. However, for balance, it should be noted that the California Air Quality Management Board rules for electric vehicles (EVs) function as drawbacks. If a car company does not sell a high enough percentage of EVs in California, the company must buy credits from a company that does. In practice, this has meant means buying the credits from Tesla Motors, which sells only electric cars. *Los Angeles Times* reporter Jerry Hirsch estimated the value of the credits to be "as much as $35,000" for each car Tesla sold in California in 2013.

By the beginning of the twentieth century electric lights were replacing kerosene lamps. However, around that time, people began driving cars with gasoline engines. Henry Ford's Model T was introduced in 1908 (Chapter 9.1.1). This gave oil a new market that turned out to be larger than the lighting market. The history of the worldwide per-person consumption of hydrocarbons was shown in Figure 1.6. Consumption grew rapidly until the Iranian Revolution in 1979, but it has been relatively flat since then. It was 1.0 toe/p in 2017.

Figure 5.5 shows the per-person hydrocarbon consumption broken out between the OECD countries and the non-OECD countries. The OECD consumption was 2.8 toe/p in 2017, 4 times the non-OECD consumption of 0.7 toe/p. This ratio has been declining slowly. It was 5 in 2007. To put the consumption in perspective, a student at a California university on a junior year abroad in Copenhagen would consume 0.7 toe in airplane fuel in one round trip during the year. This pattern is different from coal, where the consumption for OECD and non-OECD countries has been converging to 0.5 toe/p (Figure 4.4).

Over time, other hydrocarbons that were produced from oil wells began to find markets. These include natural gas, which is primarily methane (CH_4), and the natural gas liquids: ethane, propane, butane, and pentane. Figure 5.6 shows the natural-gas fraction of hydrocarbon production. It has risen from 22% in 1930 to 42% in 2017. Natural gas gradually replaced town gas for heating. A major

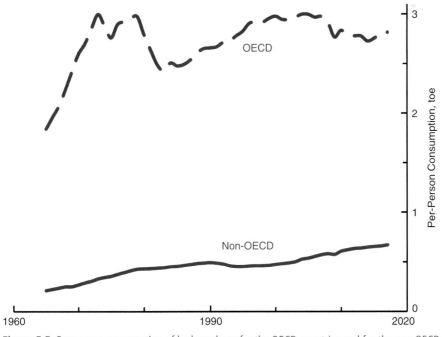

Figure 5.5 Per-person consumption of hydrocarbons for the OECD countries and for the non-OECD countries. The data come from the BP *Statistical Review*.

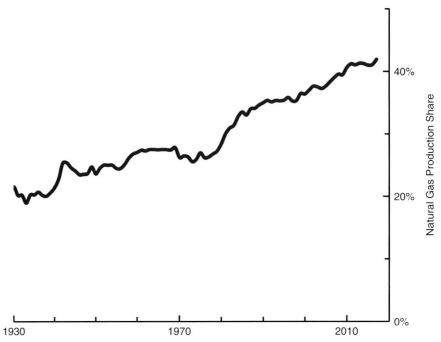

Figure 5.6 History of the natural gas fraction of world hydrocarbon production on an energy basis. Oil and natural gas before 1950 from Brian Mitchell, *International Historical Statistics*, various editions. Oil and natural gas from 1950–1959 from the United Nations, 1976, *World Energy Supplies, 1950–1974*. More recent production from various editions of the BP *Statistical Review*.

advantage of natural gas over town gas is that methane is not poisonous, while town gas contains carbon monoxide, which is. Methane has also become indispensable in fertilizer production. In addition, natural gas competes with coal in electricity generation. Natural gas plants are less expensive to build than coal plants, and they have higher efficiency and lower pollution levels. They have an important role in electricity grids with high levels of wind and solar generation because natural gas generators can adjust power levels more quickly than coal generators. Ethane (C_2H_6) is the most important feedstock for the plastics industry. Propane and butane are gases at atmospheric pressure, but can be liquefied at reasonable pressures. This allows them be to be transported conveniently as liquids in pressurized tanks. They can be burned in stoves and heaters like natural gas, but without a pipeline distribution system.

5.1.3   Oil and War

Historically oil has had a critical influence on military strategy and operations. The advantages of oil were recognized for warships early on by the United Kingdom and the United States. Previously battleships had been powered by coal. Oil has twice the energy density of coal and can be supplied to engines with pumps rather than stokers with shovels. This allowed the ships to have a greater range and smaller crews. The UK laid down the keel for its first oil-fired battleship, the *Queen Elizabeth*, on October 21, 1912 and the US followed with the *Oklahoma* five days later. At the beginning of World War I, at the First Battle of the Marne when the fate of Paris was in doubt, the French commander, General Joseph Gallieni, commandeered 500 gasoline-powered Renault taxis. The taxis transported 5,000 soldiers to the front at a vital moment. In World War II, oil fields themselves were important campaign objectives. The Japanese army attacked Sumatra to obtain oil for its armed forces. The attack was successful and the Japanese were able to operate the oil fields. However, Sumatra is 6,000 kilometers away from Tokyo. American submarines sank the Japanese tankers and prevented the oil from getting to Japan. Japanese navy and air force training and operations later in the war were crippled by lack of oil in the home country. In Europe, a primary objective of the German army in invading the Soviet Union was to capture the Baku oil fields. Baku is 3,000 kilometers from Berlin and in the push to the oil fields, the German army became over-extended. The Russian army successfully counter-attacked at Stalingrad (Volgograd today). The Battle of Stalingrad was the turning point of the war in Europe.

Oil has also been a factor in more recent wars. After the Iranian Revolution in 1979, Iraqi President Saddam Hussein judged that the Iranian military was sufficiently weakened that he could capture the Iranian province of Khuzestan that borders Iraq. Most of Iran's oil production comes from this province. Iraq invaded in September, 1980. By 1981, Iranian oil production had fallen to 1.3 Mb/d, compared with 5.3 Mb/d in 1978. The war became a major conflict, lasting eight years. Hundreds of thousands of people were killed. Then in 1990, Saddam Hussein came into conflict with Kuwait. Iraq owed Kuwait 14 billion dollars that Saddam Hussein did not want to pay. Another source of tension was the Rumaila oil field. This field is primarily in Iraq, but the tip extends into Kuwait. In this situation, it is common to negotiate a sharing agreement. The Saudi–Kuwaiti Neutral Zone is an example. There was no such agreement for the Rumaila oil field. Saddam Hussein complained that Kuwaiti production from the Rumaila oil field was too large.[3] In August, Iraq invaded and occupied Kuwait. The following year, in February, US President George H. W. Bush led a multi-national force that ejected Iraq from Kuwait. Oil was involved in almost every aspect of these two wars. Oil fields were Iraqi military objectives. Oil money was the basis of the loan from Kuwait to Iraq. The tanks and military aircraft for all three countries, Iraq, Iran, and Kuwait were purchased with oil money.

---

[3] Saddam Hussein also accused Kuwait of drilling slant wells to steal oil from the Iraqi side of the border. It is easy with logging equipment to determine if a well has a substantial slant. The Iraqi government had ample opportunity to prove its claim of theft when it occupied Kuwait, but it did not. Without further evidence, the charge should be considered unsubstantiated.

The oil business has been dominated by booms that alternate with crashes. There would be times when new discoveries were limited and oil companies would get high prices. Often during these periods some geologists would announce that the world was running out of oil and many people would pay attention to them. In recent years there has been a social movement associated with this concern called *peak oil*. This situation is different from coal, where there has never been a concern about the supply at the world level. Coal is a rock, and the seams can often be easily mapped from outcrops. However, there is nothing easy about going underground to mine coal. Hydrocarbons, on the other hand, are hard to find. They are usually not obvious from the surface, and they migrate from the source rocks where they were formed. However, once discovered, they are relatively easy to produce with wells. Pressure is your friend in hydrocarbon production. When there are new oil discoveries, production surges, and prices collapse. Then people criticize the geologists who had said that the world was running out of oil.

5.1.4   The Texas Railroad Commission and OPEC

A prime example of a boom was the development of the East Texas Oil Field that was discovered in 1930. The production reduced Texas oil prices to pennies per barrel. In response the Texas state government gave its Railroad Commission the authority to set oil production. The approach they developed was called *prorationing*. Each month, the Railroad Commission would make an estimate of the supply inside and outside the state of Texas. In addition, they would be provided with an estimate of the demand from the federal Bureau of Mines in Washington. Based on these two estimates, each oil company operating in Texas was assigned a number of days of the month that they were allowed to pump. The goal was to match the total supply to the demand. The policy effectively limited companies to a fixed proportion of their production capacity, hence the name prorationing. Prorationing was a radical idea. There were legal barriers to prorationing because American business markets were relatively free of regulation at the time. Some oil above the prorationing limits, called "hot oil," was produced and refined. In time these obstacles were overcome. The Railroad Commission found that the hot oil could be detected by monitoring refinery production. One factor that helped was that the state of Texas controlled its own land. Texas had been an independent republic and it joined the United States as a state rather than becoming a federal territory first. As a result, the federal government is not a significant land owner in Texas, in contrast to other states, where the federal government owns much of the land. A legal theory supporting prorationing based on avoiding damage to the oil fields was developed that was upheld by the United States Supreme Court. The advocates for prorationing persuaded the court that if the wells produced oil too quickly, the oil field production in the long run would suffer.

Figure 5.7 plots world crude-oil prices against production with time as the independent parameter. The era of the Texas Railroad Commission lasted from 1930 to 1972. During this time, world crude-oil production increased by a factor of 13. Prices were relatively stable and restrained, falling in a range from $10/b to $20/b, expressed in 2017 dollars. There were several factors in the success of the Texas Railroad Commission in establishing a stable environment for oil production and prices for 43 years. The state succeeded in stopping the hot oil. The demand estimates were set by a federal agency that remained independent of both the oil companies and commission. The commission solicited inputs from oil companies,

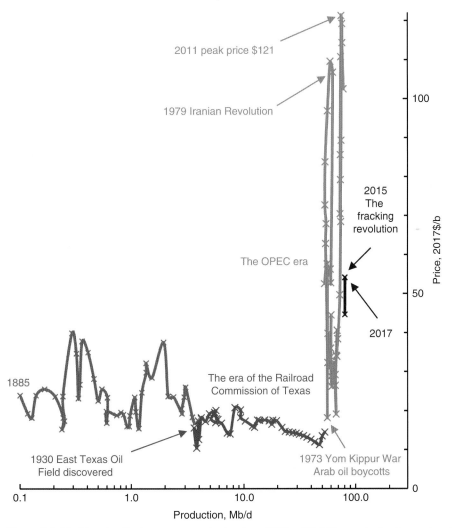

Figure 5.7 World crude-oil prices versus production from 1885 to 2017. The production is on a logarithmic scale. The inflation-adjusted prices are taken from the BP *Statistical Review*. Production before 1980 from Brian Mitchell, *International Historical Statistics,* and from 1980 on from the EIA, "International Energy Statistics." Crude oil excludes the natural gas liquids.

but for the most part it maintained its independence from them. In addition, it was not involved with foreign policy. Perhaps most importantly, Texas produced a significant share of world crude-oil production, so when the commission limited production, it mattered. Figure 5.8 shows the Texas share of world production over time. The average from 1930 to 1972 was 15%. It reached its all-time high of 24% in 1944, during World War II. However, by 1972, it had fallen to 6%, below Saudi production.

The OPEC era began in 1973 with the Yom-Kippur War and the Arab Oil Embargo. In percentage terms, the expansion during the 42 years of the OPEC era through 2014 was modest, only 40%. Prices were higher and much less stable than during the time of the Texas Railroad Commission, ranging from \$18/b to \$121/b. Several factors contributed to this. Saudi Arabia has been the largest producer in OPEC, but it has not consistently focused on stabilizing prices. It increased production during the 1990s when the price was low. It maintained production during the rise of tight-oil production in the United States in recent years. The Saudi average production share during the OPEC era was lower than the Texas share from 1930 to 1972. OPEC reserves accounting has not been transparent and it has been common for OPEC countries to produce more than their quotas.

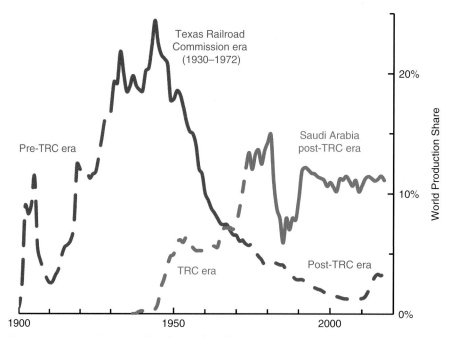

Figure 5.8 Texas and Saudi Arabia shares of world crude-oil production from 1900 to 2017. The source of the Texas oil production data through 1934 is personal communications from Jean Laherrere. The later Texas production data come from the Texas Railroad Commission. Saudi production from 1937 to 1979 from Brian Mitchell, *International Energy Statistics*. Later Saudi production from EIA, "International Energy Statistics."

OPEC countries have also participated in boycotts. The oil companies in OPEC are national oil companies, so they are subject to the oil ministries that regulate them. There is no independent agency comparable to the US Bureau of Mines to estimate demand. Finally, there have been wars between OPEC members, Iran versus Iraq, and Iraq versus Kuwait.

In Figure 5.8, there is a rise in Texas production in recent years. This increase is from the combination of two technologies. The first is called hydrofracturing, *fracking* for short, or less commonly, *fracing* or *fraccing*. The process consists of pumping water at very high pressure into the oil-bearing formation. This fractures the rocks. The idea goes back to the early days of oil production when explosives were sometimes set off down in wells to increase flow. In fracking, sand is usually included in the water to keep the cracks from immediately snapping shut. This increases the permeability of the rock so that the oil can flow. Oil rocks with initial low permeability are called tight reservoirs, and this type of production is called *tight-oil* production. Hydrofracturing has been used for more than a hundred years, but it has become important in recent years through the efforts of George Mitchell, who developed the Barnett Shale natural gas field in Fort Worth, Texas. The second technology is horizontal drilling. Reservoirs typically have a flattened shape, and drillers can greatly increase the well reservoir contact by drilling horizontally through the reservoir. Again, the technology has an older basis. Previously

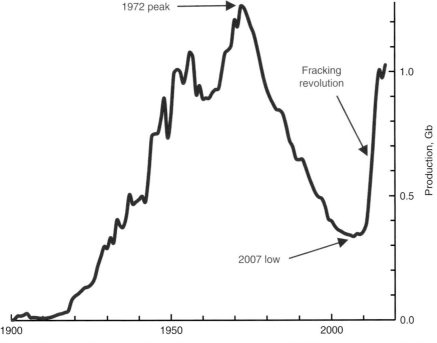

Figure 5.9 Texas oil production history. The data source through 1934 is personal communications from Jean Laherrere. The later data come from the Texas Railroad Commission.

wells were sometimes drilled on a slant. This allowed access to a reservoir under an area where there was no surface access. It was also sometimes used to drill into a neighbor's reservoir illegally. This happened in the East Texas Oil Field in 1962 when hundreds of slant wells were drilled. The wells were discovered and the Texas Rangers shut them down.

Figure 5.9 shows the production history for Texas crude oil. The fracking revolution is associated with a sharp rise in production after 2007. Figure 5.10 shows the same data as a Hubbert linearization. There was an established trend that gives a projection for ultimate production of 63 Gb. The new fracking technology caused a dogleg in the plot. The cumulative production through 2017 was 67 Gb, so it has already passed the earlier projection.

In this chapter, we will start with a discussion of the origins of oil and gas. Then we will consider the history of the concerns about peak oil, with a critical look at the work of King Hubbert. Following this we will discuss the role of new technology in producing hydrocarbons, focusing on George Mitchell's development of fracking. We conclude with a projection of ultimate hydrocarbon production to go along with the projection of ultimate coal production in Section 4.7.3. In contrast to coal, the projections for ultimate hydrocarbon production are larger than the original reserves.

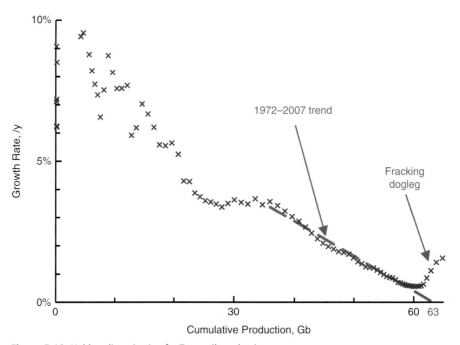

Figure 5.10 Hubbert linearization for Texas oil production.

5.2   Origins

In the previous chapter, we associated coal with the accumulation of organic material on land. In contrast, oil and gas come from organic material in the sea. In the ocean organic material is constantly raining down – plankton, parts of plants, dead animals and feces. Ordinarily this material would decay, its carbon ultimately oxidized by respiration. However, under some circulation conditions the dissolved oxygen at the bottom of the sea becomes depleted. Then organic material accumulates in the sediment along with mud, sand, and shells. Much of the organic material will initially be in chains of 20 carbon atoms or more. These compounds are usually waxy solids at room temperature. The next step in forming oil and gas is burial. This can happen at a river delta where massive amounts of mud and sand are dumped into the sea. Figure 5.11 shows the growth of the Atchafalaya delta over a thirty-year period. Compare the two photographs carefully to find the places where the delta has grown. These piles of sediment can bury organic material thousands of meters below the surface. As the depth increases, the temperature rises. This *geothermal gradient* is about 2.5 K/100 m. Once the organic material is deep enough, the heat breaks the carbon chains in a process called *cracking*. When the depth is greater than 2,300 meters, corresponding to a temperature of 80 °C, the carbon fragments become short enough to form a liquid, oil.

Oil contains many different compounds. The majority are *saturated* hydrocarbons, where the carbon–carbon bonds are single. These include the *alkanes*, made up of carbon chains and branched carbon chains, and the *naphthenes* with carbon

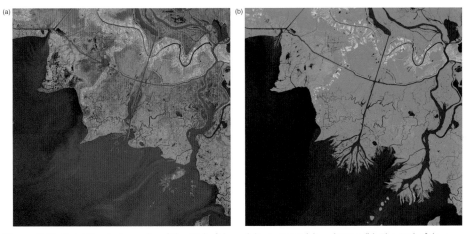

Figure 5.11 Satellite photographs of the Atchafalaya delta in 1984 (a) and 2014 (b). The Atchafalaya River is a distributary, or branch, of the Mississippi River. The width of the photograph encompasses 32 km. Credit: US National Aeronautics and Space Administration (NASA).

rings. The alkanes can be written in the form C_nH_{2n+2}. For example, octane, C_8H_{20}, is an important fraction of gasoline, and dodecane, $C_{12}H_{26}$, is a significant component of diesel fuel. Octane has a lower boiling temperature than dodecane, and this allows them to be separated by distillation in a refinery. It is interesting that the US and Europe have historically split a barrel of oil, with the gasoline fraction going to the US and the diesel portion to Europe. This is the result of a difference in emphasis, with the US focusing on minimizing air pollution and Europe on maximizing fuel mileage. Air pollution from cars was first a problem in the United States, particularly in Los Angeles. This led the US to favor gasoline engines for cars, because it is easier to control the emissions in spark-ignition gasoline engines. In diesel engines the combustion is initiated by compression and the fuel burns at a higher temperature than in a gasoline engine. This makes it easier to design diesel engines that are more efficient than gasoline engines. On the other hand, the higher temperatures in diesel engines increase nitrogen-oxide emissions, and diesel engines have more trouble with particle emissions than gasoline engines. Europe has historically had high fuel taxes that encourage drivers to buy small cars with efficient engines. In contrast, the most popular vehicle in the US for many years has been the Ford F-series pickup truck. The pickups have lower fuel mileage than European sedans, but they can carry cargo and pull heavy trailers.

5.2.1 The Oil Window

If the organic material is pushed down below 4,600 meters, corresponding to a temperature of 130 °C, the molecules are cracked all the way to methane, CH_4, which is the major component of natural gas. In the early days, companies were much more interested in oil than natural gas because oil sells at higher prices than natural gas on an energy basis. Natural gas also requires pipelines, while oil can be moved by trucks and trains. In analyzing an oil prospect, the geologists needed to assure that the organic material had been deeper than 2,300 m but not deeper than 4,600 m. The range from 2,300 m to 4,600 m is called the *oil window*. Much of the natural gas that was produced was actually associated with oil wells. This *associated gas* occurs as a cap on top of the oil because of buoyant forces. The gas pressure actually provides a drive that helps oil production. Sometimes oil-field operators pump carbon dioxide down into a field to provide an artificial gas drive. This is called *enhanced oil recovery*, or EOR. In recent years, where a pipeline network is available, people have come to appreciate natural gas as a clean, flexible fuel that can be distributed to homes and businesses. Wells are now drilled specifically for natural gas. The fracking revolution is at least as much about gas as oil, and the production of natural gas in the US is larger than either oil or coal in energy terms.

There are also intermediate compounds that are recovered primarily from gas wells. These are called *natural gas liquids* (NGLs). They are primarily ethane (C_2H_6), propane, (C_3H_8), and butane, (C_4H_{10}). In interpreting oil and gas production statistics, one needs to be aware that the natural gas liquids are sometimes included in oil production, sometimes in gas production, and sometimes they are treated as a third category. The BP *Statistical Review* includes NGLs with oil. When people talk about *crude oil* production, they are excluding the natural gas liquids. The oil price indexes are for crude oil. *Wet gas* includes NGLs with the natural gas, *dry gas* excludes them. One should be aware of these details because it is easy to make plots that inadvertently count the NGLs twice or leave them out. The NGLs are valuable. Ethane is the foundation for the plastics industry. It is converted to the non-saturated alkene, ethylene (C_2H_4), which is then made into polyethylene. Polyethylene consists of long saturated carbon chains that can be represented by the chemical formula $(C_2H_4)_n$. It is the most important plastic, with world production approaching 100 Mt/y. Propane and butane are widely used as fuel in heaters, stoves, and refrigerators.[4] They are gases at room temperature and atmospheric pressure, but they are liquids at modest pressures, and they are sold as liquids in pressurized containers. At room temperature the vapor pressure of propane in a tank is eight atmospheres while for butane it is two atmospheres. Propane and butane are also sold in a mixture called *liquefied petroleum gas* (LPG). These fuels are burned as gases. A heater or stove needs a pressure of only a fraction of an atmosphere, so a regulator is attached to the tank to reduce the pressure to the appropriate level. Propane and butane are important when a natural gas pipeline system is not available. LPG has a special significance in lower-income countries where poor families burn wood indoors for cooking and heating. The smoke is a major health hazard. In addition, in some areas burning wood fuel is a major reason for deforestation. As family incomes rise, they can switch to LPG, which greatly reduces the smoke hazard and the deforestation pressure. This is one of the major benefits of the increasing income in lower-income countries in recent years (Figure 1.19).

When oil drops form by cracking the longer carbon chains, surface tension will tend to hold them to the surface of pores in the rock. However, if enough oil is formed, the drops can coalesce and flow. The liquid has a lower density than the original organic material and this provides pressure to induce the oil to move. Oil has a lower density than water and when water is present, there is a buoyant upward force on the oil, which causes the oil to migrate towards the surface.

---

[4] These are *absorption* refrigerators, rather than the usual *compression* refrigerators. An absorption refrigerator burns a fuel like propane to boil a solution containing the refrigerant. The refrigerant could be ammonia and the solvent water. This liberates ammonia as a gas, and the water is returned with a condenser. Absorption refrigerators do not have a compressor, so they are quiet. Electricity use is low, limited to the control circuits and the display. This makes them useful in recreational vehicles (RVs) where the refrigerator can be close to a bed and the electricity may come from a battery with a limited capacity.

Near the surface the oil is attacked by bacteria. What's left after the bacteria have had their dinner and the volatile organic compounds have evaporated are natural asphalt lakes, or tar pits. "Asphalt" is the American English. The British English is *bitumen,* as it is for geologists. Asphalt is viscous, and animals are often trapped in tar pits. The skeletons are preserved in large numbers. More than a million bones have been recovered from animals in the La Brea Tar Pits in Los Angeles. Incidentally *brea* means asphalt in Spanish.

Asphalt is also produced in refineries as a residue after the more volatile compounds have been separated out by distillation. The terminology is confusing to people who do not work with the material professionally. In the chemical industry, the word "tar" is reserved for this residual distillation product. Tar can also be produced from coal and wood. Asphalt is not ordinarily burned for fuel.[5] However, it is very useful. It is mixed with sand and gravel to make durable roads. Asphalt shingles are the most common roofing material in the United States. Under the heat of the sun the asphalt softens to form a seal between the shingles. They are inexpensive to install and they hold up under the sun for decades, even in sunny places. They are usually constructed with a fiberglass base to add strength and embedded mineral granules on the surface to protect against the sun.

Natural gas also tends to move toward the surface. The methane may vent or it may dissolve in aquifers. The vents are the sources of the eternal flames that are sometimes found in nature. These can be ignited by lightning and can burn for thousands of years. Methane is common in water wells and often makes its presence known by fizzing. It is not poisonous. We have bacteria in our digestive system that produce methane through anaerobic digestion, and this make flatus flammable. In water wells, methane is a fire hazard and there should be a vent to get the gas safely into the atmosphere. Coal seams also contain methane produced by bacteria and if a well penetrates a coal seam, the methane will get into the water. The methane can be released even if the well does not reach the coal seam. When a well draws water, it reduces the local hydrostatic pressure, and that pressure change can release the methane from the coal. The methane in coal is dangerous for miners because the methane can explode. Today wells are often drilled into the coal seams to capture the methane. This eliminates the explosion risk, and the methane is distributed through the regular natural gas pipeline system. This *coalbed methane* is about 5% of US natural gas production.

Where oil and gas wells pass through an aquifer, cement is pumped around the casing to seal off the aquifer. Sometimes these cement seals leak. The leaks are

---

[5] When asphalt burns, it produces terrible fumes. However, it can be upgraded to a synthetic oil. Asphalt has many linked carbon rings and it is upgraded by breaking carbon–carbon bonds between the rings and replacing them with carbon–hydrogen bonds. This is done for much of the Canadian tar sands production.

more common in natural gas wells than in oil wells because gas diffuses more readily than liquid. In the media, the leak risk is often linked to the new fracking technology. However, the risk of a poor cement job at the aquifer is present in all oil and gas wells that pass through aquifers, whether the wells are fracked or not. It is not a new risk. A chemical analysis on the water can usually determine whether oil and gas operations are responsible or if the cause is natural. If the methane is *biogenic*, that is, produced by bacteria, there will not be any propane or butane. If the methane is *thermogenic*, produced by the natural cracking of large organic molecules, propane and butane will be present. This can be an important analysis for a family that owns a water well with methane contamination, because it determines whether the family would be compensated for the contamination by oil and gas companies working in the area.[6]

5.2.2 Hydrocarbon Traps

It is possible that oil and gas may be blocked in its migration toward the surface so that it accumulates in a *reservoir*. An example is shown in Figure 5.12. In the figure, oil is produced in the *source rock*, and pressure and buoyancy forces cause

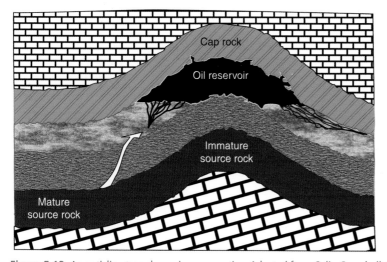

Figure 5.12 An anticline trap shown in cross section. Adapted from Colin Campbell, 2005, *Oil Crisis*, Multi-Science Publishing Company, Ltd.

---

[6] The most famous of these is the family of Mike Markham of Weld County, Colorado. Markham set his water on fire in the 2010 movie *Gasland*. Two years earlier the Colorado Oil and Gas Conservation Commission had analyzed his water. The chemical analysis of his water showed propane and butane were not present in the water, so the contamination did not come from oil and gas operations. The movie maker, Josh Fox, did not let the viewers know about the exculpatory analysis. On the contrary, viewers were left thinking that the Markham family problem resulted from oil and gas operations. A good starting point for investigating this further is the statement by the Colorado Oil and Gas Commission at www.energyindepth.org/wp-content/uploads/2018/02/GASLAND-DOC.pdf

it to migrate upward. Tar pits are a sign of a nearby source rock and people often drill near tar pits hoping to find a reservoir. Los Angeles is known for Hollywood now, but it first became a major city when wildcatters found oil in 1890 by drilling near the La Brea tar pits. In the early 1900s the Los Angeles oil fields produced a quarter of the world's oil. The reservoir should be a porous rock like sandstone or limestone. There must be an impermeable *cap rock* above the reservoir rock. The cap rock must make a seal. The rock itself must be impermeable and there cannot be a fracture path through the rock. The seal must be very good. A leak of only a liter per hour will cause a loss of a billion barrels over 20 million years. Mineral layers that form by evaporation like salt can make good seals because they are ductile and they flow under pressure to close cracks. However, the most common cap rocks are *mud rocks*. Mud rocks are sedimentary rocks made of fine mud particles. Examples are *shale*, which is made up of thin layers, and *mudstone*, which lacks the layers. There is plenty of mud rock. It is the most common sedimentary rock.

In the figure, the immature source rock has not been deep enough to enter the oil window. The organic precursor to oil in the immature source rock is called *kerogen*. Kerogen is a waxy solid. With additional heat, it could become oil. The most important kerogen formation is the Green River Basin in Wyoming. There were large programs started during the Carter Administration in the late 1970s to try to produce oil from the Green River formation. The projects all failed and they were shut down. The process was extremely expensive and it generated enormous amounts of waste rock. It does not appear to be competitive with the current alternatives to fossil fuels and there is little sign that the programs will be started again. These kerogen formations are properly classified as occurrences rather than resources or reserves.

In Figure 5.12, the layers have buckled to make a ridge. This structure is called an *anticline*. A spectacular example of an exposed anticline is shown in Figure 5.13. An anticline can trap oil and gas. The largest oil field in the world, the Ghawar field in Saudi Arabia, has an anticline trap. Its ultimate production will likely exceed 100 Gb. Anticlines are often visible from the surface. Many early oil fields were discovered simply by drilling into anticlines. For comparison, the Ghawar source rock is 150 million years old.

5.2.3   Dad Joiner and the East Texas Oil Field

There are many different types of traps and a trap may give no sign at the surface. One that is historically significant is the trap for the East Texas Oil Field, the second most productive oil field in the US after Prudhoe Bay, with a cumulative production of 5 Gb. The East Texas Oil Field was discovered by Columbus Marion "Dad" Joiner (Figure 5.14). Joiner acquired his nickname "Dad" in recognition that he was the father of the field. The story is one of great drama. It is not easy to distinguish the myth from the reality because the principals were part oil men and part con men. This account mostly follows that given in *The Last Boom*, by James

Figure 5.13 The Blue Mountain anticline exposed by a highway cut in Juniata County, Pennsylvania. The highway is US22/US322. Credit: Scott A. Drzyzga, PhD, GISP, and Remote Pilot.

Clark and Michel Halbouty. It is listed in the references at the end of the chapter. Joiner was born in 1860 in Alabama. He was a state legislator in Tennessee. Then he moved to Oklahoma, where he made and lost a fortune in land deals. In Oklahoma he became interested in wildcatting. A *wildcatter* is someone who drills wells on land where no oil has yet been found. He teamed up with A. D. "Doc" Lloyd, born Joseph Durham, a self-taught mining engineer, who chose the drilling spots. Joiner and Lloyd came within a whisker of discovering the Seminole Oil Field when they ran out of money and stopped drilling a few hundred feet above the reservoir. Joiner then moved to Texas, where he set up an office in Dallas in 1925. One of his first oil leases in Texas was the farm of a widow, Daisy Bradford, near the town of Henderson. Joiner then set out to raise money to drill a well. By all accounts, he was a charming and persuasive man and he was strongly supported by the local citizens. Doc Lloyd wrote a geology report for investors and Joiner added a cover letter that predicted that oil would be found in the Woodbine Sandstone that started beneath the Austin Chalk at 3,550 feet. Professional geologists scoffed at Lloyd's report. Joiner persisted nevertheless. He was limited by poor equipment. His first well, in 1927, Daisy Bradford number 1, reached 1,100 feet and then the pipe stuck in the well. In 1928, Joiner changed drillers and started a second well. By March, 1929, the well reached 2,500 feet, but then a piece of pipe broke off and

Figure 5.14 The most famous photograph in the history of the petroleum industry. The two men shaking hands are Dad Joiner, on the left, and Doc Lloyd, on the right. The man with the straw hat and cigar, third from the right, is H. L. Hunt, who bought Joiner out. The man to Hunt's right with his hat off is Ed Laster, the driller. This picture was taken in 1930 at the time of a successful earlier test before the casing was cemented in. From the Prints and Photographs Collection, the Dolph Briscoe Center for American History, the University of Texas at Austin.

blocked the well. Undaunted, Joiner hired a third driller, Ed Laster, and spudded (oil speak for "started") Daisy Bradford number 3 in May 1929. In July 1930 Laster hit sandstone at 3,500 feet. He retrieved a core sample and took it home to inspect it in secrecy. Oil! The drilling crew cemented in a casing and added storage tanks to prepare the well for production. On Sunday, October 5, 1930, in front of thousands of onlookers, the crew swabbed out the well. The earth trembled and Laster ordered cigarettes to be put out. Up went a black geyser. The drilling crew diverted the oil into a tank and measured the flow rate – 6,800 barrels per day.

The trap for the East Texas Oil Field is shown in cross section in Figure 5.15. There is nothing at the surface that indicates its presence. This is why oil companies were dismissive of Joiner and Lloyd's efforts. The oil reservoir in the Woodbine Sandstone is shown in red. The Woodbine Sandstone formed 90 million years ago during the Cretaceous Period from a river delta. Next it was tilted by tectonic activity. Then the eastern part eroded away, and the Austin Chalk layer was deposited on top. The angle of the eroded lower sandstone layer is shifted from that of

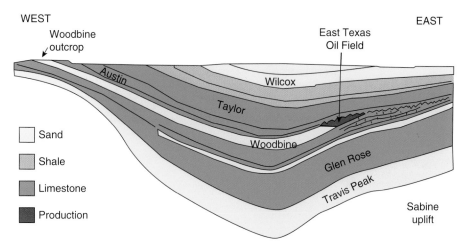

Figure 5.15 The East Texas Oil Field trap. For scale, the distance from the red oil production area to the Woodbine outcrop is 200 km. From Michel Halbouty, 2000, "Exploration into the New Millennium," at the Second Wallace Pratt Memorial Conference, American Association of Petroleum Geologists.

the upper limestone layer. Geologists call this an *angular unconformity*. The tilt shown in the figure is exaggerated. The dip is actually 0.5°. The Austin Chalk layer forms the cap rock. To the west, the Woodbine sandstone outcrops at the surface. For this reason, the western section could not form an oil reservoir. The East Texas reservoir is superb – 80% of the original oil in place has been recovered. This is the highest recovery efficiency of any major oil field in the world. The maximum thickness of the oil zone was 38 meters. The *porosity* of the rock, which is the fraction occupied by oil and water, is 25%. The *permeability*, which is a measure of the ability of the oil to flow through the rock is measured in darcies, abbreviated D. The East Texas reservoir rock has an outstanding permeability of 2.1 D. In contrast, the permeability of the Austin Chalk cap rock is less than one millidarcy.

There is water below the oil in the sandstone. This is shown in Figure 5.16. The pressure from the water helps drive the oil into the wells. This is called *water drive*. As the oil is withdrawn, the contact level between the oil and water rises. In addition, water is produced with the oil. Water from this depth contains dissolved minerals and is not suitable for drinking. This presented a problem for disposal. In addition, there was concern that the water pressure was dropping and becoming less effective in oil production. There are other problems if the original pressure is not maintained. The land surface subsides and causes damage to structures. In seismically active areas, earthquakes can be induced. The solution was to drill wells to inject the water back into the sandstone reservoir. Water injection started in 1938 and continues to this day. The field currently produces 100 barrels of water for every barrel of oil. This water injection to improve oil production is called *secondary recovery*. It is distinguished from gas and steam injection, which is *tertiary recovery*. Nowadays

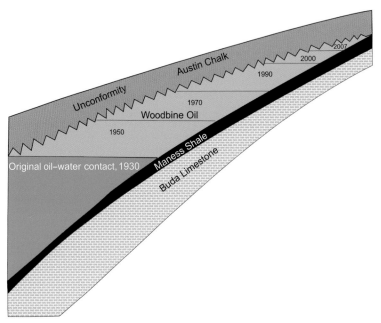

Figure 5.16 Water encroachment over time in the East Texas Oil Field. Redrawn from Fred Wang, 2012 "Development Strategies for Maximizing East Texas Oil Field Production," RPSEA Onshore Production Conference: Technological Keys to Enhance Production Operations, Midland, Texas.

water injection is a routine but critical part of oil field management from the beginning.

Figure 5.17 shows the production history for the East Texas Oil Field. There was an early peak in 1933 at 216 Mb, followed by a gradual decline. In the late 1960s and early 1970s, the Texas Railroad Commission relaxed the prorationing limits. Production increased for a while, but then resumed its decline.

The Laherrere linearization is shown in Figure 5.18. In the early years from 1931 to 1933, production is increasing, so the points would not be used in a fit for the x-intercept. The projection for the years from 1934 to 1955 was for an ultimate production of 5.4 Gb. For the years after prorationing was lifted from 1972 to 1986, there is a trend with an intercept of 6.3 Gb. But this was a mirage caused by the lifting of the prorationing limits. Since then, the locus has reverted back to the earlier trend. The current cumulative production is now 5.4 Gb, equal to the early projection. Production in the East Texas Oil Field continues, even though the output is 99% water. The output was 2.6 Mb in 2017.

Dad Joiner's faith and determination were remarkable, particularly considering his age. At the time of the discovery of the East Texas Oil Field, he was 70 years old. However, he had no clue of the size of the oil field that he had discovered, and the flow from the well was sporadic. Moreover, Joiner had greatly oversold the interest in his wells, so that when the well came in, he owed much more than he

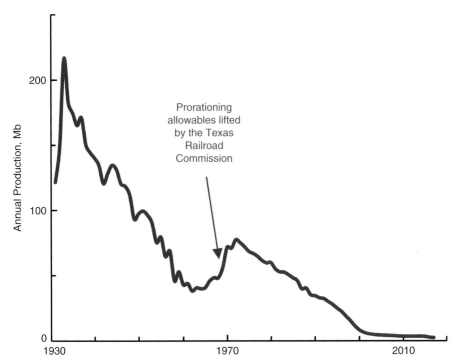

Figure 5.17 Production history of the East Texas Oil Field. The source of the production data through 1992 is personal communications from Jean Laherrere. The later data come from the Texas Railroad Commission.

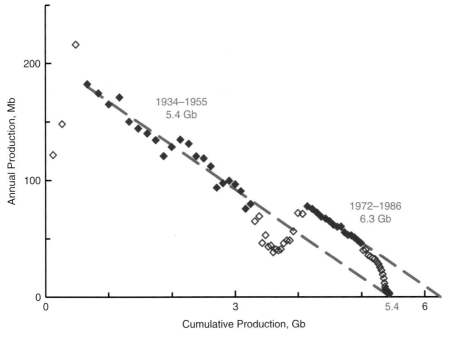

Figure 5.18 The Laherrere linearization for the East Texas Oil Field.

made. This made him a target of lawsuits. As it happened, there had been a visitor to Joiner's well the month before, Haroldson Lafayette "H. L." Hunt, who appears in the historic photograph, Figure 5.14. Hunt made an inspired call that the well was on the eastern edge of a substantial oil field. After an intense negotiating session, he bought Dad Joiner's leases and his legal troubles for $1,335,000. Hunt was occupied for years in settling lawsuits, hundreds of them. But his instincts were correct. The East Texas Oil Field was a monster, extending 70 km north to south and 20 km east to west. The leases were the foundation of an enormous family fortune. Joiner, on the other hand, went to Juarez in 1933 to divorce his wife and marry his 25-year-old secretary Dea England. He continued wildcatting and died poor at age 87.

5.3   Drilling for Oil and Gas

Oil and gas well drilling is a highly evolved technology. Early wells were vertical, but the trend has been towards horizontal wells that give much greater contact with the reservoir. Although horizontal wells are more expensive than vertical wells, they have higher ultimate production and shorter lifetime than vertical wells. This makes it easier to hit a time of high prices. In addition, with horizontal wells it is possible to drill many wells from the same pad. This reduces the impact on the surrounding community. Oil and gas wells can be drilled in urban areas where it would be out of the question to mine coal. A pad with eight wells is shown in Figure 5.19. In the foreground are seven gas wells. There is an eighth well, not visible, to the right. The wells produce from the Barnett Shale. The light blue box in the background is a compressor to separate the natural gas liquids. These are stored in the four tanks in the background on the right. There is also a compressor, not visible, off to the right to remove the produced water, along with tanks to store the water. The gas is removed by pipeline in the background on the left. The natural gas liquids and water are taken away with trucks. Originally the land formed part of a small horse ranch for teaching children to ride horses and it could be returned to a similar purpose when the wells are shut down. The main impacts of these wells are the compressor noise and the truck traffic. The truck traffic is significant at the fracking stage, but drops once the wells are in production.

The wells pass through aquifers, and these aquifers must be protected from contamination. The contamination risk is usually not from the production reservoir directly, which is typically thousands of feet below the deepest fresh-water aquifer. However, on its way to the production reservoir, the well can pass through other rock layers containing hydrocarbons and coal seams with methane. In addition, it should be recognized that the contamination risk is present in all oil and gas wells, not just those that have had hydraulic fracturing. Figure 5.20 shows how

Figure 5.19 A well pad in the city of Fort Worth, Texas. The area of the pad is comparable to a football field. Photograph by the author.

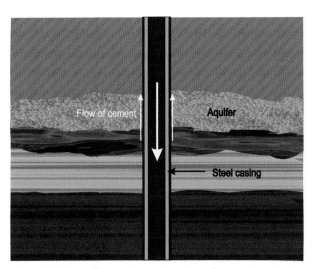

Figure 5.20 Installing and cementing the surface casing to isolate an aquifer. In practice there would be additional cemented casings inside the surface casing that are not shown in the figure.

the isolation is done. The drilling proceeds until the well has passed through a fresh water aquifer. Then a steel pipe, called the *surface casing*, is lowered into the well. Cement is pumped down the inside to the bottom of the casing and forced back up the outside of the casing. The cement and casing prevent the walls from collapsing. Drilling can proceed after the cement plug inside the casing is drilled out. The cement forms the barrier isolating the aquifer. It is not foolproof. A cement job can be botched and the cement barrier can degrade over time. The cement is a better barrier for oil than for natural gas, which diffuses more easily through cement than oil. There are times when contamination occurs and oil and gas companies have to pay compensation for the damage.

5.3.1 Fracking Technology

The next step is to create the horizontal section of the well. A downhole drilling motor is lowered into the well. The motor is powered hydraulically with drilling fluid. The motor assembly has a bend that allows it to drill on a curve to enter the reservoir horizontally. The horizontal part of the well may be several kilometers long. In Figure 5.21 steel casing is installed and cemented to stabilize the well. The casing is perforated with explosive charges to allow the oil or gas into the well. It is now common to increase the permeability of the reservoir rock by *hydraulic fracturing, or fracking.* In fracking a water solution is pumped in at high pressure to fracture the rock. The fracturing solution includes a *proppant*, typically sand grains a half millimeter across, to keep the fractures from sealing shut. The process is designed to take place entirely inside a reservoir that is initially relatively impermeable. This reduces the chances of contamination from the fracking process itself.

The hydrofracturing process does increase the reservoir pressure when water is injected. Increasing or reducing the pressure in a reservoir creates a risk of earthquakes. However, the pressure drops back down as the water and hydrocarbons are

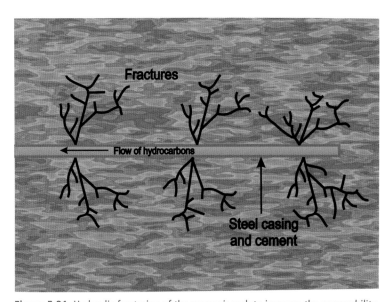

Figure 5.21 Hydraulic fracturing of the reservoir rock to increase the permeability.

Table 5.2 Earthquake risks from energy production in the United States.

	Number of projects	Number of events	Maximum magnitude
Steam geothermal (water injection)	The Geysers, California	300–400/y	4.6
Secondary recovery (water injection)	108,000 wells	18 sites	4.9
Tertiary recovery (gas injection)	13,000 wells	none known	na
Hydraulic fracturing	35,000 wells	1	2.8
Hydrocarbon withdrawal	6,000 fields	20 sites	6.5
Wastewater disposal	30,000 wells	9	4.8

Information from Table S.1 in National Research Council, 2013, *Induced Seismicity Potential in Energy Technologies*, National Academies Press.

produced, so it is a temporary risk. Table 5.2 shows a comparison of the seismic activity that has been associated with different energy technologies. A critical factor is whether the wells are being drilled in an area that is already prone to having earthquakes. It is often not possible to map the faults, and they may be discovered only after the earthquake has occurred. In making comparisons of risk, one also needs to consider the number of wells involved and the magnitudes of the observed earthquakes. People who live in earthquake country often have wood-framed houses with stucco walls. The wood frame sways, but is not damaged in even large earthquakes, and stucco cracks are easily patched. Earthquakes with magnitudes less than 3.0 are usually not perceptible unless one is in a special situation, like reading in bed or sitting in an office in an upper story of a building. Earthquakes of magnitudes from 4.1 to 4.9 will usually be felt. They do minor damage, like cracks in walls and windows. An earthquake of a magnitude 6.5 kills people.

From the table, the Geysers geothermal electricity field in California has been associated with an earthquake of magnitude 4.6, secondary recovery, magnitude 4.9, and waste water disposal wells, magnitude 4.8. Since this table was compiled, there was a magnitude 5.8 earthquake at Pawnee, Oklahoma in September, 2016. The movement occurred on a fault that had not been previously recognized. There were water disposal wells in the area and the state government shut them down after the earthquake. Fortunately, there was only one minor injury. The largest earthquake in the table was associated with a conventional oil field. This was a magnitude 6.5 earthquake in 1983, eight miles northeast of the town of Coalinga, California. The quake injured 94 people. By comparison with these earthquakes, the seismic activity associated with fracking is small, with one magnitude 2.8 quake for 35,000 wells.

5.3.2   Enhanced Oil Recovery (EOR)

If an oil reservoir is near the surface, the oil can be attacked by bacteria, and volatile organic compounds evaporate. The liquid that remains is quite viscous, and it is called *heavy oil*. An example is the Kern River Oil Field, near Bakersfield, California. This is the fifth largest oil field in the United States, with a cumulative production of 2.3 Gb through 2017. Oil was discovered at a depth of 70 feet in 1899. In the initial phase, production peaked in 1903 at 15 Mb, and then faded away to 2.6 Mb in 1954 (Figure 5.22). By 1961, cumulative production had reached 360 Mb. Production was hampered by the fact that the oil moved extremely slowly towards the wells. In 1962, the operators tried injecting steam into the reservoir to reduce the viscosity by heating the oil. It is done in a cycle called *huff and puff*. In the first stage steam is pumped into the well for a number of days. Then there is pause for a few days to let the hot water heat up the oil. Finally, there is a production phase to produce the now less viscous oil. This was successful, and production increased to 52 Mb by 1985. However, since then production has fallen steadily, reaching 22 Mb in 2017. To date 84% of the production has been in the huff and puff phase. The water produced with the oil is treated and used for irrigation. Figure 5.23 shows a photograph of the oil field. This is a heavily industrialized site that would not be available for any other purpose for a very long time.

Injecting a gas like steam into a well to improve recovery is called *tertiary recovery*. It is considered a form of *enhanced oil recovery*, or EOR. It could be

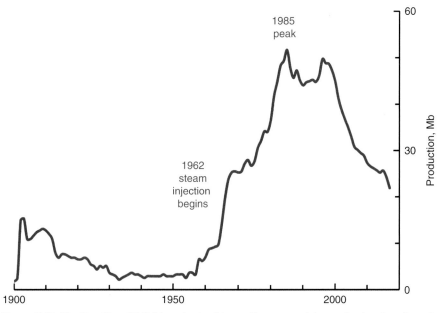

Figure 5.22 The Kern River Oil Field production history. The source of the production data through 1976 is personal communications from Jean Laherrere. The data from 1977 through 2017 are from the California Department of Conservation.

Figure 5.23 The Kern River Oil Field, near Bakersfield, California. Credit: halbergman/iStockphoto.

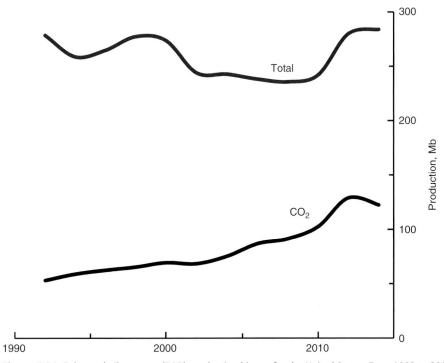

Figure 5.24 Enhanced oil recovery (EOR) production history for the United States. From 1992 to 2014, the *Oil and Gas Journal* did an EOR survey every other year and published it in April at the beginning of the month. These are the data from the surveys.

compared to secondary recovery, where water is injected into the well. At times when there are concerns about oil supplies it is common to propose that EOR can be used to greatly increase production. Historically the contribution has been of some significance in the US, but it has not been growing. Figure 5.24 shows the production history. The production in 2014 from EOR was 284 Mb. This was 7% of US production for that year. In later years, there has been a growing share of EOR with CO_2 injection. Carbon-dioxide EOR produced 122 Mb in 2014. Carbon-dioxide EOR is also sometimes proposed as a way to bury carbon dioxide

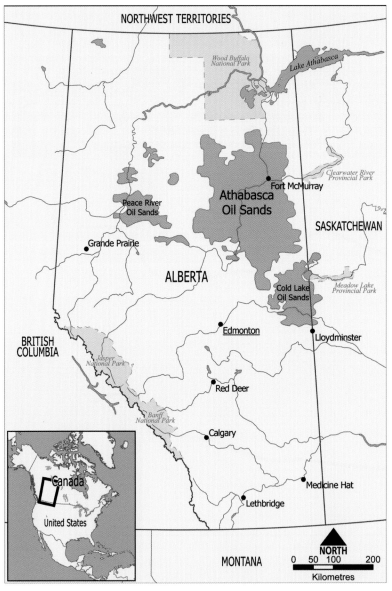

Figure 5.25 Map of the Canadian tar sands. Credit: Norman Einstein, public domain.

to reduce CO_2 emissions. The problem with this is that more carbon atoms are brought up in the oil that is produced and burned than are sequestered in the oil field. This means that on balance that CO_2 EOR increases CO_2 emissions rather than decreases them.

While EOR is not particularly significant in the United States, it is important for producing the bitumen in the Canadian tar sands. These reserves were 163 Gb in 2017, 10% of the world total. Figure 5.25 is a map of the tar sands fields. They are in northern Alberta. This area is boreal forest, savagely cold in the winter, with a record low temperature in Fort McMurray of −53 °C. Nevertheless, tar sands production has been growing rapidly. Figure 5.26 shows the production history. The 10-year annualized growth rate through 2017 is 8.5%/y. Tar sands production is now 55% of Canadian oil production, and 2.9% of world production.

The Canadian bitumen production has been criticized because it has a higher carbon coefficient than conventional oil. For example, here are comments by influential climate scientist Jim Hansen on the proposed Keystone XL Pipeline that would take Canadian bitumen to American Gulf Coast refineries.

"What would the effect be on the climate?" [Bill] McKibben asked. [Jim] Hansen replied, "Essentially, it's game over for the planet."[7]

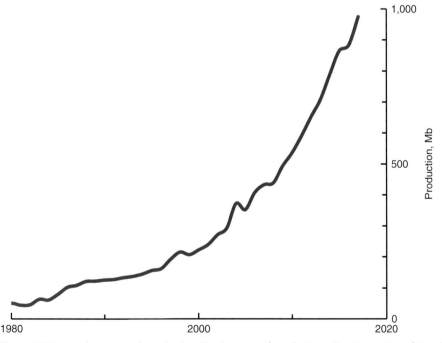

Figure 5.26 Canadian tar sands production. The data come from the Canadian Association of Petroleum Producers, *Canadian Oil Producer's Handbook.*

---

[7] Quoted in Jane Mayer, 2011, *The New Yorker*, "Taking it to the Streets."

The issue is that hydrogen and heat are needed to turn bitumen into a liquid fuel and these are supplied by local natural gas. In Hansen's accounting, the carbon-dioxide emissions from this natural gas are assigned to the bitumen. It should be noted that natural gas is a premium fuel, and it is likely that the gas would be produced and burned in any event.

The tar sands production is of strategic importance to the United States. In 2017, Canada provided 43% of US oil imports, up from 18% ten years earlier. Unlike the Arab countries, Canada has never boycotted the US. At the same time, the US is a vital customer for Canada. The United States has refineries that can handle bitumen, and virtually all of Canada's oil exports, 98%, go to the US.

5.4 Impacts

5.4.1 The First Gulf War

The greatest environmental disaster associated with oil and gas production occurred during the First Gulf War in early 1991. When the Iraqi Army was expelled from Kuwait, the retreating soldiers set six hundred oil wells on fire under orders from Saddam Hussein (Figure 5.27). The fires produced an enormous amount of smoke that affected the entire Gulf Region. It is likely that a billion barrels of oil were burned. The Iraqi Army had laid mines around the wells, and this made it difficult to put out the fires. This was a deliberate destructive act

Figure 5.27 Oil well fires in Kuwait set by retreating Iraqi soldiers in 1991. Credit: Everett Historical/Shutterstock.

by an authoritarian government without any military justification whatsoever. Kuwait's fires were put out by November and production recovered fully by 1993. We turn next to an unintended oil well problem that was mitigated by technology, land subsidence.

5.4.2   The Long Beach Subsidence

Long Beach is a bustling coastal city just south of Los Angeles with a population of 460,000 people in 2010. The port of Long Beach is the second busiest in the United States, with the busiest being the neighboring port of Los Angeles on the other side of the Dominguez Channel. Long Beach sits on top of the Wilmington oil field, which is the fourth most productive field in the US. Production began in 1900, and the cumulative production through 2017 was 2.8 Gb. Figure 5.28 shows oil rigs at the port around the time of World War II.

However, it became apparent that the land was sinking. We let the Long Beach Oil and Gas Department tell the story.

Figure 5.28 Oil rigs at the Port of Long Beach, California. The following information comes from Chris Berry, Senior Electronic Communication Specialist for the port. "The oil well photo is along Seaside Boulevard, probably in the very late 1930s or early 1940s. This part of Long Beach has been completely redeveloped, with landfill and new roads added; none of the structures or oil wells in the photo are there anymore, although there are still oil wells in the Port." Courtesy of the Port of Long Beach.

In 1951, the rate of subsidence exceeded two feet per year. By 1958, the affected area was 20 square miles and extended beyond the Harbor District. Total subsidence reached 29 feet in the center of the "Subsidence Bowl". The ocean inundated wharves, rail lines and pipelines were warped or sheared, while buildings and streets were cracked and displaced. Ninety-five oil wells were severely damaged or sheared off by underground slippage.[8]

Figure 5.29 shows flooding caused by the subsidence. The problem was that the oil withdrawals reduced the pressure and caused the reservoir to compact. It took some time to figure out that the oil production was causing the subsidence. However, once this became clear the City of Long Beach responded admirably to the challenge. The solution was to drill water wells to pump water into the reservoir to maintain the pressure. This required centralizing the control of the oil production, the water injection, and the monitoring of the ground level. The water injection operations have been able to stabilize the ground level to a few millimeters.

Figure 5.29 Subsidence at the Port of Long Beach in 1956. Note the oil rigs in the background. The building in the background is the port administration building. Courtesy of the Port of Long Beach.

---

[8] From www.longbeach.gov/lbgo/about-us/oil/subsidence/

To this day the Long Beach City Council receives surveyor reports periodically on the changes in the land elevation.

In this case, the oil companies wanted to solve the problem as badly as the city did. The city had restricted further drilling because of the subsidence. Once drilling resumed, there were jobs for oil workers, profits for investors, and taxes for the city. While the water injection was able to stop the ground from subsiding, it was not able to undo the earlier compaction. So a program was established to backfill to raise the level of the land. Figure 5.30 is a picture of the preparation for backfilling. The fire hydrant has been raised first. The length of the pipe shows how much dirt will be added. The picture evokes the eternal relationship of dogs with fire hydrants.

Figure 5.30 The classic photograph from Long Beach in the subsidence era showing the preparation for backfilling. The picture was taken in 1959 by Roger Coar. The dog's name is King. Courtesy of the Long Beach Historical Society.

Figure 5.31 shows the Long Beach port area as it is today. This is a completely industrial coastline. Nothing is natural about it. The only sign that the subsidence ever occurred is that Long Beach has the deepest harbor of any major port in the United States. The harbor itself requires ongoing maintenance with costs that are much larger than those incurred in fixing the subsidence problem. We can also compare the 9 meters of subsidence with the background sea level rise rate. The Los Angeles tide gauge station is across the Dominguez Channel from Long Beach. It is on stable ground, and it has been measuring sea level since 1924. The rise rate at the station has been steady at 1.0 mm/y, or 0.1 m per century.[9]

Long Beach is an extreme example, but other coastal cities around the world also manage significant subsidence that greatly exceeds the background sea level rise. In Tokyo and Houston, water wells are the cause of the subsidence. Coastal cities tend to grow by landfill, sea level rise or no. Most of the first World Trade Center in New York City was built on landfill, as well much of Stuyvesant Town. In Boston, Massachusetts landfill has more than doubled the land area

Figure 5.31 The container ship Fabiola is pushed by tugboats into Pier T at the Port of Long Beach in 2012. The Fabiola is operated by the Mediterranean Shipping Company (MSC), headquartered in Geneva, Switzerland. The Fabiola has a capacity of 6,000 of the common forty-foot containers (12,562 TEU). Courtesy of the Port of Long Beach.

---

[9] The data come from the Permanent Service for Mean Sea Level (PSMSL), which gives the uncertainty for the Los Angeles sea level rise rate as ±0.3 mm/y, calculated as a 90% confidence interval, including a correction for first-order auto-correlation.

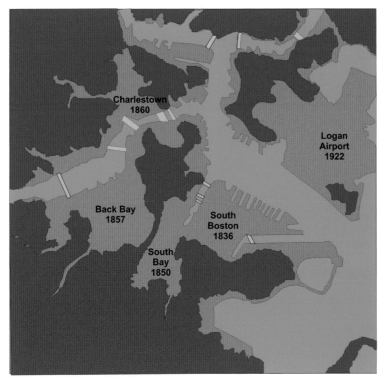

Figure 5.32 Map of Boston, Massachusetts, showing the original land in green, the landfill in tan, and the current water in blue. Adapted from a figure by Professor Jeffery Howe of Boston University.

(Figure 5.32). The names sometimes give a clue to the history. One area is called Back Bay, even though it now is dry land.

5.5 King Hubbert and Peak Oil

In Section 4.5, we analyzed Stanley Jevons' calculation of how long British coal would last. His insight was that exponential growth shreds a static lifetime calculation based on the R/P ratio. Oil is much harder to find than coal, and because of this there have been concerns about oil supplies from the very beginning, starting with Jevons himself. The most influential thinker on oil supplies was an American geophysicist named Marion King Hubbert (Figure 5.33). King Hubbert was born on a farm in San Saba, Texas in 1903. He moved to Weatherford, a small town 40 km west of Fort Worth, Texas and attended Weatherford College for two years. At the suggestion of the college president, he then began studies at the University of Chicago, graduating in 1926 with degrees in geology and physics. Later he received Master's and Doctor's degrees from the university based on scientific papers that he published. After graduating, he worked on reflection

Figure 5.33 King Hubbert (1903–1989). Hubbert was a leading American geophysicist who publicized the idea of Peak Oil. Courtesy of the Hubbert Tribute, www.hubbertpeak.com/hubbert/tribute.htm

seismology at the Amerada Petroleum Corporation. In the 1930s, he taught at Columbia University. He then worked at the Shell Laboratory in Houston, Texas, for 20 years. After reaching the company retirement age, he became a research geologist for the US Geological Survey. Even without his papers on oil supplies, King Hubbert would have been remembered as a leading geophysicist for contributions in hydrology, petrology, fault mechanics, scaling laws, and hydraulic fracture. Francis Pettijohn, a colleague and a prominent geologist in his own right, characterized Hubbert in this way, "He turned out to be something of an iconoclast, a sharp critic with an excellent analytical mind and skill in mathematical and physical analysis."

Hubbert also had a strong interest in theories of government. While he was at Columbia, he was a leader in the Technocracy movement. This was the idea that the society should be governed by scientific and technical people. King Hubbert and Howard Scott founded an organization called Technocracy, Inc. The movement emphasized evening study courses, and Hubbert wrote the book for the classes. The movement rejected a central role in society for prices in favor of energy as an organizing principle. Seen from the vantage point of American society in the twenty-first century, technocracy would be considered authoritarian, with central planning and restrictions on personal economic freedom. For perspective it should be recognized that at this time totalitarian states were developing in Europe – fascism in Germany and communism in the Soviet Union, and both regimes had supporters in the United States. But technocracy advocates never developed a real-istic plan for winning elections. To have political influence in the United States, a movement must win elections in many states at the same time, and the tech-nocracy movement never did this. However, in the United States today, scientists do have a strong influence in federal agencies in making diet recommendations, setting pollution standards, and in determining climate policy, and courts have generally deferred to their decisions.

Hubbert was best known for his paper "Nuclear Energy and the Fossil Fuels." This paper was written in 1956 for a presentation at a meeting of the American Petroleum Institute in San Antonio, Texas. A link to the paper is given in the ref-erences at the end of the chapter. In the paper, Hubbert noted that because fossil fuels are not renewable, there would be a peak in crude-oil production. This is a mathematical certainty, but Hubbert went further. He made a prediction for when the peak would occur. Figure 5.34 shows his key graph. In the figure, Hubbert plots the history of US crude oil production. He also made an estimate of ulti-mate production of 150 Gb based on his geological analysis. At the time, Alaska was not a state, and there had been no discoveries of oil there yet, so people take this estimate to be for the lower 48 states. Hubbert also took note of a survey at the time of other geologists' opinions of what the ultimate production would be. The high prediction was from the Dallas firm of DeGolyer and McNaughton, who proposed 200 Gb. Everette DeGolyer had achieved fame as the discoverer of Mexico's Tampico oil field while he was taking a break from being an *undergrad-uate* at the University of Oklahoma. He had also been a principal at Amerada, where Hubbert had worked. Hubbert took 200 Gb as a high variant, recognizing that recovery factors might improve in the future. He modeled production as a

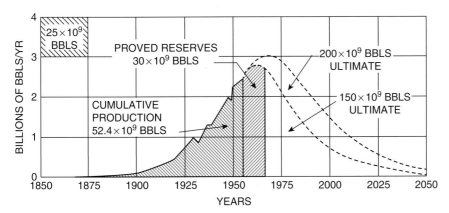

Figure 5.34 Hubbert's peak prediction for US crude oil. BBLS is an abbreviation for barrels. The graph is from King Hubbert, 1956, "Nuclear Energy and the Fossil Fuels," Shell Development Company, Houston, Texas, publication 95. Courtesy of the Shell Oil Company.

bell-shaped curve. He sketched representative production curves for the two ultimates. Hubbert argued that the peak year was not very sensitive to the shape of the bell curve, and he proposed the peak years of 1965 for the 150-Gb ultimate and 1970 for the 200-Gb ultimate.

Hubbert's prediction was not well received. The Shell management tried to persuade him not to distribute his paper, and he was criticized both by oil companies and by the US Geological Survey. Nevertheless, US oil production did peak in 1970. This was the beginning of a long slide in production that was not arrested by the end of prorationing in Texas nor by the development of the Alaskan oil fields. It meant the end of the era of the Texas Railroad Commission with its low, stable prices and steady increases in production (Figure 5.7). It ushered in the OPEC era, with its boycotts, wars, unstable prices, and relatively low production increases. The American production decline continued for 38 years. Finally, in 2009 production started to rise again because of the one-two technology punch of hydraulic fracturing and horizontal drilling.

How have Hubbert's 1956 estimates for ultimate production held up? His personal estimate was 150 Gb. He also considered a higher ultimate, 200 Gb. Expressed as a percentage range, this would be ±14%. Today we would consider these ultimates to be too low for several reasons: the development of near substitutes, new hydrocarbon fields, and new technology. Natural gas liquids can substitute for crude oil in many applications, and today people typically include NGLs in oil production numbers. They are much more significant now than in Hubbert's time. In 1956, the crude-oil production was 10 times NGL production. In 2017 the ratio was 2.9. The Prudhoe Bay production started in 1976, after Hubbert wrote his paper. Alaskan production reached a maximum share of US crude-oil oil production of 21% in 1988. However, by 2017, it had fallen to 4%. Tight-oil production based on fracking technology has recently become important. Tight oil reached 36% of the total in 2017, up from 6% ten years earlier. Figure 5.35 shows a history of US oil production, including NGLs, Alaska crude, and tight oil. The cumulative contributions are 38 Gb for NGLs, 18 Gb from Alaska, and 10 Gb from tight oil. With these additions, cumulative oil production was 263 Gb in 2017 and annual production was rising steeply. Without them cumulative production was 198 Gb, and annual production was flat.

It would have been best if Hubbert had stopped at this point, but he did not. Figure 5.36 shows a peak year prediction for world crude-oil production. For this graph, he took the ultimate to be 1.25 Tb and he proposed a peak year of 2000. Hubbert's ultimate was way too low. World cumulative crude-oil production passed 1.25 Tb in 2013 and production has not peaked as of 2017 (Figure 5.7). There was simply not enough information available in 1956 to make credible estimates for ultimate world oil production.

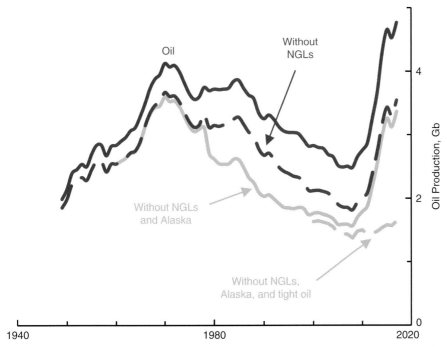

Figure 5.35 US oil production and its components. US oil production is taken from various editions of the BP *Statistical Review*. NGLs, tight-oil production, and Alaska crude-oil production from 1981 on come from the EIA. Earlier Alaska production is from the Alaska Department of Revenue, Tax Division.

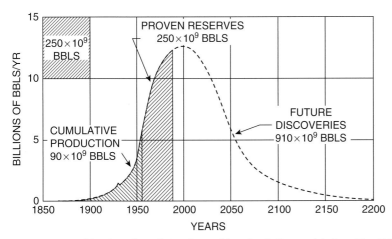

Figure 5.36 Hubbert's peak prediction for world crude-oil. The graph is from King Hubbert, 1956, "Nuclear Energy and the Fossil Fuels." Shell Development Company, Houston, Texas, publication 95. Courtesy of the Shell Oil Company.

5.6 Natural Gas

Natural gas has a growing share of overall hydrocarbon production (Figure 5.6). It is primarily methane, CH_4. Natural gas has historically been cheaper on an energy basis than oil. Some applications for natural gas use the fact that it is both a source of heat and a source of hydrogen. Upgrading bitumen to synthetic oil is an example (Section 5.3.2). This is like using coal in steel making, where it provides heat as well as carbon for chemically reducing the iron ore. However, the most important application of this type for natural gas is in ammonia synthesis for nitrogen fertilizer (Section 6.2). Natural gas provides heat to drive the reaction and hydrogen atoms to synthesize ammonia, NH_3. It has been estimated that half of the nitrogen atoms in our bodies derive from this reaction. We will see in Chapter 7 that natural gas generators have particular advantages in an electrical grid with large solar and wind generation. This is because most gas generators include a combustion turbine, where the hot exhaust directly turns the blades. This is different from a coal generator where the burning coal heats a boiler to make steam to drive a turbine. When the wind picks up and wind turbines begin to produce electricity, a combustion turbine can be throttled down quickly to accommodate the wind production. Jet aircraft engines are also combustion turbines. A coal power plant is much slower in reducing power. Natural gas is clean enough to be burned even in urban areas with extremely strict pollution regulations. In particular, sulfur can be separated from the gas before it is distributed to users. In contrast, washing coal removes only about half its sulfur. The rest is chemically embedded in the coal. Natural gas is also finding new applications. As an example, the shipping company TOTE Maritime is developing container ships that are powered by liquefied natural gas (LNG). The *Isla Bella* was launched in 2015. It is rated at 3,100 twenty-foot equivalent units, or TEUs. The typical container is 40 feet long, so a ship with this rating would actually be carrying about half this many containers at the rated limit. TOTE claims that compared with conventional diesel ships burning bunker fuel oil, emissions of sulfur dioxide are reduced 97%, particles 98%, nitrogen oxides 60%, and carbon dioxide 72%.

5.6.1 George Mitchell and the Barnett Shale

The development of shale gas is the most important event in energy production in the 21st century. It is a remarkable story, and it is primarily associated with one man, George Mitchell (Figure 5.37). George Mitchell was born in Galveston, Texas, the son of Greek immigrants. His father's name was Savvas Paraskevopoulos. He took the name of a paymaster, Mike Mitchell, who complained that Paraskevopoulos was too long. George Mitchell attended Texas A&M University, graduating with a degree in petroleum engineering. He and his wife Cynthia sponsored the Mitchell Prize, a $100,000 paper prize on the transition to a sustainable society. Mitchell

Figure 5.37 George Mitchell (1919–2013), the father of the shale gas revolution. Credit: Mitchell Family Corporation.

incorporated his ideas on sustainability in a development near Houston, Texas, called Woodlands. The Woodlands development preserves trees and access to ponds and a lake. Today 100,000 people live there.

Mitchell also founded Mitchell Energy, which focused on natural gas production. He became interested in the idea that natural gas could be produced from the Barnett Shale in Fort Worth, Texas. This was not a conventional gas play. The shale is the original source rock. There is no separate reservoir and no cap rock. The gas never went anywhere because the permeability of the shale is so low. The low permeability also had made it impossible to make money with a gas well in the shale because the flow of natural gas was too low to pay off the well. Mitchell's staff worked on improving hydraulic fracturing techniques to artificially increase the permeability of the shale. The major oil companies were skeptical of his efforts. Even Mitchell's own geologists resisted. But he persisted, almost driving his company into bankruptcy. Finally, his engineers got the recipe right, and they started to see significant gas production from the shale. However, his company did not

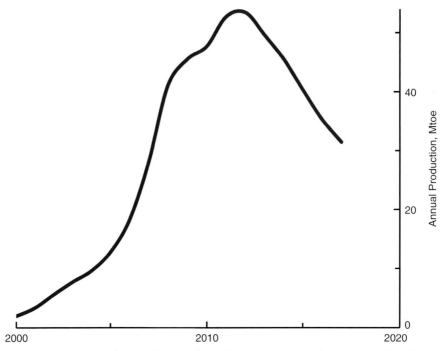

Figure 5.38 Barnett Shale natural gas production from 2000 to 2017. The data come from the Texas Railroad Commission.

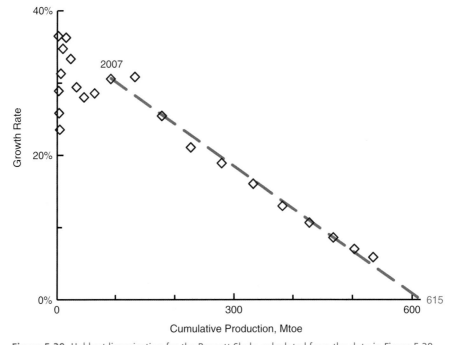

Figure 5.39 Hubbert linearization for the Barnett Shale, calculated from the data in Figure 5.38.

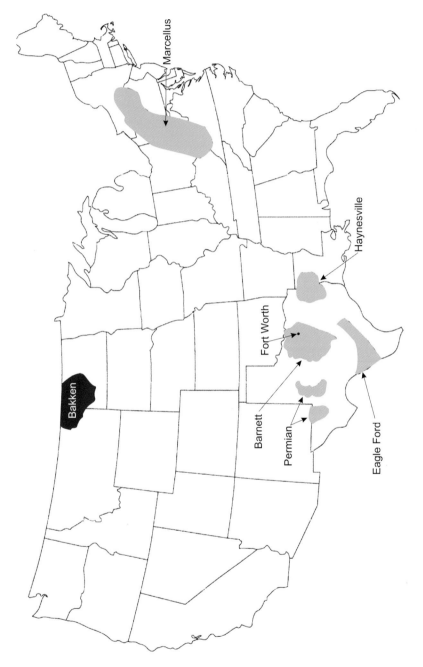

Figure 5.40 The most important shale gas plays in the United States. The Bakken field, which is mainly tight oil, is also shown. This figure is adapted from the EIA map, "Lower 48 states shale plays."

have expertise in drilling horizontal wells, so Mitchell Energy was bought by Devon Energy, which did. Figure 5.38 shows the history of the Barnett Shale production. The production is primarily from fracked horizontal wells. At its peak in 2012, the Barnett Shale produced 9% of the natural gas in the United States. The production cycle is much faster than for the other fossil-fuel fields that we have considered.

Figure 5.39 shows the Hubbert linearization for the Barnett Shale. The projection for ultimate production from the x-intercept is 615 Mtoe. The cumulative production in 2017 was 535 Mtoe, so the exhaustion factor e is 87%. We can compare the projection for the ultimate with the EIA's 2016 original reserves, 931 Mtoe. The projection is 66% of the original reserves.

5.6.2 American Shale Gas Production

After the Barnett, the fracking revolution spread to other areas. Figure 5.40 shows the most important American shale gas fields. The Marcellus Shale in Pennsylvania and West Virginia has been the most productive. Other important fields are the Eagle Ford Shale (pronounced as one word, like "Weatherford") and the Permian, in Texas, and the Haynesville, in Texas and Louisiana. The Eagle Ford and the Permian also have significant tight-oil production. The Bakken (rhymes with "rockin'") in North Dakota was the first important tight-oil field.

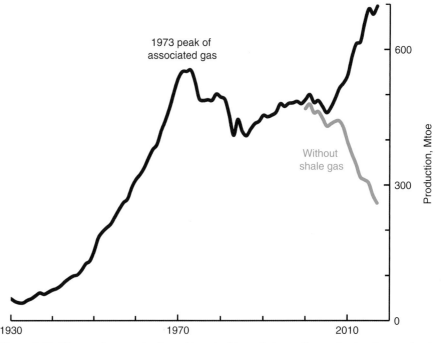

Figure 5.41 US natural gas production, with and without shale gas. The production data are for dry production and come from the EIA.

Figure 5.41 shows the history of natural gas production in the United States. Much of the earlier production was associated gas from oil wells. The early production peaked at 554 Mtoe in 1973, three years after the peak in oil production. Production then declined until 1980. After 2005 gas production rose sharply because of shale gas. At the same time the non-shale gas production has fallen, hitting 260 Mtoe in 2017. One can get a perspective on the importance of shale gas by plotting the shale gas share of the production. This is shown in Figure 5.42. It reached 63% in 2017.

5.6.3   Natural Gas Infrastructure

To produce and consume natural gas, a country must build and manage a national pipeline system. There are many *stranded* gas fields around the world where a pipeline for gathering the gas has not been built. Without gas distribution pipelines, buildings typically use more expensive fuel oil or liquefied petroleum gas for heating. Figure 5.43 shows the natural gas pipeline system for the United States.

One of the important applications for natural gas is in heating buildings and for this reason consumption peaks in the winter. Storage may be needed to accommodate the increased winter consumption. Figure 5.44 shows monthly storage levels of natural gas in the United States. Gas is withdrawn in the winter and injected in the other seasons. The storage facilities can be depleted oil and gas fields or salt formations where caverns are formed by removing salt in brine. A significant amount of gas is needed to pressurize the facilities. This *base gas* is not removed

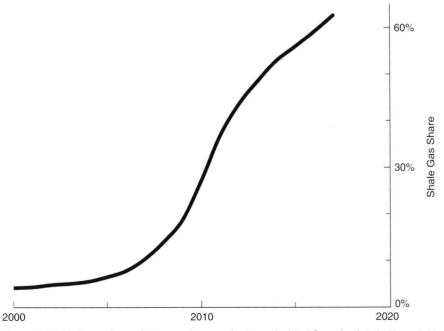

Figure 5.42 Shale gas share of US natural gas production, calculated from the data in Figure 5.41.

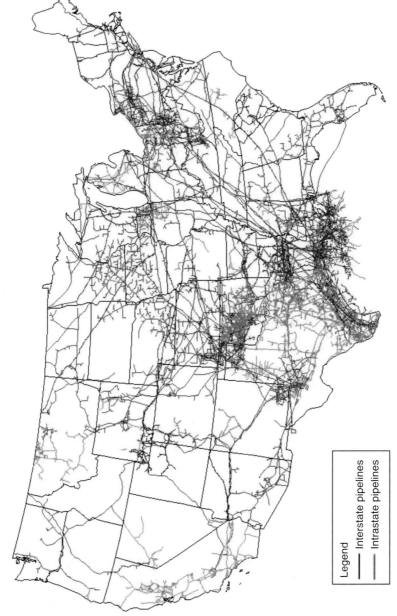

Figure 5.43 US natural gas pipeline system. Credit: EIA Office of Oil and Gas, Natural Gas Division, Gas Transportation Information System.

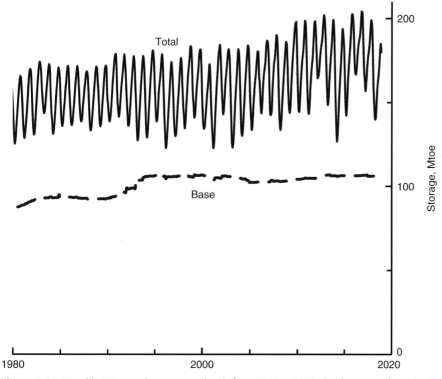

Figure 5.44 Monthly US natural gas storage levels from 1980 to 2018. The data come from the EIA *Natural Gas Monthly.*

during the year. In 2018, the base gas was 106 Mtoe. The *working gas* is the difference between the total storage and the base gas. Working gas is at a maximum in the late fall. In November 2018 it was 74 Mtoe. The working gas drops to half this level in the early spring. The working gas in March 2018 was 34 Mtoe. For comparison the total consumption during 2017 was 640 Mtoe. The base storage corresponds to two months of consumption, while the working gas storage is a month.

5.7   Projections

We will conclude the chapter with projections for U and t_{90} for world hydrocarbons. We will compare them with the corresponding numbers for coal from Section 4.7.3. Figure 5.45 shows the production history for world hydrocarbons. This analysis will be global because the market for hydrocarbons is global – 56% of hydrocarbon production in 2017 was exported, compared with 17% for coal.[10]

---

[10] The BP *Statistical Review* tracks hydrocarbon trade flows and production, and this allows us to calculate an export fraction. A good source for the coal export total is the IEA's annual *Coal Information.*

Figure 5.46 shows the logit transform linearization for $U = 764$ Gtoe. There is a kink in the linearization at the time of the Iranian Revolution in 1979. This is interpreted as a shock that set us on a slower production path with t_{90} increasing from 2043 to 2082.

Figure 5.47 shows a comparison of the evolution of the ultimate production projections with original reserves. The range of the projections is 578 to 766 Gtoe ($\pm 14\%$). The original reserves are less than the projections for the ultimate. This is different from coal, where the original reserves are much larger than projections for ultimate production (Figure 4.46). Note that the original reserves for hydrocarbons have been increasing steadily. However, a major part of the increase is from OPEC reserves. These have arbitrary jumps and are not adjusted downward to account for production (Figure 3.16). It is not clear how much significance to attach to this rise. There will also be new reserves for tight oil and shale gas. How much? The US EIA made the estimate in its 2013 report "Technically recoverable shale and shale gas resources" that tight oil increased oil resources by 11% and natural gas resources by 47%. This gives an indication, but the truth is that we do not know yet.

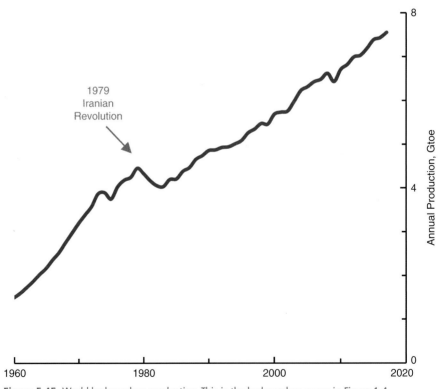

Figure 5.45 World hydrocarbon production. This is the hydrocarbon curve in Figure 1.4.

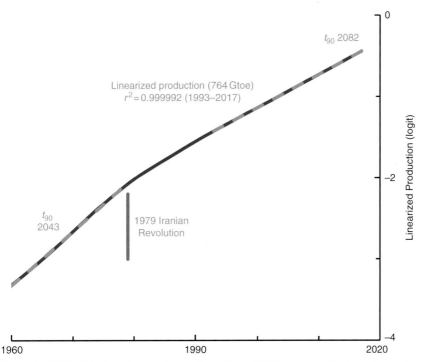

Figure 5.46 World hydrocarbon production from Figure 5.45 linearized with the logit transform.

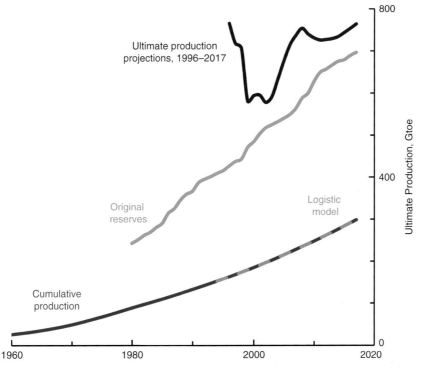

Figure 5.47 Comparing the evolution of the projections for ultimate production for world hydrocarbons with the evolution of reserves taken from the BP *Statistical Review*. The cumulative production and the logistic model are shown for comparison.

Table 5.3 Summary of the projections for world hydrocarbons and coal.

	Cumulative production q, Gtoe (2017)	Ultimate production projection U, Gtoe	t_{10}	t_{90}	Original reserves CR_o, Gtoe
Hydrocarbons	299	764	1977	2082	697
Coal	188	394	1947	2065	610
Fossil fuels	486	1,158	1970	2077	1,306

The calculations underlying these results are available in the Excel workbook fossil fuels at energybk.caltech.edu

Table 5.3 gives a summary of the projections for world hydrocarbons and coal. Note that the ultimate production projection for hydrocarbons is twice that for coal. This reflects that the fact that hydrocarbon production has been higher than coal production since 1956 and there is nothing to suggest that this will change in the future. Hydrocarbons and coal do compete in some markets, mainly electricity and heating of buildings, and this makes it appropriate to consider the combination. The projection for the ultimate production of all fossil fuels is 1,158 Gtoe. The fossil-fuel original reserves are close, only 13% higher. This is the result of an accidental cancellation. The original reserves for coal are higher than the projection for ultimate production, while original reserves for hydrocarbons are lower.

The projection for t_{90} for all fossil fuels is 2077. We should keep in mind that t_{90} is calculated from current trends. A shock could change t_{90}, as happened in 1979, the year of the Iranian Revolution. We interpret t_{90} as giving a time frame for the transition to alternatives. By then, there would need to be a substantial contribution from alternatives or a large drop in overall energy consumption. We can make one final observation. The 10%-to-90% lifetime t_l is 107 years. From the perspective of thousands of years of human history, the fossil-fuel era is likely to be short.

The shale gas revolution did not have to happen for decades, maybe not ever. The breakthrough in the Barnett Shale came from a small private gas company, Mitchell Energy. The company had creative engineers and outstanding geologists, and access to excellent workers and equipment. The drilling took place under a major city, Fort Worth, Texas (Figure 5.48). Fort Worth is an old oil town, where children are raised on tales of fortunes won and lost in the oil fields. Thousands of landowners accepted the disturbance of drilling and fracking in exchange for royalty payments. In addition, the State of Texas provided a supportive regulatory environment. It is perhaps not an accident that the two states with the largest shale gas production are Texas and Pennsylvania. Their state agencies have more than a hundred years of experience regulating oil and gas production and refining, with appalling disasters alongside wealth beyond avarice. All of these factors contributed to the success of shale gas. However, in the end, the most important factor was a pigheaded company owner.

Figure 5.48 Inside the blue square is the drilling pad shown in Figure 5.19. The drilling pad is next to the West Fork of the Trinity River, surrounded by the city of Fort Worth, which had a population of 740,000 in 2010. Imagery ©2018 Google, Map data ©2018 Google.

Concepts to Review

- The first Pennsylvania oil well
- Oil in lighting
- Importance of oil in wars
- The era of the Texas Railroad Commission
- The development of hydraulic fracturing (fracking) and horizontal drilling
- The formation of oil
- Dad Joiner and the East Texas Oil Field
- Hydrocarbon production technology
- Seismic risks
- Enhanced oil recovery (EOR)
- The Canadian tar sands
- The burning of the Kuwaiti oil fields
- The subsidence from oil production in Long Beach

- King Hubbert and peak oil
- George Mitchell and the fracking revolution
- Projections for U and t_{90} for world hydrocarbons and coal

Problems

Problem 5.1   Footprint for natural gas electricity generation

a. Calculate the area of drilling pads in square kilometers that would be needed to produce gas to generate all of the electricity consumed by the United States. You can take the Barnett gas pad with eight wells (Figure 5.19) as the reference. Its dimensions are 125 m × 137 m. Assume that an average of 2 billion cubic feet is recovered from each of the eight wells and take the time until reclamation is complete to be 30 years. Use 50% for the power-plant efficiency.

b. Use Google maps to locate the 550-MW El Segundo power plant shown in Figure 5.49. There are two generators associated with this plant, and you will need to make sure that you have the correct ones. There are other generators at the same site that should not be included in calculating the area. If we take the

Figure 5.49 The two generators of the El Segundo Power Plant, located near the Los Angeles Airport. Courtesy of Siemens AG.

capacity factor to be 50%, what is the area needed for natural gas power plants sufficient to generate all of the electricity consumed by the United States?

Problem 5.2   Propane Combustion

Propane (C_3H_8) is a fuel for cooking and heating that burns more cleanly than wood. It is sold as a liquid in cans under pressure. The liquid generates 8 atmospheres of vapor pressure at room temperature, so with a regulator, it can provide propane gas for a stove or furnace. Propane is important in rural areas that do not have natural gas pipelines.

a. Propane combustion is a chemical reaction with molecular oxygen (O_2) that produces carbon dioxide (CO_2) and water vapor (H_2O). Write the balanced chemical equation for the combustion of propane.
b. Enthalpy is a measure of the heat that is produced in a chemical reaction. The enthalpies of the constituents of the reaction are given in Table 5.4 below. Calculate the enthalpy change for the combustion of a mole of propane by subtracting the enthalpies of the reactants from the enthalpies of the products.
c. The heating value (GJ/t) is the enthalpy change for the reaction expressed on a mass basis. Calculate the heating value for propane.
d. The carbon coefficient (tC/toe) is a measure of how much carbon dioxide is produced when a substance is burned. Calculate the carbon coefficient for propane.

Problem 5.3   Hubbert's paper

Write an essay, no more than one printed page, discussing King Hubbert's analysis of future US energy production in his 1956 paper, "Nuclear Energy and the Fossil Fuels." Which parts of his analysis for the US turned out to be correct? Which parts of his analysis turned out to be wrong?

Table 5.4 Standard enthalpies of formation from Sara McAllister, Jyh-Yuan Chen and A. Carlos Fernandez-Pello, 2011, *Fundamentals of Combustion Processes*, Springer.

Constituent	Enthalpy, kJ/mol
Propane	−104.7
Molecular oxygen	0
Water vapor	−241.8
Carbon dioxide	−393.5

The convention is to give the enthalpies relative to those of the elemental components at room temperature and atmospheric pressure. The enthalpy for molecular oxygen is zero because that is the state of oxygen at room temperature and atmospheric pressure.

Further Reading

- James Clark and Michel Halbouty, 1972, *The Last Boom*, Random House. The most detailed history of the East Texas Oil Field, including the slant drilling era.
- William Childs, 2005, *The Texas Railroad Commission.* An excellent history of the Texas Railroad Commission. Childs makes it clear how difficult it was to establish prorationing, particularly the legal challenges. It also helps in understanding why the prices were more stable in the railroad commission era than in the OPEC era.
- Energy Information Administration, 2013, "Technically recoverable shale and shale gas resources" at www.eia.gov/analysis/studies/worldshalegas/archive/2013/pdf/fullreport_2013.pdf
- King Hubbert, 1956, "Nuclear Energy and the Fossil Fuels," Shell Development Company, Houston, Texas, publication 95. This is the original Peak Oil paper. It is available at www.hubbertpeak.com/hubbert/1956/1956.pdf
- Mason Inman, 2016, *The Oracle of Oil: A Maverick Geologist's Quest for a Sustainable Future*, W.W. Norton. An excellent biography of King Hubbert.
- Paul Lucier, 2010, *Scientists and Swindlers: Consulting on Coal and Oil in America, 1820–1890*, Johns Hopkins University Press. The best source for background on George Bissell and the Drake well.
- Benjamin Silliman, Jr. 1855, *Report on the Rock Oil or Petroleum from Venango County, Pennsylvania*, J.H. Benham. This is likely the most significant consultant's report ever.
- Dan Steward, 2007, *The Barnett Shale Play*, Fort Worth Geological Society and the North Texas Geological Society. A technical history of Mitchell Energy's development of fracking technology for the Barnett Shale.
- Daniel Yergin, 1991, *The Prize*, Simon and Schuster. The classic history of oil. Mill Creek Entertainment's 1992 documentary film, *The Prize*, was inspired by Daniel Yergin's book. Excellent profiles of the development of the different oil companies and remarkable interviews with the people who started OPEC.

6 Farming and Fishing

Cannery Row in Monterey in California is a poem,
a stink, a grating noise, a quality of light,
a tone, a habit, a nostalgia, a dream

John Steinbeck
Cannery Row

6.1 Introduction

6.1.1 Norman Borlaug and the Green Revolution

For almost all of human history and for almost all people, the most important question each day has been, "Will our family have enough food to eat?" That, is until the Green Revolution. In the 1950s American agronomist Norman Borlaug (Figure 6.1), funded by the Rockefeller Foundation, developed a hybrid that was a cross between a Mexican wheat that was well adapted to a range of conditions and a short, stiff Japanese wheat that could support the weight of extra grains. This allowed farmers to apply more fertilizer without the wheat *lodging*, that is, falling over. Mexican farmers began growing Borlaug's dwarf wheats in 1961 and doubled their yield. Since then there has been a spectacular increase in world crop yields. Figure 1.21 showed the history for cereals. Borlaug also argued that increasing crop yield was the best way to reduce deforestation. This has been called the *Borlaug hypothesis*. We will see in Section 6.5.1 that there is evidence to support his idea.

6.1.2 The Role of Nitrogen in Agriculture

Nitrogen is a major limiting factor for plant growth. Plants and people must have nitrogen to grow. Chlorophyll contains nitrogen and proteins are 16% nitrogen by weight. Tantalizingly, 78% of the atmosphere is molecular nitrogen, N_2, but molecular nitrogen has a triple bond that is difficult to break. Plants cannot incorporate it directly. However, there are bacteria with enzymes that catalyze the breaking of the triple bond. Some of these are free living, but the ones important for agriculture live in the root nodules of legumes. *Legumes* are a family of plants that include vegetables like peas and beans, oilseed plants like peanuts and soybeans,

Figure 6.1 Norman Borlaug (1914–2009), shown in a Mexico wheat field around 1964, developed hybrid varieties of wheat that could take advantage of intensive nitrogen fertilizer application. Winner of the Nobel Prize. Courtesy of the Norman Borlaug Heritage Foundation.

and forage for livestock like clover and alfalfa. These bacteria convert free nitrogen to ammonia, NH_3, which the plants can use. This nitrogen is called *fixed* nitrogen to distinguish it from the atmospheric *free* nitrogen and the process of converting free nitrogen to fixed nitrogen is called *fixing* nitrogen. This is a symbiotic relationship. The plant absorbs the nitrogen, while the bacteria absorb sucrose from the plant that provides energy for the bacteria to grow and to fix nitrogen. When the plant dies the fixed nitrogen remains in the soil and is available for future plants.

Traditionally the nitrogen level in a farmer's field is maintained by rotation, where a primary crop like wheat or corn alternates with a legume that restores fixed nitrogen to the soil. Plants plowed under for their nutrient content are often called *green manure*. However, the amount of nitrogen added to the soil in this way is limited. In 1909, Fritz Haber, a chemistry professor at Karlsruhe, Germany, invented a process for fixing nitrogen that synthesized ammonia. Haber's process was a small-scale demonstration, but later Carl Bosch, a chemist at BASF (Badische Anilin und Soda Fabrik, or, in English, Baden Aniline and Soda Factory), developed an industrial process to produce ammonia in large quantities. The ammonia can be used directly as a fertilizer or converted to urea, $CO(NH_2)_2$ or ammonium nitrate, NH_4NO_3. Nitrogen fertilizer production enabled the Green Revolution. Today over half of the nitrogen in our crops, and hence in us, was fixed by the Haber–Bosch process.

Urea and ammonium nitrate are important in other contexts. Urea is the major organic component of the urine of mammals. It is formed in the kidneys by a reaction of carbon dioxide with ammonia. Ammonium nitrate is the major component of the explosive ANFO (ammonium nitrate fuel oil) made by mixing ammonium nitrate with diesel fuel. ANFO is the most important explosive used in civil engineering, with millions of tons exploded each year. Nitrogen compounds are also the basis for the production of military explosives and the propellants in rifle bullets and artillery shells. The history of nitrogen fertilizer and military explosives are completely entangled. Factories making nitrogen fertilizers can be converted to military purposes and back again.

6.1.3   The Brazilian Cerrado

The Brazilian *cerrado* offers a modern example of agricultural development. Cerrado is Brazil's bush land. In recent years, large farms of hundreds of square kilometers have been established. Figure 6.2 gives a sense of the scale of the operation. This is industrial agriculture on a massive scale. It requires huge capital investments for the equipment. In addition, there were large applications of limestone, $500 \, t/km^2$, to raise the pH of the soil to allow plants to absorb nutrients better. The farms have had extensive government research support. New fast-growing varieties have been developed that allow two soybean crops per year. In addition, researchers invented new varieties of nitrogen-fixing bacteria that are adapted to the area.

The major crop in the *cerrado* is the soybean. Soybeans have a high protein content and are mainly used for animal feed. Figure 6.3 shows the production history for Brazilian soybeans. The yield shows the same linear characteristic that we saw for cereals in Figure 1.21, increasing from $113 \, t/km^2$ in 1961 to $290 \, t/km^2$ in 2016. The land planted for soybeans has been increasing rapidly, up 50% from 2006 to 2016.

Figure 6.2 A group of more than twenty combines harvesting soybeans in the Brazilian cerrado. Credit: Alf Ribeiro/Shutterstock.com.

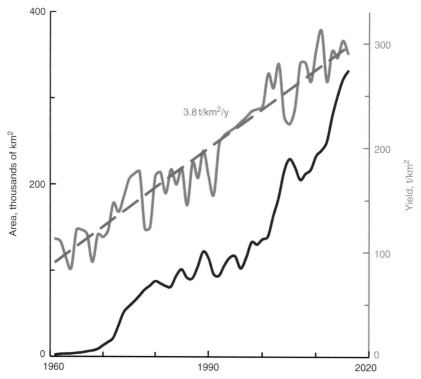

Figure 6.3 Soybean production in Brazil. The data come from the UN Food and Agriculture Organization database FAOSTAT.

6.1.4   The Food Supply

The Green Revolution has resulted in a stunning change in our food supply, shown in Table 6.1. The table shows the supply in food calories per person per day, as reckoned by the United Nations Food and Agriculture Organization (FAO). The world average has increased 31% from 1961 to 2013. The current world average of 2,884 Cal/p/d in 2013 is essentially identical to the US average in 1961. The Asian food supply was the lowest in 1961, but it is up 54% since then. Africa is now the lowest, 9% below the world average.

Table 6.1 The total food supply in food calories per person per day.

	1961	2013	Increase
US	2,880	3,682	28%
EU	2,999	3,409	14%
Asia	1,805	2,779	54%
Africa	1,993	2,624	32%
World	2,196	2,884	31%

The data come from the UN Food and Agriculture Organization database FAOSTAT. The totals include livestock and fish, which at the world level were 18% of food calories in 2013.

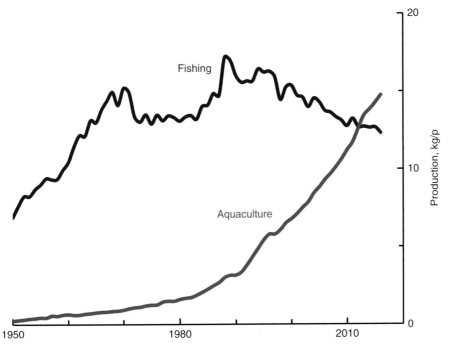

Figure 6.4 World annual fishing and aquaculture production on a per-person basis. The data come from the UN Food and Agriculture Organization online fisheries capture database and the aquaculture database. The population data come from the United Nations Population Division.

Fish provide only a modest fraction of our calories, but they are an excellent source of protein. Figure 6.4 shows annual world fishing and aquaculture production, calculated on a per-person basis. In recent years, aquaculture, or fish farming, has become increasingly important, passing fishing in 2013. Aquaculture production was 15 kg/p in 2016. The fishing catch was rising until 1970, but it has had trouble keeping up with the population increase since then. It was 12 kg/p in 2016. Even though the total catch is relatively flat, the catches in many fisheries for individual fish have varied tremendously over this time. Compared with land crops, fishing is limited to relatively small productive areas in the ocean where there are nutrients for the growth of phytoplankton that support the fish. The rest of the ocean is relatively devoid of life. Even within the productive areas the catch can be highly variable over time. The California sardine fishery made by famous by John Steinbeck in his novel *Cannery Row* peaked in 1936 at 664 kt. By 1953, the catch had fallen to 9 kt and the canneries closed down. At the time, people thought that the fishers had destroyed the fishery by overfishing. But the sardines returned and the catch reached 758 kt in 2009. Now it is thought that the rise and fall of the California sardine, while affected by fishing, are primarily associated with changes in currents and water temperatures.

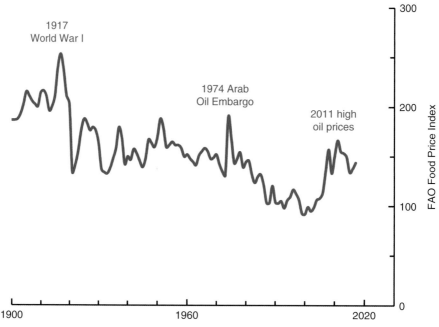

Figure 6.5 The inflation-adjusted food price index of the UN Food and Agriculture Organization. The reference is the 2002 to 2004 average, which is set to 100.

6.1.5   Oil and Food

The price history of food reflects the fact that food supplies have increased faster than the population. The Food and Agriculture Organization of the United Nations has calculated an annual food price index that starts in 1900. It is shown in Figure 6.5. The index is weighted by the prices in cereals, meat, dairy products, oil and fat products, and sugar. Overall we can say that the general trend during the last hundred years is down. Food is cheaper than it was a hundred years ago. The peak of the price index was in 1917, when it was 80% higher than in 2017. The year 1917 was during World War I. During the war there was heavy fighting in the rich farmlands of Belgium and France. This took much of the supply off the market. Farmers in other countries did respond. The largest area ever planted to corn in the United States was in 1918. There are noticeable spikes in 1974 and 2011. There was an Arab Oil Embargo following the 1973 Yom Kippur War that extended into 1974. The year 2011 set the record for the highest oil prices ever. Farmers' tractors and trucks run off diesel fuel. Another factor is that most nitrogen fertilizer is produced from natural gas. This means that the price of natural gas affects farmers' costs. In

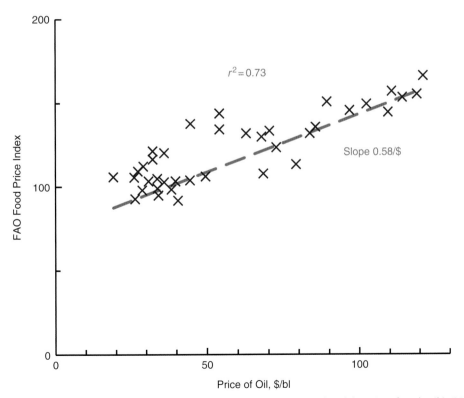

Figure 6.6 Relationship between the FAO food price index (Figure 6.5) and the price of crude oil in 2017 dollars (Figure 5.7). The data are from 1980 to 2017.

some markets, the natural gas price is indexed to the oil price. This gives another connection between the price of food and the price of oil.

We can get more insight into the relationship between oil prices and food prices by plotting them against each other (Figure 6.6). The data are from 1980 to 2017. A regression line has been added. The value of r^2 is 0.73. This means that in a statistical sense, the price of oil explains a significant part of the price of food over this time period.

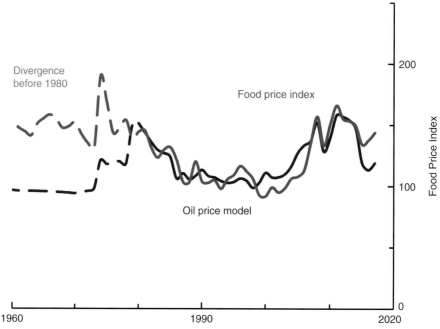

Figure 6.7 The food price index, compared with the linear regression model with the oil price as the input. The regression is calculated from 1980 on. The model fails for the years before 1980. These points are shown with a dashed line to distinguish them.

We can use the regression line to construct a linear model for the food price based on the oil price. The result is shown in Figure 6.7. The agreement from 1980 on is reasonable, but the model fails for the period before 1980. This coincides with a slope change for nitrogen fertilizer application (Figure 6.12). It is possible that this is yet another aspect of the shock of the 1979 Iranian Revolution. Recall that we saw a change in slope in per-person production of oil that year (Figure 1.6) and a change in the relationship between per-person oil production and income (Figure 1.14).

Agriculture is a significant consumer of energy, natural gas for fertilizer and gasoline and diesel fuel for farm equipment On the other hand, in recent years, agriculture has been a significant source of energy, in the form of the liquid bio-fuel ethanol and the biogas methane. Ethanol is produced by fermentation from sugar cane in Brazil and from corn in the United States. In these countries ethanol is used as a transportation fuel, either on its own or mixed with gasoline. Methane is produced through decomposition of organic material by anaerobic bacteria. In Germany, the feedstock is primarily corn and rapeseed, and the methane is mostly burned to generate electricity. In Asia, the source material is manure and plant waste, and the methane is burned in homes for cooking, heating, and lighting.

6.2 The Haber–Bosch Process

6.2.1 The Guano Islands

Livestock manure and urine contain nitrogen, as do human wastes. The nitrogen content is typically in the range of 2% to 4%. For this reason, they have traditionally been collected and spread on fields as fertilizer. However, it turns out that the *guano*[1] of seabirds has exceptionally high nitrogen content, 16%, primarily as uric acid. Seabirds mostly eat fish and squid, a high-protein diet. Ordinarily, the nitrogen compounds in bird droppings are leached away by rain. However, some islands that the birds nest on are dry, and the guano can accumulate in layers many meters thick. In the 1800s, the Chincha islands off the coast of Peru were extensively mined for guano, and it was sold overseas for fertilizer. Figure 6.8 is a contemporary etching showing the islands, with many ships waiting to be loaded. Figure 6.9 is an American advertisement for the fertilizer. The Peruvian government taxed the production heavily. The taxes supplied three-quarters of the government income in 1859. However, the guano ran out.

With fossil fuels, there is often a discussion about what it means to run out. Were the remaining resources difficult to produce? Was there a shift to another fuel? On the Chincha Islands, there was no doubt about what running

Figure 6.8 Chincha Islands, Peru, from the *Illustrated London News*, 1863. Credit: Manuel González Olaechea y Franco/CC BY-SA 3.0, https://creativecommons.org/licenses/by-sa/3.0/

---

[1] The word "guano" comes from the South American Quechua language. In Quechua, the word *wanu* means "manure."

Figure 6.9 New York advertisement for guano fertilizer. Courtesy of the Mystic Seaport Museum, Mystic, Connecticut.

out meant – the miners got down to bare rock. Mariners sailed the seven seas in search of other guano islands. The United States passed the Guano Islands Act of 1856 that allowed its citizens to claim a guano island for the United States. The most notable island that came into its possession this way was the Midway Atoll, 2,000 km northwest of Honolulu, Hawaii. Midway never became an important source of guano, but it did become a station for a telegraph line between Hawaii and Guam. The island was the focus of a savage battle between the Japanese and American navies in 1942. Today the island is a wildlife refuge, the nesting ground for hundreds of thousands of Laysan and black-footed albatrosses.

When the Chincha Islands guano ran out, attention shifted a thousand kilometers down the coast to the Atacama Desert, then divided between Peru, Bolivia, and Chile. Rain is almost unknown in the Atacama, and for this reason, minerals exist there that would have been dissolved by water anywhere else. There are large deposits of the mineral sodium nitrate, known in English as Peruvian or Chilean saltpeter, and in Spanish as *salitre*. Ordinary saltpeter is potassium nitrate. Potassium nitrate is used as a nitrogen fertilizer and for making gunpowder. Gunpowder, also known as *black powder*, is an explosive and was a propellant in muskets and cannons. It is made by mixing saltpeter, sulfur, and charcoal. The supplies of saltpeter have historically been limited, but in the 1800s, chemists developed a way for sodium nitrate to replace saltpeter in making fertilizers and explosives. Mining of sodium nitrate in the Atacama Desert began on a large scale, both for fertilizer and for explosives. In 1878, the Bolivian government attempted to raise the tax on a Chilean mining company operating in Bolivian territory, violating an earlier agreement. This set off a war from 1879 to 1883 with Chile opposing Bolivia and Peru. In English it is called the War of the Pacific, or the Nitrate War, and in Spanish, it is the *Guerra del Pacífico*, or *Guerra del Salitre*. Chile won the war and claimed the mines as its prize. The country became the leading supplier of nitrates for fertilizer and explosives.

6.2.2   Fritz Haber and Ammonia Synthesis

In 1898, William Crookes, the discoverer of thallium, was the incoming president of the British Academy of Sciences. In a speech in Bristol, he issued a remarkable challenge. Crookes argued that the Chilean nitrate mines would run out in several decades. He said, "Wheat preeminently needs nitrogen. ... It is the chemist who must come to the rescue." He noted that there is plenty of nitrogen in the atmosphere. He envisioned using electricity to synthesize sodium nitrate. Many chemists undertook Crookes' challenge, particularly in Germany, where it was appreciated that in the event of a war, the British Royal Navy would cut them off from the Chilean nitrates, and they would lose the ability to make explosives. The man who developed the modern approach to this problem was a professor of chemistry at the University of Karlsruhe, Fritz Haber (Figure 6.10). This was an exciting time for Germans. Germany was a new country and Haber was intensely patriotic. In 1909, he succeeded in producing ammonia from nitrogen and hydrogen with the reaction

$$N_2 + 3H_2 \rightarrow 2NH_3. \qquad\qquad 6.1$$

Figure 6.10 Fritz Haber (1868–1934). The German chemist who synthesized ammonia from atmospheric nitrogen. Winner of the 1918 Nobel Prize in chemistry. This photograph was taken in 1891, the year he received his doctor's degree.

Fritz Haber was appointed the founding director of the Kaiser Wilhelm Institute for physical chemistry in 1912.[2]

While Haber's process was successful as a laboratory demonstration, as an industrial process it presented severe challenges. It worked at an extremely high pressure of 200 atmospheres and it required osmium, a rare element, as a catalyst. Scaling up the process for a large output would require a different catalyst. And no one had ever built a large reactor that could take 200 atmospheres. Representatives from the BASF company visited Haber's laboratory to discuss whether it would

---

[2] The Kaiser Wilhelm Institutes were renamed Max Planck Institutes after World War II.

be possible for their company to produce ammonia with his process. The senior members of the group were skeptical and might have turned Haber down, but for a junior member, Carl Bosch. Bosch was trained as a metallurgist and he had worked in a smelter.[3] He persuaded the others that because of advances in steel making, it would be possible to build a reactor large enough to produce large quantities of ammonia. Bosch was given the assignment to develop the Haber process.

6.2.3   Carl Bosch and Ammonia Production

It could be said about Carl Bosch (Figure 6.11) and ammonia production that the greatest chemical engineer who ever lived met the most important chemical production project in history. His group solved many difficult problems to get Haber's process to work on a large scale. They developed a process for producing large quantities of hydrogen from coal. This was an extension of the wood gas process (Section 2.3.2). To replace the osmium catalyst, they invented a hybrid catalyst based on magnetite and metal oxides. At the time it was appreciated that at high pressures, hydrogen makes carbon steels brittle. The hydrogen diffuses into the metal and reacts with carbon in the steel to form methane. The pressure from the methane causes fractures and the vessels fail. Bosch invented an insert that absorbed the hydrogen and an outer jacket that took the pressure. By 1913, BASF was producing large quantities of ammonia.

August 1914 – the war came. The BASF output was diverted to make nitrates for explosives. Carl Bosch supervised the development of an enormous new plant in Leuna that was out of the range of French bombers. Fritz Haber joined the army. On his initiative, the German army developed artillery shells to deliver chlorine, a poison gas. This was in spite of the fact that Germany had ratified the Hague Convention of 1899 that proscribed the use of "asphyxiating or deleterious gases." The first German attack was near the Belgian town of Ypres in April, 1915. The French soldiers fled the chlorine clouds and a gap developed in the French line. However, the German commanders had been uneasy about the gas attack, and had not made plans to follow up the attack with a major offensive. The French quickly plugged the hole. By comparison with other battles of the war, the casualties were light because the French soldiers got out of the way. In recognition of his role in the attack, Haber was promoted to the rank of captain. He held a party to celebrate his promotion, but later that night, his wife, Clara Immerwahr Haber, herself a PhD chemist, killed herself with her husband's service pistol. After the war, many of his colleagues in other countries understandably considered Haber a war criminal. Nevertheless, he was awarded the Nobel Prize in 1918. It is worth noting that the Nobel Prize committees are Swedish and Sweden was neutral during the war.

---

[3] Carl Bosch's uncle, Robert Bosch, founded the Robert Bosch Company, which is still an important maker of car parts.

Figure 6.11 Carl Bosch (1874–1940). The German chemical engineer who developed the industrial process for synthesizing ammonia. This photograph was taken in 1908. Winner of the 1931 Nobel Prize in chemistry.

Haber's service was appreciated in Germany and he kept his job as director of the Kaiser Wilhelm Institute for Physical Chemistry. However, when Adolf Hitler became chancellor in 1933, the situation changed because Haber had a Jewish background. He had been raised Jewish, but he had converted to Christianity as an adult. He left Germany and died the following year in Switzerland.

Like Haber, Carl Bosch received the Nobel Prize. In 1924, he helped found the chemical conglomerate I.G. Farben, which included BASF as one of its member companies, and he served as its first director. The 1920s were one of those times when there was a concern about how long oil would last, and Bosch started a program at the Leuna plant to synthesize gasoline from coal. With the discovery of the gigantic East Texas Oil Field in 1930, the concern receded, but Bosch persevered.

One motivation was that Germany produced no oil itself, and in the event of a war, it would be cut off. And that is exactly what happened when war came again to Europe in September, 1939. Germany tried to capture Soviet oil fields, but failed, and there was never enough oil. Bosch's plant provided much of Germany's gasoline, particularly high-octane aviation fuel for its fighters. Although Carl Bosch died in 1940, his gasoline synthesis plant likely lengthened the war in Europe.

Nevertheless, Carl Bosch's politics were not Hitler's. He assigned his Jewish employees to the United States and Switzerland to get them out of Hitler's reach. He wrote,

I stand here in conscious opposition to those who say and demand that the individual will and must subordinate himself to the general good. ... I consider such desire absolutely irreconcilable with human nature. Man in his entire makeup and evolution is not a herd animal but a family animal. ... The purpose of the state is to make sure that the gainful employment and co-existence of individuals and nations proceeds with the least possible amount of friction.[4]

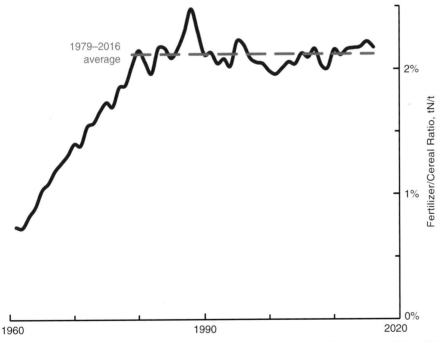

Figure 6.12 The ratio of world nitrogen fertilizer production to cereal production from 1961 to 2016. The data come from the UN Food and Agriculture Organization database FAOSTAT. The Food and Agriculture Organization (FAO) does not distinguish between fertilizer applied to cereals and other uses. For this graph, the fraction of nitrogen fertilizer applied to cereal land is taken to be the same as the cereal land fraction of arable land.

---

[4] This quotation is from Thomas Hager's superb dual biography of Haber and Bosch. The book is listed in the references section for this chapter.

We conclude this section with Figure 6.12, which shows the ratio of synthetic nitrogen fertilizer to the cereal harvested. It increased steadily until 1979. From 1979 to 2016, the ratio is flat with an average of 2.1%. You will recall that 1979 was a shock year for hydrocarbon production because of the high oil prices at the time of the Iranian Revolution and the Iran–Iraq war (Figure 1.6). This percentage is consistent with the nitrogen content of cereals, which is about 2%. It should be appreciated that there are offsetting factors. Some of the nitrogen fertilizer that is applied to the fields is not absorbed by the plants. There are also additional sources of nitrogen – manure, legumes, and nitrogen oxides entrained in rain. However, the figure does make clear that since 1979, fertilizer and cereal production are tightly coupled. Increases in cereal production will apparently need increases in nitrogen fertilizer production. At the present time, this does not appear to be a problem. There are many nitrogen fertilizer plants around the world. Most use natural gas as the source of hydrogen, but coal gasification could also be used, as Carl Bosch did in the early days.

6.3   Cereals

From time immemorial, our primary foods have been cereals. In East Asia, rice dominated, while in Europe and the Mideast, wheat was the most important crop. Sorghum is significant in Africa. Corn, or maize, comes from the Americas. Other cereals that the UN Food and Agriculture Organization (FAO) tracks are barley, buckwheat, canary seed, fonio, millet, oats, popcorn, quinoa, rye, and triticale. Rice is boiled and eaten directly. Wheat is made into bread. Barley provides the starch for beer. Corn dominates the modern food industry. In his 2006 book *The Omnivore's Dilemma*, Michael Pollan relates that the average American grocery store sells ten thousand items that contain corn. Corn is also the pre-eminent feed crop for livestock and the most important crop for biofuels. Figure 6.13 shows how production of some cereals has evolved over time, expressed on a per-person basis. Corn now dominates, with per-person production up 114%, from 183 g/p/d in 1961 to 391 g/p/d in 2016. Rice and wheat production were higher than corn in 1961, but both plateaued in the 1980s at 275 g/p/d, and corn passed them around 2000. The other cereals are fading away, together amounting to only 107 g/p/d. An important factor in the rise of corn is that it has a different photosynthetic pathway than rice and wheat. Corn is classified as a C4 plant, while rice and wheat are C3 plants. The labels C3 and C4 refer to the number of carbon atoms in the molecule that fixes the carbon in the plant. C4 plants eliminate an energy loss component in hot weather called photorespiration. This improves tolerance of hot, dry climates. Sorghum and sugar cane are also C4 crops.

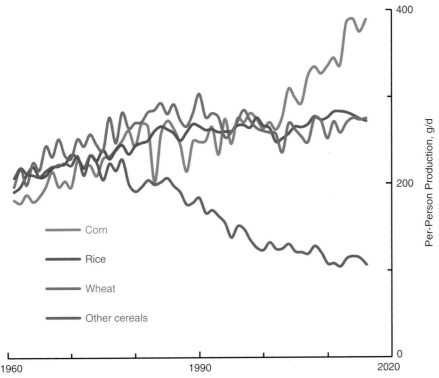

Figure 6.13 World cereal production per person from 1961 to 2016. The data come from the UN Food and Agriculture Organization database FAOSTAT. The population data are from the United Nations Population Division.

6.3.1 Corn

Corn is grown in many countries, but the largest producer is the United States, with 36% of the world production in 2016. The US Department of Agriculture has corn production data back to 1866. These are plotted in Figure 6.14. The largest area harvested with corn was 449,000 km² in 1918, during World War I. American farmers had a large demand for their crops during the war because many European farms were in areas where there was fighting. After the war, the European farms recovered, and the area planted to corn on American farms dropped. The yield was relatively steady during this time, but in the 1940s it began to pick up, slowly at first, then faster. Starting in the 1960s, the yield has increased linearly. This is like the linear plot for world cereal yield (Figure 1.21). Table 6.2 compares American corn production in 1918, when the planted area was largest, to 2018, one hundred years later. The total production in 2018 was five times larger than in 1918, from an area 25% smaller. This was made possible by a continuing rise in yield. It was not predicted. On the contrary, in 1974, prominent agronomist Lester Brown, founder of Worldwatch Institute, wrote,

In looking at future prospects for raising yields, it seems inevitable that increases in the more advanced countries – for example, in corn yields in the United States, in wheat yields in the United Kingdom and France, and in rice yields in Japan – will slow down markedly during the remainder of this century. A steadily expanding share of growth in world food supplies must come from developing countries, which have far more unrealized agronomic potential. The difficulties of getting rapid gains in rice yields anywhere and of getting further rapid gains in cereal yields in the more advanced countries do not encourage optimism about the adequacy of future supplies in the absence of a sharp slowdown in the world rate of population growth.[5]

Table 6.2 US corn production in 1918 and 2018. The data are taken from Figure 6.14.

	Area, 1,000 km^2	Yield, t/km^2	Production, Mt
1918	449	165	74
2018	335	1,108	371

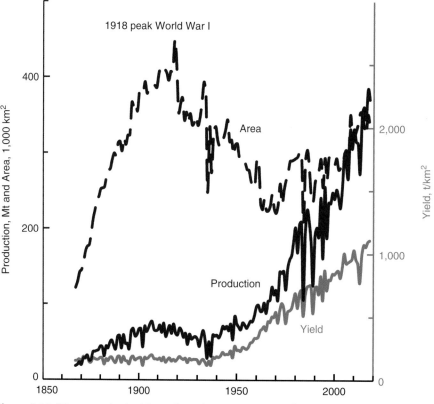

Figure 6.14 US corn production history, from the US Department of Agriculture, *Feed Grains Data Workbook*, Table 1. Production data is converted at the USDA recommended rate of 39.39 bushels/t. The area is the harvested area, calculated as the quotient of the production and the yield.

[5] Lester Brown and Erik Eckholm, 1974, *By Bread Alone*, Elsevier Science.

6.3.2   Regional Cereal Production

Next we consider several graphs that show cereal production since 1961 in different regions of the world, the United States (Figure 6.15), the European Union (Figure 6.16), Asia (Figure 6.17), Africa (Figure 6.18), and Brazil (Figure 6.19). There are no graphs in this book that are more important than these. They show the basis of our food supply. Table 6.3 summarizes the results. At first glance, the results for the United States and the European Union appear similar. The harvested area in both has dropped about 10% since 1961. This allows land to go back to forest. Both show an ongoing linear increase in yield. However, if one looks carefully, the yield in the United States is accelerating, while the yield in the European Union is decelerating. One difference between the two is that the United States has supported GMO (genetically modified organism) crops, while the European Union has banned them. Asia shows the strongest increase in yield, up 223% since 1961. However, unlike the United States and Europe, land is not being released to the forest in Asia. Harvested land is up 24%. However, in compensation, the Asian food supply has gone up dramatically, 54% since 1961, to 2,779 Cal/d (Table 6.1). On the other hand, Africa has not had the same level of success. The yield is up 91%, but the area is up 104%. The final plot, for Brazil, shows that it does not

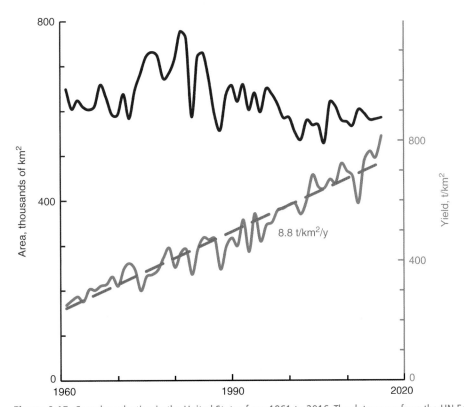

Figure 6.15 Cereal production in the United States from 1961 to 2016. The data come from the UN Food and Agriculture Organization database FAOSTAT.

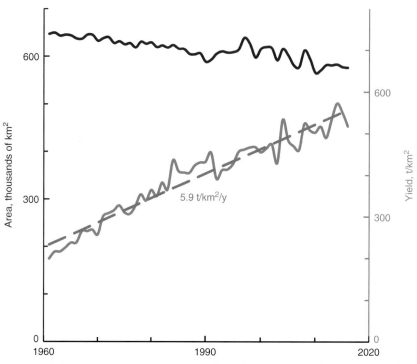

Figure 6.16 Cereal production in the European Union from 1961 to 2016. The data come from the UN Food and Agriculture Organization database FAOSTAT.

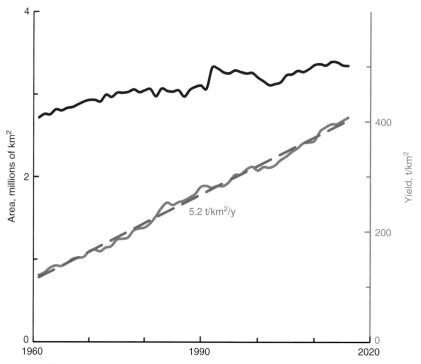

Figure 6.17 Cereal production in Asia from 1961 to 2016. The data come from the UN Food and Agriculture Organization database FAOSTAT.

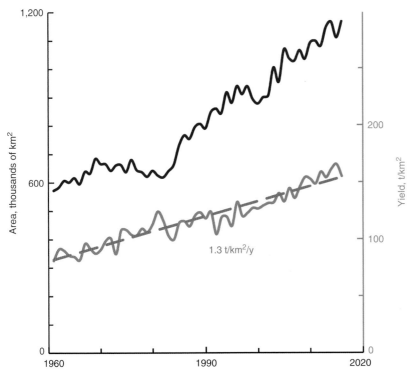

Figure 6.18 Cereal production in Africa from 1961 to 2016. The data come from the UN Food and Agriculture Organization database FAOSTAT.

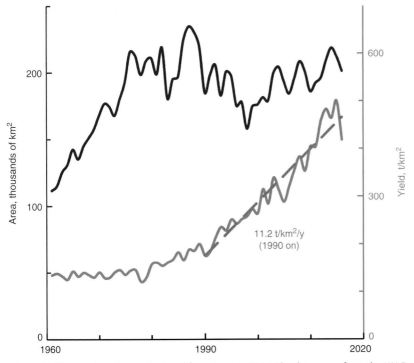

Figure 6.19 Cereal production in Brazil from 1961 to 2016. The data come from the UN Food and Agriculture Organization database FAOSTAT.

Table 6.3 Cereal production summary. Data from figures Figure 6.15 through Figure 6.19.

	2016 Area, 1,000 km²	Increase from 1961	2016 Yield, t/km²	Increase from 1961
US	585	−10%	814	223%
EU	576	−11%	517	160%
Asia	3,347	23%	408	236%
Africa	1,167	104%	155	91%
Brazil	201	80%	418	211%
World	7,181	11%	397	193%

have to be this way. Before 1990, Brazilian yield was growing slowly, like Africa. However, after 1990, the yield took off, growing at $11.2 \, t/km^2/y$, compared with $1.3 \, t/km^2/y$ for Africa. As we saw at the beginning of this chapter, Brazil did not become an agricultural powerhouse with small farms. It succeeded through large-scale private capital, government research, and chemical fertilizers. The planted area for Brazil was growing quickly in the early years when the yield stalled. However, since 1990, the planted area has been relatively flat. The large increase in yield appears to have restrained deforestation, as Norman Borlaug argued it would. A measure of the shortfall in Africa is that it uses twice the land the United States does, but produces 62% less.

We conclude this section with two open questions. Why does the yield increase linearly in time? In the figures for the world (Figure 1.21) and for the different regions in this section, the yield in each case is linear, in Brazil's case after 1990. However, the slopes for the different regions are quite different. The largest slope, Brazil's, is nine times Africa's. It does appear that the linear improvement is a consistent characteristic of yield, but we do not know why. The second question is, "How long will the trend continue?" Again, we do not know. There is likely to be no problem scaling up nitrogen fertilizer production as needed for future yield improvements (Figure 6.12). For additional insight, consider Figure 6.20, which shows that the US yield has been close to twice the yield in the rest of the world ever since 1961. Note that this relationship means that while the American yield and the yield outside the US are both increasing, they are moving farther apart rather than closer together. This suggests that when the yields do approach well-defined saturation levels, the limits may be different in different regions.

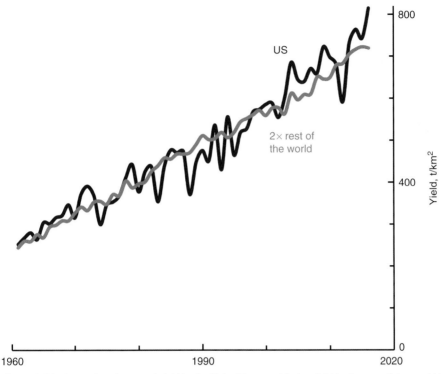

Figure 6.20 Comparing the cereal yield in the United States with the yield in the rest of the world from 1961 to 2016. Since 1961, US yield has been twice that in the rest of the world.

6.4 Livestock

Meat and dairy products contribute greatly to our eating enjoyment. They also make it easy to get a good supply of protein. Livestock production is a significant part of agriculture. It has been estimated that a quarter of the world's crops are grown as feed for animals.[6] Corn and soy beans are particularly important. From an energy perspective it would be more efficient for people to eat the corn and soy beans directly. However, it would be oversimplifying to view livestock simply as competing with people for food. Dairy cows and their calves typically graze on grass. We could not digest the grass ourselves. For beef cattle, ranchers often aim at a certain percentage of fat in the meat. They can achieve this by raising the calves in pastures, and then finishing in a feed lot. The more time the cattle spend in the pasture, the leaner the beef. Legumes like alfalfa and clover provide feed for cattle and at the same time they increase the

---

[6] There are large uncertainties in this type of calculation. The number here comes from Emily Cassidy, Paul West, James Gerber, and Jonathan Foley, 2013, "Redefining yields: from tonnes to people nourished per hectare," *Environmental Research Letters*. They calculate that 24% of the crops go to feed on a weight basis, 36% on a calorie basis, and 53% on a protein basis. Available at http://dx.doi.org/10.1088/1748-9326/8/3/034015

soil nitrogen for the next stage of the crop rotation cycle. Livestock manure is itself a significant source of nitrogen. When corn is made into ethanol, the starch in the kernels is separated out and fermented. The remainder, called distillers grains, is high in protein, and is used as a feed supplement for animals. The name "distillers grains" comes the fact that they originally came from distilleries and breweries.

The following figures, Figure 6.21 through Figure 6.25, show the production history for milk, eggs, chicken, pork, and beef. There are different ways to express how efficiently feed is converted to an animal product. We will use the feed ratio, which is the ratio of the weight of feed to the weight of the food product, whether it is milk, eggs, or the live animal. Smaller feed ratios are better. Our source for feed ratios is Vaclav Smil's 2000 book, *Feeding the World*, which is listed in the references. For milk, the feed ratio is 1.0. A cow produces 1 kg of milk for each kilogram of feed, which is remarkable. This is the best ratio for land animal products, but aquaculture can achieve comparable ratios. Note that the energy density of the input and output are not the same, so in principle the ratio could be less than 1. From Figure 6.21, the European Union has a 60% larger milk production than the United States, but the EU population is also 60% larger. China has become a significant milk producer only recently. Milk in China is mainly for children. Most Chinese people lose the ability to digest milk well as they grow older and that limits the consumption.

Figure 6.22 shows the egg production history since 1961. The feed ratio is 2.5. The EU and US are comparable. In 1961, China had a lower egg production than either the US or the EU, but production has grown rapidly since 1980, and it is

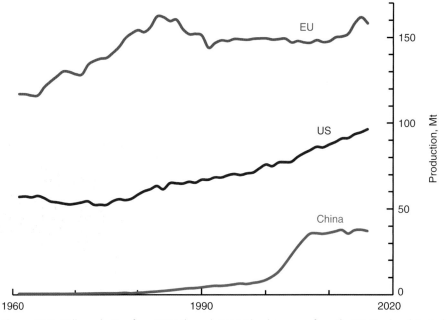

Figure 6.21 Milk production from 1961 through 2016. The data come from the UN Food and Agriculture Organization database FAOSTAT.

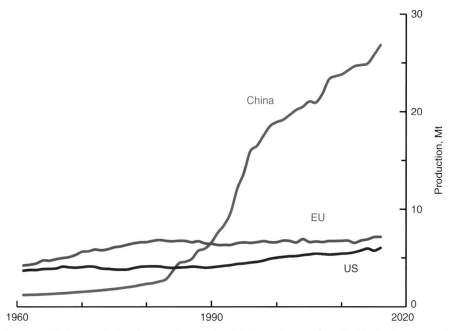

Figure 6.22 Egg production from 1961 through 2016. The data come from the UN Food and Agriculture Organization database FAOSTAT.

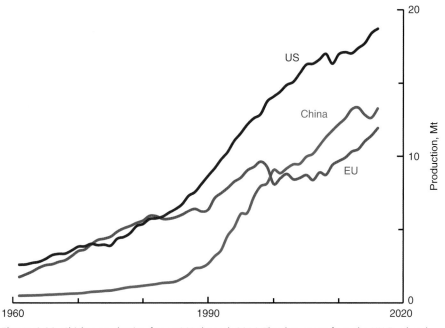

Figure 6.23 Chicken production from 1961 through 2016. The data come from the UN Food and Agriculture Organization database FAOSTAT.

now four times the US production. This is close to the population ratio, so the per-person production in China and the US are similar.

Figure 6.23 shows the chicken production history. The feed ratio is 2.5, the same as for eggs. The production of chicken in the US, China, and the EU have all been rising steadily, with US production the largest. On a per-person basis, chicken consumption in the US is a multiple of that in the EU and China. For decades, the US government has recommended that people eat chicken in preference to beef, based on the idea that fat in the diet is harmful, and the fact that chicken is leaner than beef. In recent years the idea that fat is bad for health has been challenged, and the government is wavering. At this point, it would be premature to judge how this debate will end.

Figure 6.24 shows the pork production history. The feed ratio is 4.0. Again Chinese production started off lower than in the US and the EU, but now has surpassed both. On a per-person basis, production is now comparable in China and the EU.

Finally, Figure 6.25 shows the beef production history. The feed ratio is 8.0, double that of pork. Cattle have a higher metabolism than pigs. This is mitigated somewhat by the fact that milk is a by-product of the cattle industry and milk production is quite efficient. Nevertheless, beef ends up being more expensive than pork and chicken. US production is the largest. In recent years, Chinese beef production has risen rapidly, and is now comparable to the EU production.

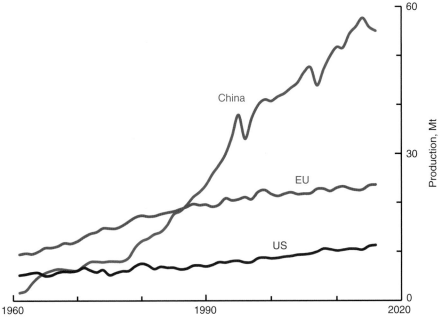

Figure 6.24 Pork production from 1961 through 2016. The data come from the UN Food and Agriculture Organization database FAOSTAT.

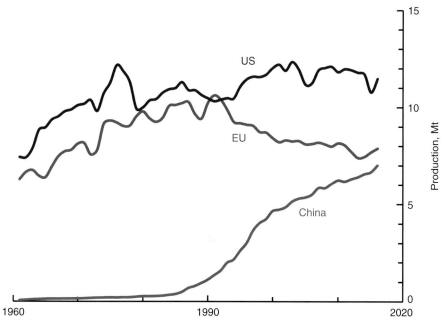

Figure 6.25 Beef production from 1961 through 2016. The data come from the UN Food and Agriculture Organization database FAOSTAT.

Table 6.4 Food supply in food calories per person per day from animal products, including seafood, which at the world level was 7% of animal calories in 2013.

	1961	2013	Increase
US	1,010	984	−3%
EU	822	978	19%
Asia	111	451	306%
Africa	156	215	38%
World	338	514	52%

The data come from the UN Food and Agriculture Organization database FAOSTAT.

Table 6.4 shows the food calories per person provided by animal products. The United States has been flat near 1,000 Cal/p/d since 1961. The European Union started out 19% lower than the US in 1961, but is now close to the American level. The world average has risen 52% since 1961, with the increase concentrated in the lower-income countries. The supply for Asia has quadrupled to 451 Cal/p/d since 1961. Africa shows a much smaller rise, 38%, which is comparable to its 32% rise in the overall food supply.

6.5   Impacts

No human activity has as great an impact on the earth as farming. Vast forests have vanished before the plow. Cropland is utterly unlike any natural environment. Fields are planted with a single crop. Fertilizers and pesticides are added. Some of the fertilizer goes elsewhere in the runoff. This can result in algal blooms in the sea that consume the oxygen in the water and cause die-offs. Insects evolve resistance to pesticides. And then there is water. Growing a kilogram of wheat requires more than a tonne of water. Irrigation water can bring up salts from underground. Diverting water for irrigation has reduced great seas to sheets of salt. Some of the most painful choices societies must make involve allocating water for agriculture. The water loss in agriculture is fundamental. In photosynthesis light provides the energy to synthesize glucose. The reactants are carbon dioxide from the air and water from the ground. The stomata on a leaf that open to let carbon dioxide in also let water vapor out by transpiration.

6.5.1   Agricultural Land

Table 6.5 shows the areas for arable land and permanent crops, pastures and meadows, and forests. The world total for arable land and permanent crops is 16,000,000 km², comparable to the size of the Russian Federation. Pastures and meadows is twice as large, 33,000,000 km². For pastures and meadows, we cannot be as categorical about agricultural impacts as we were with cropland. There are large areas, like the steppes in Asia, the savannahs in Africa, the pampas in South America, and the Great Plains of North America, where large wild animals graze on grass. Our cattle descend from the aurochs of Eurasia. In many places

Table 6.5 Land areas and changes from the UN Food and Agriculture Organization database FAOSTAT.

	Land, millions of km²			Annualized growth rate, /y		
	Arable land, permanent crops	Pastures and meadows	Forests	Arable land, permanent crops	Pastures and meadows	Forests
EU	1.2	0.7	1.6	−0.3%	−0.6%	0.3%
US	1.6	2.5	3.1	−0.8%	0.4%	0.2%
Asia	5.7	10.8	5.9	0.1%	−0.1%	0.3%
Africa	2.7	8.6	6.3	1.1%	−0.4%	−0.5%
Brazil	0.9	2.0	4.9	1.4%	0.0%	−0.3%
World	15.8	33.2	40.0	0.2%	−0.2%	−0.1%

The areas are for 2014, and the annualized growth is calculated for the 10-year period from 2004 to 2014.

pastures are similar to natural grassland and may be shared by wild animals. The largest area in the table is forests, 40,000,000 km². There is movement among all of the categories. In some places, forests are cut down to make cropland and pastures. In others, cropland and pastures may return to forest. At the world level, deforestation and reforestation nearly cancel out and the change in forest area is small, –0.1%/y.

In the European Union, the United States, and Asia, the forest area is increasing. In the EU and the US, this is a result of farms being abandoned to the forest. One factor in the Asian forest growth is an aggressive program to plant trees in China. On the other hand, in Africa and Brazil, forest land is being converted to cropland. In Africa, this reflects a slower increase in crop yield than in other regions (Figure 6.18), and a faster growing population (Table 1.3). However, the ratio of arable land and permanent crops to forest land in Africa is still only 0.4, a smaller ratio than in the US (0.5), the EU (0.7), and Asia (1.0). The situation in Brazil is different from Africa. The deforestation in Brazil can be attributed to crop exports. The yield has improved enormously and Brazil has become a major food exporter. In any event, the ratio in Brazil is only 0.2, far below the ratios for the other regions. The deforestation in Africa and Brazil is often criticized, but it is difficult to argue that it is different from what Europe, the US and Asia did earlier.

6.5.2   The Destruction of the Aral Sea

In addition to these global changes, agriculture can have an enormous local impact. The Aral Sea extends across the border between Uzbekistan and Kazakhstan in Central Asia. Formerly it was completely inside the Soviet Union. In Soviet times it was the fourth largest lake in the world, after the Caspian Sea,[7] Lake Superior, and Lake Victoria. The Aral Sea supported a thriving fishing community. It is fed by two rivers, the Amu Darya and Syr Darya. The word *darya* means "river" in the Kazakh language. These rivers are supplied with water from glaciers high in the Hindu Kush and the Tian Shan ranges. The Aral Sea is *endorheic*, meaning that there is no outlet. The water was brackish rather than salty like two other notable endorheic lakes, the Great Salt Lake and the Dead Sea. When the Soviets came to power, they developed an extensive irrigation system for cotton, an export that brought in cash to the central government. Figure 6.26 shows the history of the withdrawals. They increased steadily from 16 km³ in 1922 to a peak of 123 km³ in 1985, shortly before the collapse of the Soviet Union. Since then, the withdrawals have been lower, reaching 92 km³ in 2010.

---

[7] The Caspian Sea was connected to the world's oceans in relatively recent geologic times and is salt water. From a geological perspective its structure is more like an ocean basin than a lake in the middle of a continent.

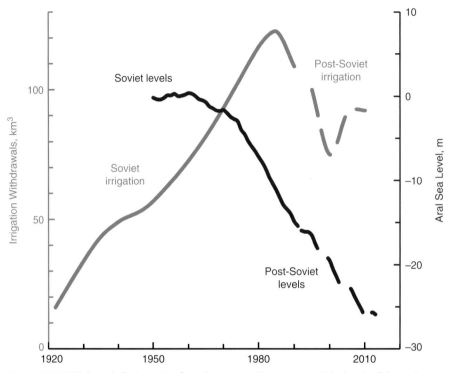

Figure 6.26 Withdrawals for irrigation from the Amu and Syr rivers, and the levels of the Aral sea. The reference for sea level is the average from 1950 through 1960. The data are from Philip Micklin, N. V. Aladin, and Igor Plotnikov, editors, 2014, *The Aral Sea, The Devastation and Partial Rehabilitation of a Great Lake*, Springer-Verlag.

In an endorheic lake, the incoming river water is balanced by evaporation losses. If water is diverted for irrigation, the area shrinks until the evaporation losses rebalance. In the figure, the lake level started falling in the 1960s. This continued for the remaining Soviet years and through the post-Soviet years. By 2010, the level had dropped 26 meters. Figure 6.27 shows a 2014 satellite photograph. The 1960 shoreline is indicated by a thin gray line. Instead of a single large lake, there are now three separate lakes, the Little Sea, Lake Tscherebas, and the Great Sea. The Kokaral, Barsa-Kelmes, and Vozrozhdeniya islands are now connected to the mainland. Vozrozhdeniya Island was a bioweapons laboratory. The Little Sea at the north is in Kazakhstan. It is fed by the Syr River, and there is a dike that releases some water into the Great Sea basin. The Little Sea is brackish rather than salty. Fish can live in it and there is still some fishing there. The large southern portion, the Great Sea, and the Amu Darya delta are in Uzbekistan. The Great Sea has been reduced to the green remnant on the west side. It is now so salty that no fish survive. The salt in the dry lake bed is kicked up by the wind and is now a source of air pollution. Figure 6.28 is a haunting photograph of a fishing boat stranded on the lake bed. The Aral Sea and the river deltas had formed an important refuge for

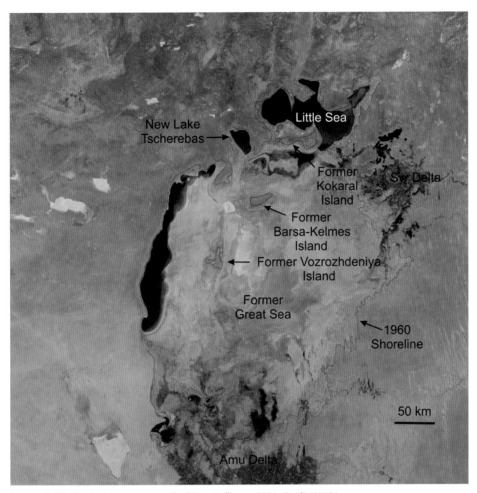

Figure 6.27 The Aral Sea photographed by satellite in 2014. Credit: NASA.

wildlife. Migrating birds visited the lake and the deltas. Most of that refuge has been destroyed.

The Soviet planners did realize that diverting the water for irrigation would reduce the size of the Aral Sea and its fishing industry, but they judged that the increase in cotton production was worth it. It is important to appreciate the underlying incentives. The central planners got credit for the cash that the cotton sales provided to the central government, while the fishing families that were injured by the policy were a local problem. The same incentives are present today in Uzbekistan.

6.5.3 The American Endangered Species Act

In the United States, the most important legislation that protects wildlife is the Endangered Species Act, enacted in 1973 during the Nixon Administration. This law gives the Fish and Wildlife Agency and the National Marine Fisheries Service

Figure 6.28 A fishing boat stranded by the destruction of the Aral Sea. Courtesy of Lukasz Kruk.

the authority to designate animals and plants as *threatened* and in more critical cases, *endangered*. The number of animals and plants listed is large. As of October 26, 2018, there were 2,344 animals and plants listed. Once there is a designation, government agencies gain extraordinary power to regulate land and water use to encourage the designated plants and animals. Agriculture requires large amounts of water. Solar power plants and the lumber industry requires large amounts of land. Workers in these industries can be greatly affected by the Endangered Species Act, particularly in the western United States, where the federal government is also a major landowner and water supplies are often limited. It is uncommon for a population to recover; only 2% have been taken off the list. Perhaps the most notable recoveries are gray whales and nine populations of the humpback whale.

The legislation refers to species, but in practice the law is applied to distinct populations. The populations may be distinct species, subspecies, or distinct populations of a single species. The traditional definition of a species is a population with members that mate and produce fertile offspring. Recall from Section 2.1.1 that when a mare and a donkey jack mate, the hybrid offspring is a useful animal, the mule. However, horses and donkeys are different species and mules are sterile. Subspecies are populations of a species that occupy distinct ranges and that can be distinguished in some way. For example, the Northern Flicker is a woodpecker with two subspecies, one with yellow feathers in its wings in the eastern United States and the other with red wings in the western United States. In the Great

Plains there are intergrade birds of intermediate color. However, in practice the distinction between species and subspecies is not always clear. Western Gulls and Glaucous Gulls are considered two species. However, they often mate and produce fertile hybrid offspring.

Biologists may not agree on whether two populations are different species or different subspecies. They also change their minds. The red-shafted flicker and the yellow-shafted flicker used to be considered two species. The change can also go the other way. The desert tortoise of the southwestern United States and northern Mexico used to be considered one species, the Desert Tortoise. Now it is two. The Sonoran Desert Tortoise is on the Arizona side of the Colorado River, and the Mojave Desert Tortoise is on the California side. The Sonoran Desert Tortoise is not endangered, but the Mojave Desert Tortoise is. The Ivanpah Solar Thermal Power Plant is on the California side and it is under a strict court allowance. If more than nine Mojave Desert Tortoises are killed, the 2-billion-dollar project will be shut down. The plant employed fifty biologists to move the tortoises to safety.[8] The difficulty in managing the tortoise is that they spend 95% of their lives underground, so no one knows where they are. There is more discussion of the Ivanpah plant in Section 7.8.2. We will consider two examples, the Northern Spotted Owl in the Pacific Northwest and the Delta Smelt in the Sacramento River Delta in California.

The Northern Spotted Owl, shown in Figure 6.29, is a subspecies of the Spotted Owl. The range of the Northern Spotted Owl is extensive, from southern British Columbia, through Washington, Oregon, and down into California. To a bird watcher, the Northern Spotted Owl would not be considered particularly rare or difficult to find. The birds tend to nest in old Douglas Fir trees. They hunt birds and small mammals.

The Northern Spotted Owl was declared threatened in 1990. The government asserted that the limit on the bird's population was the area of forest that had never been logged. In earlier times the government had reserved 12,000 square kilometers in Oregon and Washington to be off limits for logging. However, there was another 12,000 square kilometers of land owned by the government that had never been logged. The Forest Service had been releasing this land slowly to loggers. The plan was that as trees were logged, trees would be replanted so that the production would be maintained. This would provide steady employment for loggers and a continuous supply of wood for housing and wood products.

To help the Northern Spotted Owl, the government changed its plan and blocked logging on most of the rest of the government forest land that had never been logged. The result of the policy is shown in Figure 6.30. The plot shows the lumber production in Oregon on government land and on private land from 1962 to 2015.

---

[8] This information is from a meeting at the site between the author and the Ivanpah plant manager in October, 2015. One tortoise had been killed by a biologist's truck at that time.

Figure 6.29 Northern Spotted Owl (*Strix occidentalis caurina*) near the McKenzie River in central Oregon. Credit: United States Fish and Wildlife Service.

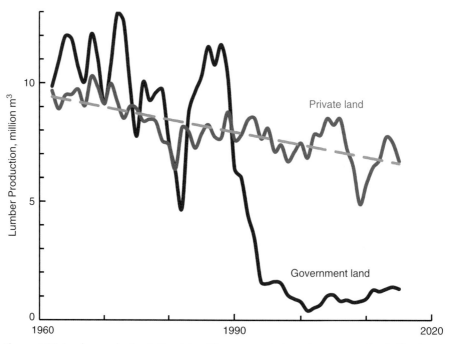

Figure 6.30 Lumber production in the state of Oregon on private and government lands. The data come from the Oregon Department of Natural Resources.

The logging on private land shows a slow decrease over this period. The trend line decreases from 9.4 million cubic meters in 1962 to 6.6 million cubic meters in 2015. For government land there is a sharp drop after the Northern Spotted Owl was listed as threatened in 1990 and a further decline through the 1990s. The 1962 to 1989 average was 10.0 million cubic meters while the average from 1998 to 2015 was 1.0 million cubic meters. Many people lost their jobs and lumber communities were wrecked. There is still much resentment of the government decisions in rural Oregon. People there perceive that their livelihoods and families are less important to people in the city than the owls. And the incentives do work that way. Government agencies receive credit for trying to preserve the Northern Spotted Owl, but the damage to the lumber communities is a local problem.

From the beginning, the rural people argued that the government model was shaky. They claimed that the birds often hunt outside the old-growth forests. Moreover, the Douglas Fir is a *seral* tree species. This means that it is part of a succession rather than a *climax* tree. Douglas Firs start in disturbed areas, for example, following a fire that allows light to reach the ground. Even if the Douglas Firs are not logged, they die in other ways, from insects, wind, and disease. As they die, they are replaced by climax species like the Grand Fir and Western Hemlock. In any event, the government solution did not work. Surveys indicate that the number of spotted owls in the Pacific Northwest has declined 3%/y since the bird was listed. One current theory for the decline is that a closely related species, the Barred Owl, is displacing the Spotted Owl.

Now we turn to the Delta Smelt. The Sacramento River is the most important source of water in California (Figure 6.31). Its origins are in the north near the city of Redding. It flows south, joining the Feather River and the American River along the way. Below the city of Sacramento, the river forms a delta. The San Joaquin River also flows into the Sacramento Delta, but from the south. Originally the delta was a wetland, but in the early 1900s, much of it was drained for farming. The fields are still productive today, but they need constant attention. Underlying the fields is a layer of peat twenty meters thick. As the peat dries out it compacts. The subsidence is severe. The fields are now several meters below the water level in the delta and they are enclosed by levies to keep them from flooding. In addition, tides push water from the San Francisco Bay into the delta. There must be a constant flow of water from the delta out to the bay to keep salt from the bay away from the delta farms.

Within the Sacramento Delta is a small fish called the Delta Smelt, shown in Figure 6.32. The Delta Smelt lives only in the Sacramento Delta, and it has endangered status. The California Department of Fish and Wildlife does a survey during the year to look for the fish. In 2018, only seven Delta Smelt were found in 800 tows of a net. One problem for the Delta Smelt has been that several species of bass have been introduced from the eastern United States. The introductions have been

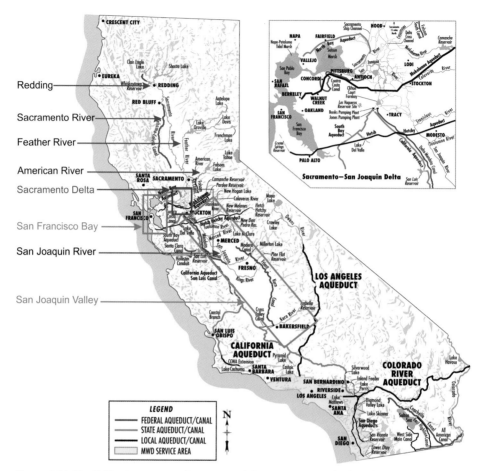

Redding

Sacramento River

Feather River

American River

Sacramento Delta

San Francisco Bay

San Joaquin River

San Joaquin Valley

Figure 6.31 The California water supply system and the Sacramento Delta (inset). Arrows and boxes have been added for the features mentioned in the text. Underlying map courtesy of the Metropolitan Water District of Southern California.

successful and there are now fishing tournaments to catch the bass. However, the bass prey on the Delta Smelt. It should be noted that the Delta Smelt will not go extinct. The University of California at Davis maintains thousands in tanks.

Rain and snowfall in the Sacramento Delta watershed change dramatically from year to year. During the last hundred years the water available for the delta has varied from a high of 76 km^3 in 1983 to a low of 9 km^3 in 1931, an 8:1 range. Table 6.6 shows what this means for the users. In 2011, a wet year, the total water available was more than twice as much as in 2015, a dry year. However, the diversions upstream of the delta and for the delta farms were little changed, only a 7% drop. The majority of the water use is classified as environmental. Environmental here means that the water is allowed to flow through the delta out into San Pablo Bay.

In addition, water is exported from the delta south to the San Joaquin Valley. Farmers complain bitterly about the reduced water supply. When one drives down

Table 6.6 Water allocation in the Sacramento Delta.

	2011	2015
Upstream and delta farms	13.8	12.8
Environmental	32.6	7.6
Exports to the San Joaquin Valley	8.1	2.4
Total	54.5	22.8

The units are cubic kilometers. The year 2011 was a wet year and 2015 was a dry year. The data come from the Public Policy Institute of California, 2016, "The Sacramento-San Joaquin Delta."

Figure 6.32 The Delta Smelt (*Hypomesus transpacificus*). Credit: Peter Johnsen, United States Fish and Wildlife Service.

the valley it is common to see signs like the one in Figure 6.33. The Delta Smelt story has been framed in the same terms as for the Northern Spotted Owl. The farmers believe that people in cities think that a small fish is worth more than their livelihood. In the case of the Northern Spotted Owl, it was easy to find the impact in the production data because lumber production on public lands in Oregon collapsed. For the Delta Smelt, the story is turning out differently. It is a tale of almonds.

California grew 62% of the world's almonds in 2016. Figure 6.34 shows the production history. Yield has tripled since 1995 and the harvested area has doubled. Farm labor has become expensive in California and one advantage of growing almonds is that the production steps can be mechanized. Trees also tolerate more variation in water supplies from year to year than annual crops can. The trees begin to produce nuts in their third year, and they are productive for thirty years.

Figure 6.33 Sign in the San Joaquin Valley in 2010 in response to the shutoff of water from the Sacramento Delta. Photograph by the author.

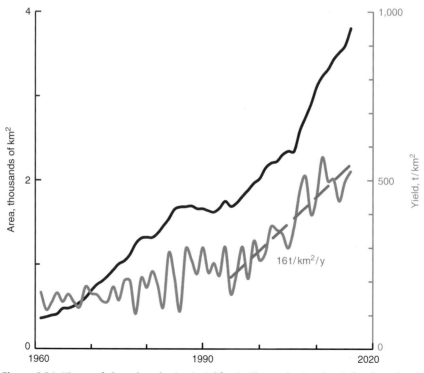

Figure 6.34 History of almond production in California. The production data is for almonds with shells for the US from FAOSTAT. California is the only state with significant commercial almond production in the United States.

Table 6.7 The ratio of water to food by weight for the United States. In addition, cotton is included.

	Water–food ratio
Corn	489
Rice	1,275
Wheat	849
Milk	695
Eggs	1,510
Chicken	2,389
Pork	3,946
Beef	13,193
Cotton	5,733

From A. K. Chapagain and A. Y. Hoekstra, 2006, "Water footprints of nations, Volume 1, Main Report," UNESCO-IHE. They cover the period from 1997 to 2001. The calculations are complex and this paper should be consulted for the details. It is available at http://waterfootprint.org/media/downloads/Report16Vol1.pdf

When a new generation of trees is planted, farmers can take the opportunity to upgrade from flood irrigation to the more efficient drip irrigation. Many farmers also have access to ground water during dry years and others have allocations from the Sierra runoff. While some individual almond farmers have undoubtedly been injured by the reduced water allocations during the dry years, it is difficult to argue from the data that the reductions have hurt state production as a whole. On the contrary, it is a spectacular success story.

6.5.4   Agriculture and Water

For agriculture, large reductions in water use are hard to come by. Modest improvements can be made by good engineering of irrigation systems and diligent maintenance. Table 6.7 shows the water requirements for food and for cotton. The huge number for cotton was the death knell for the Aral Sea. Corn, a C4 plant, is more efficient than rice and wheat, which are C3. Corn uses 62% less water than rice and 42% less water than wheat. The low water demand of corn goes a long way to explain the shift by farmers to corn in this century (Figure 6.13). The numbers for livestock include the water consumed by the animal and the water to grow the feed. Beef cattle are less efficient than pigs and chicken for meat. The most important conclusion from the table is that water use in agriculture is enormous. A teenager's hearty breakfast of eggs, sausage, toast, and milk requires a tonne of water.

We began this section on agricultural impacts with a discussion of land use changes, but the underlying issue is often water. Table 6.8 shows the components of the water demand in the United States. In practice, river water can be used more than once on its way to the ocean, so the demand is calculated on a net basis as

Table 6.8 Water demand in the United States in 2005. The units are km³/y.

	Demand
Irrigation	124
Thermal power plant cooling	14
Residential	14
Commercial and industrial	11
Aquaculture	3
Livestock	2
Mining	2
Total	169

Demand is calculated on a net basis as withdrawals minus discharges to fresh water. Ocean withdrawals by power plants are excluded because this water is separate from the rest of the supply. The data come from the US Department of Energy Lawrence Livermore Laboratory energy flow charts program.

the difference between withdrawals and discharges to fresh water. Irrigation completely dominates demand, with a 73% share. Thermal power plant cooling and residential demand are next with 8% shares. As wind turbines and solar panels replace fossil-fuel generation there is a corresponding reduction in cooling water demand. To reduce the residential demand on fresh water supplies, it is feasible for coastal communities to build desalination plants. Currently desalting ocean water by reverse osmosis requires $3 \, \text{kWh/m}^3$. Desalination technology is an area of intense research because the theoretical limit based on thermal physics is considerably lower, less than $1 \, \text{kWh/m}^3$.

6.6   Fish

Coastal communities have fished and crabbed and clammed for millennia. Seafood is delicious and it is an excellent source of protein. Table 6.9 shows the seafood contribution to the world food supply. Per-person production has doubled from 1961 to 2013. In earlier times, the ocean fisheries were viewed as inexhaustible. By *fishery*, we mean a regional stock of a fish. Thomas Huxley, known as Darwin's bulldog for his defense of the theory of evolution, said in 1883, "I believe then that the cod fishery, the herring fishery, pilchard fishery, the mackerel fishery, and probably all the great sea fisheries are inexhaustible: that is to say that nothing we do seriously affects the numbers of fish."[9]

---

[9] Thomas Huxley, 1883, inaugural address, Fisheries Exhibition, London. Available at http://aleph0.clarku.edu/huxley/SM5/fish.html

Table 6.9 Seafood supply in food calories per person per day.

	1961	2013	Increase
US	20	35	75%
EU	26	48	85%
Asia	15	37	147%
Africa	9	20	122%
World	17	34	100%

Only animal products are taken into account in this table; seaweed is neglected.
The data come from the UN Food and Agriculture Organization database FAOSTAT.

Huxley was wrong. The stock in almost every significant fishery has been reduced to less than half of the original level. Many oceanic birds, like petrels, shearwaters, and albatrosses, eat fish and squid and their numbers have been correspondingly reduced. The same is true for the marine mammals, like otters, seals, and dolphins. Some fisheries, notably the incomparable Newfoundland cod fishery, have been destroyed by overfishing. There are several contributing factors to this damage. One is technology development. Diesel trawlers can handle enormous nets that extend over many kilometers. Sonar allows a trawler captain to find *every* school of fish over large areas. A second problem is that many fishing fleets are too large for their fisheries. Often the over-building has been encouraged by government subsidies. A third factor is that much of the ocean is not under the control of a single country. Traditionally a nation's authority extended only three nautical miles from land. Ships from any country could pass freely and fish outside this range. This was convenient for freighters and warships. However, the narrow zone of control left no way to prevent overfishing. Now most countries claim an economic zone that extends out 200 nautical miles and they exclude foreign trawlers from this zone. In addition, large areas of ocean, totaling more than ten million square kilometers, larger than the land area of the United States, have been set aside as marine reserves. This allows some fish stocks to increase, as well as the birds and mammals that feed on them.

World seafood production since 1950 is shown in Figure 6.35, characterized as either fishing, where the catch is wild, or aquaculture. Aquaculture includes a wide range of activities, like fish pens, seaweed grown on ropes, and oyster farms. From 1950 to 1990, fishing production rose steadily, but since then it has been relatively flat at around 90 Mt/y. On the other hand, aquaculture production has been rising steadily, with a 10-year annualized growth rate of 6%/y. Aquaculture was 110 Mt in 2016, compared with 92 Mt for the wild catch. One should appreciate that often there is not a clear distinction between fishing and aquaculture. Many trout hatch from eggs in hatcheries. The juveniles are released into the wild, where they grow

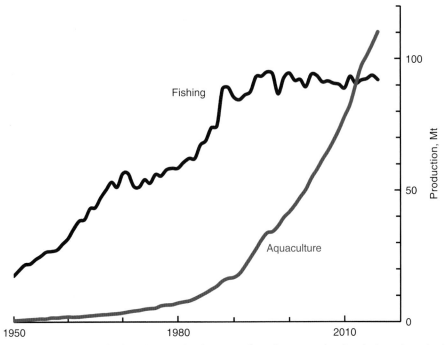

Figure 6.35 World seafood production. The data come from the UN Food and Agriculture Organization online fisheries capture and aquaculture databases.

and are then captured as adults. On the other hand, tuna are often captured young and then raised in pens. Finally, much of the wild fish catch, particularly anchovies and sardines, is used as fish meal in fish farms.

Most countries try to manage the fisheries in their economic zones thoughtfully. There may be limits on the type of equipment that is used. For example, the holes in nets may be set large enough to let young fish escape. Colored streamers may be added to fishing lines to keep birds away from the hooks. The catch may be controlled by licenses or by limiting the period of time when fish may legally be caught. The advantage of a licensing system is that it can reduce over-investment in fishing boats. In addition, the catch for each boat can be relatively large, and this helps the income of the remaining fishers. The advantage of setting a limited fishing season is that both the rich and the poor can participate. Also, fishing can be prevented during times when the fish are especially vulnerable, like the spawning season.

In the early days, it was thought that a scientific model could determine a *maximum sustainable yield*. This was defined as the maximum catch level that would allow the stock to be stable in the long run. However, over time it became recognized that there are fundamental difficulties in determining a maximum sustainable yield. One problem is that the life cycle of a fish can be extremely complex. Unlike birds and mammals, most fish species are not fed and protected by their parents. The young fry may develop in estuaries that are hundreds of kilometers away from the areas where the adults live. Fish that are top predators when they are adults are prey for even small fish when they are young. In addition, the natural variations from year to year in fish populations can be extremely large. The foundation of ocean life is *phytoplankton*, microorganisms that grow by photosynthesis. The word *plankton* means life that drifts with the currents. The phytoplankton depend on upwelling currents that pull the nutrients up from the depths. These upwellings occur only in specific places. Figure 6.36 shows a map of the sea surface temperature in the North Atlantic in March 2010. The area in the small box in the middle of the map is the Grand Bank. The Grand Bank is part of the continental shelf surrounding Newfoundland. The sea floor drops away quickly south of the box. This area is where the cold waters of the Labrador Current coming from the north meet the warm waters of the Gulf Stream coming from the south. The boundary between the two currents is called an *ocean front*, analogous to a weather front in meteorology.

The upwelling occurs at the front. Figure 6.37 is another satellite map that shows chlorophyll concentrations. This is the area inside the box in Figure 6.36 on an enlarged scale. The presence of chlorophyll is an indication of photosynthetic activity that the birds and the dolphins and the trawler fleets ultimately depend on. The front is associated with an extraordinary concentration of photosynthetic activity. The color changes from red to blue in just a few kilometers. This is a change in chlorophyll density of a factor of 10,000. Note that the challenge for the birds and the dolphins and the trawler fleets is that most of the ocean is almost devoid of life because of the lack of nutrients for the phytoplankton. Compared with the land, life in the ocean is concentrated in small areas. It is notable that

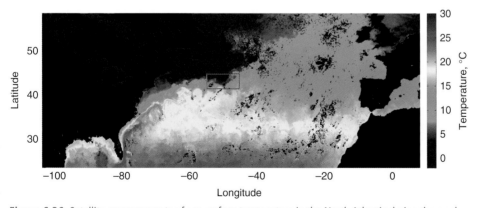

Figure 6.36 Satellite measurements of sea surface temperature in the North Atlantic during the week of March 14–21, 2010. The color scale on the right is the Sea Surface Temperature in °C. One degree of latitude is 111 km. Credit: NASA. From a 2013 talk by John Taylor, Lecturer in the Department of Applied Mathematics at the University of Cambridge given at the California Institute of Technology, "Turbulence, submesoscales, and phytoplankton at ocean fronts."

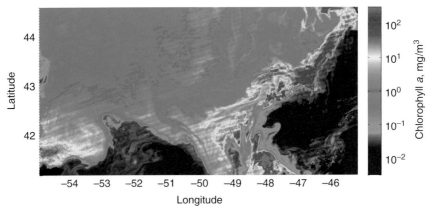

Figure 6.37 Optical measurements by satellite of chlorophyll concentrations in the North Atlantic measured by satellite. This is the area shown in the box in Figure 6.36 on an enlarged scale. The scale on the right is the chlorophyll concentration in mg/m³. One degree of latitude is 111 km. Credit: NASA. From a 2013 talk by John Taylor, Lecturer in the Department of Applied Mathematics at the University of Cambridge given at the California Institute of Technology, "Turbulence, submesoscales, and phytoplankton at ocean fronts."

seabirds typically nest on islands where there are no mice or rats to eat their eggs. These islands may be at great distances from the feeding grounds. There will be more discussion of this in Section 9.6.

6.6.1 The Destruction of the Newfoundland Cod Fishery

Most of the important fisheries, like the Newfoundland cod, Peruvian anchovies, and California sardines are associated with upwellings. These depend on the geometry of the sea floor and on currents that vary from year to year. In these situations, the ability of models to make predictions may quite limited. It may be beyond our ability, even in principle, to create a model that will correctly predict a maximum sustainable yield. The greatest failure of all was the Newfoundland cod fishery.

It seems likely that Basque fishermen discovered the cod on the Grand Banks around the time Columbus sailed to America. Fishers then and now consider the location of fishing spots as trade secrets, so Columbus got the credit. Before ships carried freezers, the cod were preserved by drying them on land and packing them with salt to stop the growth of bacteria. Fishing villages were established on the coast of Newfoundland. Figure 6.38 is a photograph from 1895 of Petty Harbour on the east coast. Most of the Newfoundland fishing villages could only be reached by boat until after World War II. The growing season is short and the land is rocky, so there are limited possibilities for raising crops. The photograph does show livestock grazing.

Traditionally cod were caught with hook and line from dories with one or two fishermen. The dorymen then brought the fish back to the mother ship. Figure 6.39 is a photograph on board the mother ship on the Grand Banks in 1949. One can get

Figure 6.38 Petty Harbour, Newfoundland, 1895. Newfoundland became a province of Canada in 1949 after World War II. Courtesy of the Centre for Newfoundland Studies (Coll – 137, 07.04.004), Memorial University of Newfoundland, St. John's/CC BY-NC-ND 2.5, CA/https://creativecommons.org/licenses/by-nc-nd/2.5/ca/

Figure 6.39 Fishing for cod on the Grand Banks, 1949. Credit: National Film Board of Canada/PA-110814.

a feel for the fishing life in those times from the 1937 movie, *Captains Courageous*. The extras were real cod fishermen. There are also dramatic scenes of schooners racing to the home port with all their sails set. Fishing then and now is a dangerous job. In the United States, fatalities among fishers were 100 per 100,000 in 2017, the highest of any job. In comparison, the average worker faced a fatality risk of 3 per 100,000.

Canada declared a 200-nautical-mile exclusive economic zone in 1977. The cod catch that year was 235 kt. In the following years, the government subsidized the construction of a large fishing fleet. The catch rose dramatically, reaching a peak of 511 kt in 1982. But too many fish were taken. Figure 6.40 shows the history. Starting in 1990, the catch collapsed. In 1992, the Canadian government closed the fishery. In 1995, the catch was only 12 kt. There it has remained. The catch in 2016 was still only 18 kt. And the Grand Banks fishery has remained closed. The collapse was a calamity for Newfoundland and many people moved to other provinces. The population in 2016 was 9% lower than in 1991.

For comparison the figure also shows the Norwegian cod catch. The Norwegians fish for cod off their west coast in the Norwegian Sea and to the north in the Barents Sea. They received a scare at the same time the Canadians fell into difficulties. From 1982 to 1990, the Norwegian catch fell from 344 kt to 125 kt. In 1989, the government restricted the catch. The Norwegians were lucky. Their fishery did recover. The catch in 2016 was 413 kt.

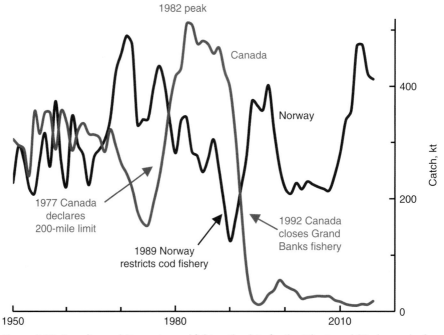

Figure 6.40 Canadian and Norwegian cod fishing. The data for the Atlantic cod (*Gadus morhua*) come from the UN Food and Agriculture Organization online fisheries capture database.

6.6.2   The Pacific Sardine and the Pacific Decadal Oscillation (PDO)

A fishery can be destroyed by overfishing, but changes in temperature and currents in the ocean can also ruin a fishery. A notable example is the California sardine fishery. Sardines are an oily fish that grow to a length of 20 cm. They are also called *pilchards*. They are bait fish that are eaten by larger fish, birds, and marine mammals. Sardines migrate in large schools (Figure 6.41). They are fished with nets called purse seines that completely envelop the schools. Most of the sardine catch today is ground up to make fish meal, which acts as a protein supplement in fish farms and for livestock. Before World War II, the sardine fleets supplied canneries in Monterey, California. These canneries were made famous by John Steinbeck's wonderful 1945 tale *Cannery Row*. The canneries are all gone today and Cannery Row is a tourist attraction. The centerpiece is the Monterey Bay Aquarium.

Figure 6.42 shows the history of the Pacific Sardine catch for the United States and Mexico. There are different common and Latin names for this fish. The United Nations Food and Agriculture Organization (FAO) calls it the California Pilchard. US fisheries managers call it the Pacific Sardine. The FAO uses an earlier Latin designation, *Sardinops caeruleus*, while US fisheries managers follow the more recent designation as one of four subspecies of the Pacific sardine and write the Latin name as *Sardinops sagax caerulea*. In the early years, the United States

Figure 6.41 A school of Pacific Sardines (*Sardinops sagax*). Credit: Andrea Izzotti/iStockphoto by Getty Images.

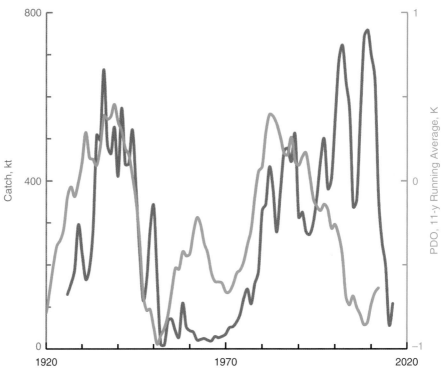

Figure 6.42 The catch history for the Pacific Sardine (*Sardinops sagax caerulea*). The catch from 1926 to 1949, for California only, is from the State of California Marine Research Committee, 1961, *CalCOFI* (California Cooperative Oceanic Fisheries Investigations), vol. 8. The data from 1950 on for US and Mexico for the California Pilchard are from the UN Food and Agriculture Organization online fisheries capture database. The Pacific Decadal Oscillation (PDO) temperature index is computed as a spatial component of sea surface temperatures and the temperature unit, K, has been retained. The PDO is plotted as an 11-year running average, so the series stops a few years earlier than the catch. The PDO data are available from NOAA at www.ncdc.noaa.gov/teleconnections/pdo/

dominated the catch. The US share in 1950 was 96%. In recent years Mexico has become the most important producer. The Mexican share in 2014 was 88%. The United States closed its sardine fishery starting in July 2015.

The catch pattern appears to be an oscillation with a period of around sixty years. The catch was high in the 1930s and 1940s, low from the 1950s through the 1970s, and then high from the early 1980s to the 2000s. The highest catch during the early cycle was 664 kt in 1936. The highest catch in the later cycle was 758 kt in 2009. However, the catch crashed again to 59 kt in 2015. Also shown in the figure is the Pacific Decadal Oscillation (PDO) temperature index.[10] The PDO is a spatial

---

[10] In spite of the word "decadal" in the name, the apparent period of the PDO is around sixty years. When we consider sea level rise in Section 10.3, we will study the Atlantic Multidecadal Oscillation, where the period is similar, but the impact is quite different.

climate pattern characterized by alternating warm and cold phases. In the warm phase, the water temperatures are warmer along the western American coasts and cooler in the North Pacific. In the cool phase, these temperatures are reversed. The index is positive in the warm phase and negative during the cool phase. It appears that the high sardine catches are associated with the warm phase and the low catches with the cold phase, with the catch lagging the PDO by a few years.

The last fishery we consider is different from the Grand Bank cod fishery or the California sardine fishery in that the catch has been stable, so far. This is the Alaska Pollock fishery. This fish is also called Walleye Pollock (Figure 6.43). The Alaska Pollock is in the same genus, *Gadus*, as the Atlantic cod. The Latin name is *Gadus chalcogrammus*. Alaska Pollock is the largest wild fishery in the world for fish that people actually eat.[11] If you have eaten a McDonald's fish-filet sandwich, you have eaten Alaska Pollock. Alaska Pollock are also used to make a fish paste called *surimi* in Japanese. One popular surimi product is imitation crab.

Alaska Pollock are fished in the North Pacific, with about half of the catch going to Russia and half to the United States. The sweet spot for catching Alaska Pollock in the US zone is the upper slope off the Bering Sea shelf, shown in red in Figure 6.44. It is about a thousand kilometers long. The fishing fleet is based at Dutch Harbor in the Aleutian Islands.

The American Alaska Pollock fishery got its start when the US adopted a 200-nautical-mile exclusive economic zone in 1979. The history of the American

Figure 6.43 Alaska Pollock (*Gadus chalcogrammus*). Adults reach a length of 1 m. Credit: Σ64/CC BY 3.0, https://creativecommons.org/licenses/by/3.0/deed.en

---

[11] Most years the Peruvian anchovy catch is larger than the Alaska Pollock catch, but the anchovy catch is mostly ground up for fish meal.

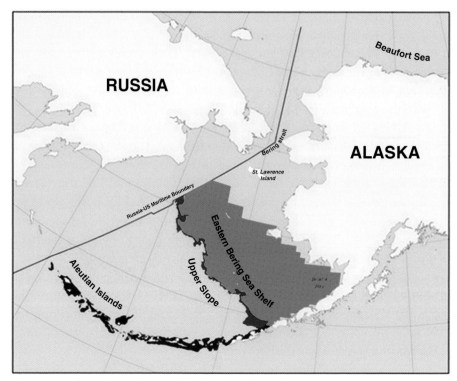

Figure 6.44 The Bering Sea, home of the Alaska Pollock. Credit: NOAA.

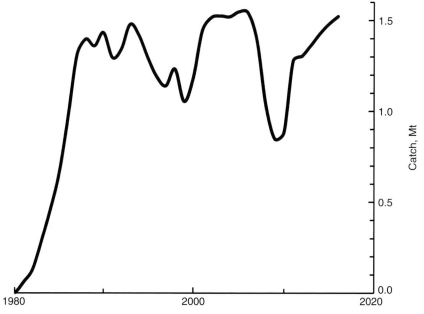

Figure 6.45 US Alaska Pollock (*Gadus chalcogrammus*) capture history. The data come from the UN Food and Agriculture Organization online fisheries capture database.

catch is shown in Figure 6.45. Starting from almost nothing in 1980, production reached 1 Mt by 1986. It has been in the range from 1.0 Mt to 1.6 Mt since. For thirty years the Alaska Pollock fishery has had relatively steady production. One factor that may help stabilize the Alaska Pollock fishery is that there are many rugged underwater canyons that act as refuges. There are also thousands of ship-wrecks that can snag a net that drifts too close to the bottom.

6.6.3 Fish Farming

For some fish, a transition is underway from wild catch to fish farming. Figure 6.46 shows the history of salmon capture compared with the production from salmon farming. The wild catch has been steady at around 1 Mt per year, while farmed salmon production has been rising quickly, with a ten-year annualized growth rate of 5%/y. The production was 2.4 Mt in 2016. Figure 6.47 shows salmon pens in the Faroe Islands. The pens are 50 m across. Each pen in the photograph holds thousands of salmon. The Faroese now make more money farming salmon than they do from fishing.

Asia completely dominates aquaculture, accounting for 92% of the production in 2016. Figure 6.48 shows the world aquaculture production history. Two impor-tant components of Asian aquaculture are carp and seaweed.

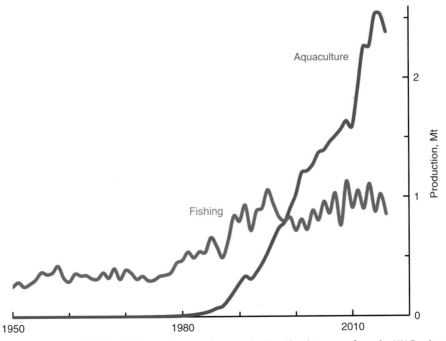

Figure 6.46 World salmon fishing and aquaculture production. The data come from the UN Food and Agriculture Organization online fisheries capture database and the aquaculture database.

Figure 6.47 Salmon pens outside the town of Vestmanna on the island of Stremoy, in the Faroe Islands. The pens are 50 m across. Photograph by the author.

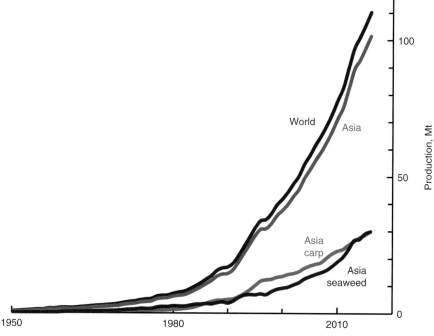

Figure 6.48 Aquaculture production. The data come from the UN Food and Agriculture Organization online aquaculture database. The carp data here are for ciprinids, and the seaweed data are for all aquatic plants.

Carp farming goes back more than two thousand years in both Asia and Europe. The Romans kept carp pens. Because of this long history, carp should be considered domestic animals, like cattle. Carp tolerate a high percentage of plants in their diet and they can live in crowded tanks. Figure 6.49 shows photographs of carp farms.

Figure 6.49 (a) A carp farm in Shanghai. Credit: Ivan Walsh CC BY 2.0, https://creativecommons.org/licenses/by/2.0/ (b) Carp feeding frenzy. Credit: st-palette/iStock by Getty Images.

Figure 6.50 Farming seaweed on ropes in Indonesia. Credit: pradeep_kmpk14/Shutterstock.

The popularity of Japanese sushi has made people aware of the seaweed *nori*, which forms the wrapper for the rice. Nori is dried and sold in sheets. Seaweed is grown on a large scale on ropes (Figure 6.50). Some of the seaweed is eaten directly like nori, but most is used in the food and drug industries.

What does this mean for the world seafood supply? History also tells us that despite best efforts by the managers, it is likely that some wild fisheries will decline. Others, like the Pacific sardine fishery, may collapse through natural changes like the Pacific Decadal Oscillation (PDO). The wild catch may never exceed 100 Mt/y. On the other hand, aquaculture production has been rising steadily and no fundamental limits have appeared.

6.7 Biofuels

For many years, agriculture has been a major consumer of energy, principally methane for fertilizer production, but also diesel fuel and gasoline in farm equipment. In recent years, agriculture has also become a source of fuels, including methane, ethanol, and diesel fuel. These are generally called *biofuels*, and more specifically *biomethane, bioethanol, and biodiesel*. Sometimes people use the word "biofuels" to mean only liquid fuels, and not *biogas*, but in this book we will not make that distinction.

Biodiesel is produced from plant and animal oils by reacting them with methanol or ethanol. The oils are made primarily of triglycerides that contain three carbon chains and the reaction products are single carbon chains. Europe is the leader in biodiesel production. The most important oil plant for biodiesel is rapeseed (Figure 6.51). Biodiesel is typically burned as a blend called B5, 5% biodiesel and 95% petroleum-based diesel. Incidentally, the variety of rapeseed oil that is used in cooking was developed at the University of Manitoba. This is recognized in North America, where it is called Canola, short for "Canadian oil."

Biomethane is produced by bacteria from a feedstock of plant and animal material. The reactions are quite complex and they involve steps by different bacteria. In the overall reaction glucose, $C_6H_{12}O_6$, is converted to methane and carbon dioxide.

$$C_6H_{12}O_6 \rightarrow 3CH_4 + 3CO_2. \qquad\qquad 6.2$$

This mixture of methane and carbon dioxide is called *biogas*. Because of the carbon dioxide, the energy density of biogas is lower than for natural gas. A representative value is 0.5 Mtoe/Gm^3 compared with 0.9 Mtoe/Gm^3 for natural gas. The chemical process is called *anaerobic digestion*. Anaerobic means that oxygen is not one of the reactants. The technique is not new. Anaerobic digestion has been used in treating sewage sludge for a hundred years. The process reduces the amount of waste material and it kills disease microorganisms through the heat that is generated in the reaction. Germany has been the leader in biogas production. Their most important feedstocks are corn and rapeseed.

Figure 6.51 A rapeseed field in the county of Cumbria, England. Photo by Robb Rutledge.

Bioethanol is produced by yeast through alcohol fermentation. In this reaction, glucose is converted to ethanol, C_2H_6O, and carbon dioxide.

$$C_6H_{12}O_6 \rightarrow 2C_2H_6O + 2CO_2. \qquad 6.3$$

The process is anaerobic, like methane production. Alcohol fermentation has been used for thousands of years to make wine and beer. Alcoholic drinks were safer than water, because the alcohol killed microorganisms that caused diseases, particularly cholera bacteria. Two countries dominate bioethanol production, the United States, using corn, and Brazil, using sugar cane. Like biodiesel, bioethanol is usually blended, typically as E10.

Collectively the processes of anaerobic digestion and fermentation are called *anaerobic respiration*. Anaerobic respiration provides microorganisms with energy for metabolism, growth, and reproduction. We distinguish anaerobic respiration from *aerobic respiration*, which does use oxygen, and produces carbon dioxide and water.

$$C_6H_{12}O_6 + 6O_2 \rightarrow 6CO_2 + 6H_2O. \qquad 6.4$$

This is the same chemical equation that would apply if we burned the glucose, but cells can do this at room temperature. Aerobic respiration provides much more energy than anaerobic respiration. Some cells use both types of respiration. Our muscle cells use aerobic respiration when oxygen is available and anaerobic lactic acid fermentation when oxygen is low.

Biofuels production can be compared with wood fuel. In energy terms, biofuels amounted to 110 Mtoe in 2014, including 30 Mtoe of biogas.[12] This can be compared to 279 Mtoe for wood fuel that year. To the extent arable land can be converted to forests and back again, biofuels and wood fuel are alternatives. There are many different factors affecting the choice. The majority of wood pellet production actually comes from the waste material associated with industrial roundwood. Much of the biogas feedstock is animal waste. Corn ethanol has a useful by-product called distillers grains that is fed to cattle as a protein supplement A forest and a farmer's field are different environments. Forests support a variety of wildlife. On the other hand, a rape field can be beautiful, as Figure 6.51 shows. The choice can come down to culture. Farms or forests? Farmers or foresters?

6.7.1   Gobar Gas

Independent journalist Michael Yon wrote a fascinating story called "Gobar Gas" in 2010 about biogas production in rural areas in Asia. A link to the story is given in the references at the end of the chapter. Gobar is the Nepali word for cow dung. In these areas biogas is produced in buried digesters. Figure 6.52 shows a cutaway digester that is used for training. Organic waste material is chopped up and put in the chimney in the background. This could be food scraps and livestock manure. The material collects under the dome. The family operating the digester

Figure 6.52 A cutaway of a biogas digester for training in Cambodia. Courtesy of Michael Yon.

---

[12] The BP *Statistical Review* tracks liquid biofuels. For biogas, we use the World Bioenergy Association publication, *Global Bioenergy Statistics*, available at www.worldbioenergy.org

must control the conditions. The material must not dry out. Antibiotics and oxygen will stop the process. The reaction works best around 37 °C. The reaction does give off heat, so the digester can be self-heating once it has started. The biogas is removed from the pipe in the center of the dome.

The biogas can be burned for heat, for light, and in a cooking stove (Figure 6.53). The advantage is that no wood fuel is burned. This greatly reduces the smoke inside the home. In addition, family members do not have to collect wood. The process of forming methane concentrates the nutrients in the residual slurry and the heat kills parasites and disease-causing microorganisms. This allows the slurry to be used as a fertilizer. In many areas toilets are connected to the digesters and this helps sanitation. For biogas of the gobar variety, there is no bad news. Indoor air pollution is reduced, pressure on forests in reduced, sanitation is improved, and fertilizer is produced as a by-product.

6.7.2 Liquid Biofuels Production History

Liquid biofuels production is dominated by the United States, Brazil, and the European Union. Together they accounted for 82% of the world production in 2017. Figure 6.54 shows the history. These are primarily blended into transportation fuel. Brazil was the early leader. The motivation was to reduce oil imports. All three curves show a sharp rise around 2005. This was the time of concerns about

Figure 6.53 A biogas stove in Nepal. Courtesy of Michael Yon.

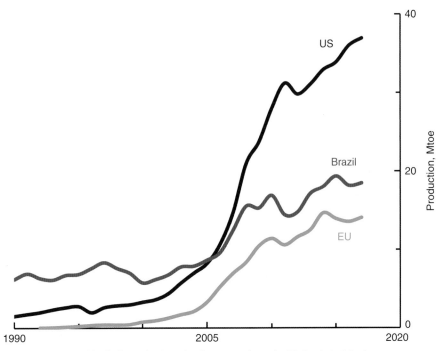

Figure 6.54 Liquid biofuels production. The data come from the BP *Statistical Review.*

Peak Oil, before the fracking revolution. The increases continue after 2010, but at a slower rate.

One concern that is often expressed with American corn ethanol is that less corn may be available for the food industry because of ethanol production. Figure 6.55 shows the history of US corn production, plotted according to whether it is used for making ethanol or not. The part that is used for ethanol shows the same pattern that we noted in the previous figure, a rapid rise from 2005 to 2010 and a slower rise afterwards. The non-ethanol corn production is flat. This makes it difficult to argue that the ethanol program has reduced American corn production for other purposes. The increase in sugar cane ethanol in Brazil has required additional land. Figure 6.56 shows the yield and area over time. From 1961 to 2016, the yield has gone up 73%. However, the planted area has increased 648%.

The production of bioethanol consumes a significant amount of energy. The reason is that yeast produces ethanol in a water solution, and the solution must be distilled to remove the water. Distillation requires heat. In Brazil, the sugar cane bagasse is burned to provide heat for the distillation. *Bagasse* is what is left after the juice has been squeezed out of the cane. However, for corn, the stover ordinarily needs to be left on the field as a soil conditioner. *Stover* is the leaves and stalks. Usually natural gas is burned to distill corn ethanol. The energy content of the natural gas used in distillation is about half the energy content of the ethanol. In 2016, the production of corn ethanol in the United States was 36 Mtoe, so the

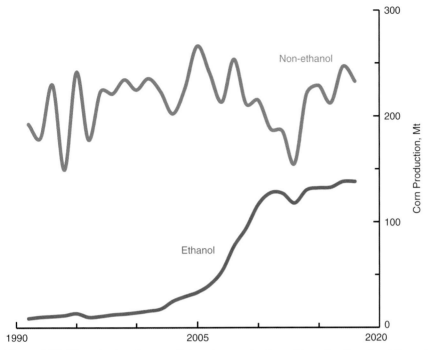

Figure 6.55 US production of corn, split into the part that is used to make ethanol and the rest. The data come from the United States Department of Agriculture, *Feed Grains Data Workbook*.

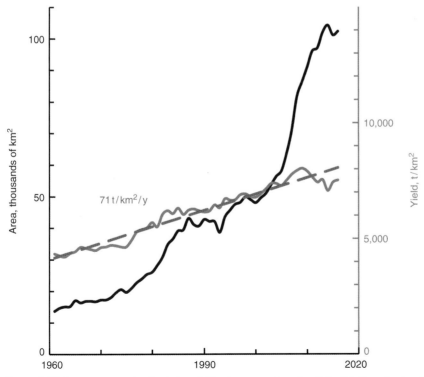

Figure 6.56 Sugar cane production in Brazil. The data come from the United Nations Food and Agriculture Organization online database FAOSTAT.

methane burned to distill American corn ethanol was comparable to the European biomethane production of 17 Mtoe that year.

Historically interest in biofuels rose when there was a concern about hydrocarbon supplies. Governments around the world began to encourage biofuels production as a supplement to oil supplies. There are many ways that governments can do this. They can subsidize the price of the biofuel. They can put tariffs on imported fuels. They may require electricity generators to have a part of their generation come from renewable energy sources. In the United States, the Environmental Protection Agency requires gasoline to be mixed with a fuel that contains oxygen to keep down the level of carbon monoxide in the exhaust. Ethanol satisfies this requirement. Figure 6.57 shows biofuels production as a share of world primary energy. There is a steep rise after 2005. Once the fear of scarcity passes, biofuels programs attract criticism and opposition. In the figure, the biofuels share starts to roll over in 2010 as the fracking revolution takes hold. Some argue that encouraging biofuels increases the price of food significantly. Returning to Figure 6.5 on the price history of food, the period from 2007 on has been a time of high food prices and it was a time when biofuels were encouraged. On the other hand, it was

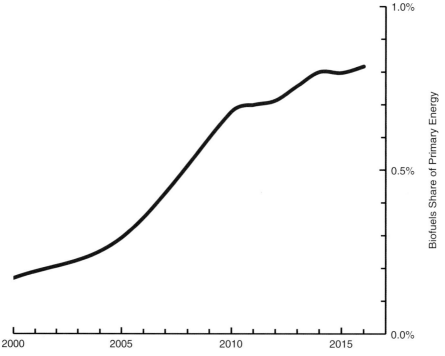

Figure 6.57 Biofuels share, including biogas, of world primary energy. The liquid biofuels production data are from the BP *Statistical Review*. Primary energy is taken from Figure 1.4. Biogas production is taken from various editions of the World Bioenergy Association, *Global Bioenergy Statistics.*

also a time of high oil prices and oil prices have been correlated with food prices, as Figure 6.6 shows. It may not be possible to separate the two effects.

The Achilles heel of biofuels is its gargantuan land requirements. The World Bioenergy Association reckons that 711,000 square kilometers were devoted to liquid biofuels in 2013. We will take the land required for biogas to be 103,000 square kilometers.[13] The total of 814,000 km^2 was 5% of the world's arable land and permanent crops area that year. The biofuels share of all primary energy was 0.8% in 2013. Absent a technical revolution, a 16% primary energy share for biofuels would require all of the world's crop land. As an example of a possible future technical revolution, the 2018 Nobel Prize in chemistry went to Professor Frances Arnold at Caltech for her work in the directed evolution of bacteria to produce liquid biofuels.

6.8 Prospects

The Green Revolution ranks with the collapse of the Soviet Union and the information technology revolution as the most significant events of the second half of the twentieth century. Even the most optimistic experts would not have predicted that world cereal yield would still be improving at the same rate that it was at the beginning in the 1960s. But it is. In the higher-income countries, the increase in yield has allowed agricultural land to be returned to forest. In the lower-income countries it has reduced hunger. It has also increased the amount of animal feed available so that more people can eat meat and farmed fish. Commercial fishing production has apparently reached its limits, but aquaculture has grown steadily so that more people can eat fish than ever before.

The outstanding aspect of the yield improvement is that it is linear in time. We do not know why. However, we can say that there are several interacting factors that contribute. One is the availability of chemical fertilizers, critically synthetic nitrogen. Another is CO_2 fertilization resulting from the increase in atmospheric carbon dioxide caused by burning fossil fuels. The Green Revolution has taken place in a warming period that has lengthened the growing season at northern latitudes. Crop varieties have been developed that take advantage of these factors. In the background is a more general increase in plant growth called *global greening*. The increase in timber density in Figure 2.22 is evidence of this. We will return to the topic in Section 10.2.

---

[13] The estimate of 103,000 km^2 for biogas is calculated on the basis of a volume yield at 5,780 m^3/ha in Peter Weiland, 2003, "Production and energetic use of biogas from energy crops and wastes in Germany," *Applied Biochemistry and Biotechnology*, vol. 109, pp. 263–274.

Farmers are enthusiastic about producing biofuels because they are an additional market for them. Aside from biogas production from organic waste in lower-income countries, it could be argued that biofuels are not appropriate at present because they require an enormous amount of land compared with alternatives like wind turbines and solar panels. However, in a post fossil-fuel world there could be a role for biomethane in combustion turbines to balance the electricity grid and for liquid biofuels as aviation fuel.

Concepts to Review

- The Green Revolution
- Nitrogen in farming
- Modernizing agriculture in Brazil
- Food prices and oil
- The use of guano as fertilizer
- Fritz Haber and Carl Bosch and ammonia synthesis
- The regional evolution of cereal yield
- The regional evolution of livestock production
- Irrigation and the Aral Sea
- The Endangered Species Act in the United States
- Agriculture and water
- The production history of fishing and aquaculture
- The role of upwellings in fisheries
- The history of the Newfoundland cod fishery
- The history of the California sardine fishery
- Salmon farming
- Biogas production in lower-income countries
- Biofuels from rapeseed, corn, and sugar cane

Problems

Problem 6.1   Desalination

A common way to desalt sea water is to pump the water through a filter with pores so fine that the salt ions are held back. The minimum work w required to overcome the osmotic pressure across the filter can be written as

$$w = kT \text{ / salt ion} = 2kT \text{ / salt molecule} \qquad 6.5$$

where k is Boltzmann's constant, 1.38×10^{-23} J/K, T is the absolute temperature at room temperature, 300 K, and there are two ions per molecule of salt.

a. Calculate the minimum energy in joules to produce 1 t of fresh water from sea water if the salt concentration of the ocean is 3.5% by mass. For this problem you may assume that the salt is entirely sodium chloride.
b. The work is done by an electrical pump. Calculate the electrical energy in kWh required to produce 1 m³ of fresh water. You may neglect losses in the pump.

Problem 6.2   Respiration

a. Aerobic respiration of glucose produces energy for cells. Enthalpy is a measure of the heat that is produced in a chemical reaction. Use Equation 6.4 to calculate the enthalpy change for the aerobic respiration of a mole of glucose by subtracting the enthalpies of the reactants from the enthalpies of the products. The enthalpies of the constituents of the reaction are given in Table 6.10.
b. Anaerobic respiration can produce biomethane. Use Equation 6.2 and the enthalpies in the table to calculate the enthalpy change for anaerobic respiration. This is the energy that the cells gain from the reaction.
c. Fermentation can produce bioethanol. Use Equation 6.3 and the enthalpies in the table to calculate the enthalpy change for fermentation. This is the energy that the yeast cells gain from the reaction.

Problem 6.3   Nitrogen fertilizer

On the download data pages of the FAOSTAT database of the United Nations Food and Agricultural Organization (FAO), find the world nitrogen fertilizer production in tN for the latest year. The item will be listed as Nitrogen Fertilizers (N total nutrients) and the Element will be Production Quantity in nutrients.

a. Taking the fundamental reaction for producing nitrogen fertilizer to be

$$3CH_4 + 6H_2O + 4N_2 \rightarrow 3CO_2 + 8NH_3 \qquad 6.6$$

find the methane consumption in tonnes through this reaction that year.
b. What fraction of the world natural gas consumption would this represent?

Table 6.10 Enthalpies of formation for the constituents of aerobic and anaerobic respiration, and fermentation.

Constituent	Enthalpy, kJ/mol
Glucose	−1,272.0
Molecular oxygen	0
Water vapor	−241.8
Carbon dioxide	−393.5
Methane	−74.9
Ethanol	−277.4

Further Reading

- *Captains Courageous*, 1937, starring Spencer Tracy, Freddie Bartholomew, Mickey Rooney, and Lionel Barrymore. This movie is based on a novel by Rudyard Kipling. It tells the story of a spoiled young man (Bartholomew) who falls overboard near the Georges Bank and is picked by a cod fisherman (Tracy).
- *The Economist*, 2010, "The Miracle of the Cerrado." The comments are also quite informative. Available online at www.economist.com/node/16886442
- Thomas Hager, 2008, *The Alchemy of Air*, Harmony Books. An outstanding dual biography of Fritz Haber and Carl Bosch. Hager also gives an excellent history of the Peruvian guano industry and Chilean nitrate mining.
- Leon Hesser, 2009, *The Man Who Fed the World*, Durban House. A biography of Norman Borlaug, the father of the Green Revolution. Hesser was a colleague of Borlaug's, and writes knowledgeably about both the scientific and political challenges.
- Philip Micklin, N. V. Aladin, and Igor Plotnikov, editors, 2014, *The Aral Sea, The Devastation and Partial Rehabilitation of a Great Lake*, Springer-Verlag. A collection of papers covering many aspects of the problem.
- Vaclav Smil, 2000, *Feeding the World*, MIT Press. This is an authoritative analysis of crop, livestock, and fish supplies, together with food needs.
- Benjamin Stout, *The Northern Spotted Owl, An Oregon View, 1975–2002*, Trafford Publishing. A collection of primary source material.
- Michael Yon, 2010, "Gobar Gas II," *Online Magazine*. *Gobar* means cow dung in Nepali. Available online at www.michaelyon-online.com/gobar-gas-ii/All-Pages.htm

7 Electricity and the Alternatives

> *I'd put my money on the sun and solar energy.*
> *What a source of power! I hope we don't have to wait*
> *until oil and coal run out before we tackle that.*
>
> *Thomas Edison*

7.1 Introduction

The electricity grid is the greatest engineering invention of all time. The grid provides light, heat, cold, mechanical power, computation, and communications to billions of people. A key aspect of the grid is that the generators are separated from the consumers by transmission lines. The electricity is generated in favorable locations – hydroelectric dams in the mountains, solar panels in the desert, wind turbines offshore, and coal plants near cooling water. We can see how electricity has grown in importance from Figure 7.1, which shows the electricity share of primary energy over time. It has increased from 9% in 1950 to 41% in 2017. The non-electrical transportation energy share has also grown, but more slowly, from 14% in 1971 to 20% in 2017. These increases came at the expense of non-electrical stationary energy whose share fell from 65% in 1971 to 39% in 2017. The electricity plot has a noticeable curvature. The rise is slowing. In the ten years from 1950 to 1960, the increase was 5.6%, but from 2007 to 2017, it was half as large, 2.8%.

The electricity share of primary energy is a critical index. A transition from fossil fuels to alternatives has two components. One is increasing the alternatives share of electricity generation and the other is increasing the electricity share of primary energy consumption. This chapter discusses the electricity grid, with an emphasis on alternatives. In the following chapters, we will consider the role of electricity in buildings and transportation.

Most of our non-electrical energy supply is based on fossil fuels, like natural gas for heating homes and gasoline for cars. There are non-electrical alternatives, most importantly wood fuel. Another non-electrical alternative would be the wind turbines that pump water for cattle on ranches. A *turbine* is a machine with vanes that converts the kinetic energy in wind or flowing water to rotational energy. One

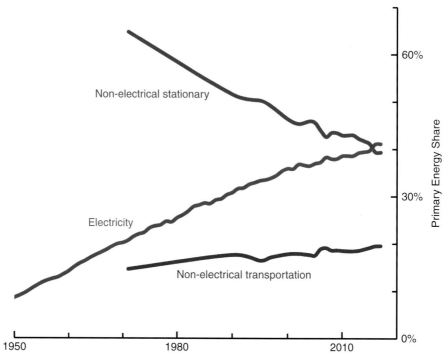

Figure 7.1 The electricity share of world primary energy consumption. Total primary energy is taken from Figure 1.4. Electricity is from Figure 1.16, re-plotted as primary energy consumption. Transportation energy comes from various editions of the IEA *World Energy Outlook*. Non-electrical stationary energy is calculated as the remainder.

can think of a turbine as a fan or propeller run backwards. Today generating electricity from wind turbines is much more important than pumping water. Similarly, solar water heaters are common in some countries, but the growth in energy production is dominated by solar panels that produce electricity. It is possible that a technical revolution could enable significant non-electrical solar power sources to be developed in the future. For example, research is underway at MIT and Caltech to develop artificial photosynthesis devices that would produce hydrocarbons from sunlight, carbon dioxide, and water.

7.1.1   Thomas Edison and the Electrical Grid

The electricity grid was the dream of one man, American inventor Thomas Edison (Figure 7.2). Edison was born in Milan, Ohio, the seventh and last child in a poor family. He attended school for only three months. Tom began work at age 12, selling food to passengers on the trains between Port Huron, Michigan and Detroit. He became almost deaf around this time, probably due to an ear infection. A turning point in his life came at age 15. A three-year-old boy was playing in the path of a boxcar that was being shunted between tracks. Tom pulled the child out of the

Figure 7.2 Thomas Edison (1847–1931), inventor of the electrical grid. This picture was taken in 1878, and it shows Edison with his phonograph. The records for this machine were cylinders, but the disk records developed later were the dominant audio recording medium during the twentieth century. Credit: Brady-Handy photograph collection, Library of Congress.

way of the car. In gratitude, the boy's father, J. U. MacKenzie, offered to teach Tom to be a telegraph operator. This was the time of the American Civil War, and telegraph operators were in great demand. When we hear Morse code today, it is usually as tones on short-wave radio. Edison likely would not have been able to hear the tones because of his deafness. However, at that time the telegraph used electromagnetic sounders that made clicks that he could perceive. Edison became an expert operator. The telegraph was his introduction to electricity. He began studying Michael Faraday's *Researches in Electricity*. Faraday's research on the relationship between electricity and magnetism laid the scientific foundation for electrical generators and motors. Like Edison, Faraday started life poor and he was self-taught. Because of this, Faraday used words rather than equations to describe his work, and Edison could understand his writing.

In the late 1800s, the best residential lighting came from town gas (Table 5.1). The advantage of a gas fuel was that it could be distributed to many lamps with pipes. Lamps could be controlled individually, but they did not have to be fueled separately. The disadvantage was that town gas was the cause of many building fires. It also contains carbon monoxide, which is poisonous. Edison hit upon the idea of wiring incandescent lamps in parallel with each lamp controlled by a separate switch. The problem with a series connection was that the lights failed as open circuits, and when one failed, they all went out. However, the resistance of existing incandescent lamps was low, only a few ohms. When connected in parallel they could not be driven efficiently. We can illustrate the problem with the circuit diagram in Figure 7.3. The generator is represented by a voltage source V with a wiring resistance R_g. A transmission line with a resistance R_t connects the generator to the load, which is a group of n lights with resistance R connected in parallel. We let the current through the transmission line be I. The power dissipated as heat in the transmission line can be written as $I^2 R_t$, while the power lost from the generator wiring is $I^2 R_g$. The current in the transmission line will divide equally between the lights, so the power in an individual light is $(I/n)^2 R$. This means we can write the fraction of the power delivered to the lights as an efficiency η given by

$$\eta = \frac{n\,(I/n)^2 R}{I^2 R_g + I^2 R_t + n\,(I/n)^2 R} = \frac{R_l}{R_g + R_t + R_l} \qquad 7.1$$

where $R_l = R/n$ is the resistance of the lamps in parallel. To keep the efficiency high, the generator and transmission line resistances should be low and the lamp resistance should be high. Through an exhaustive search, Edison and his team developed a lamp with a carbonized bamboo filament that had a resistance of $100\,\Omega$. The lamp was the critical component in his system, but his group also invented a low-resistance generator, together with regulators that allowed generators to be connected in parallel. In addition, they developed insulated cables that that could be buried underground and switches that allowed each lamp to be controlled individually. One problem that Edison did not solve was that of a low-loss, long-distance transmission line. Edison's grids extended only over a few

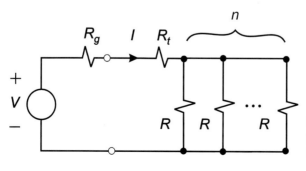

Figure 7.3 Circuit diagram for a generator, transmission line, and n lights in parallel.

kilometers, whereas modern grids span thousands of kilometers. We discuss how to solve this problem in Section 7.2.

Edison's first residential customer was the prominent banker J. P. Morgan in 1881. The generator was put in the basement of Morgan's home. In the following year, 1882, Edison set up generators in a building on Pearl Street in Manhattan. This was the first true electricity grid, with 500 customers and 10,000 electric lights wired in parallel. The system used direct current (DC). The exhaust heat from the generators was supplied to nearby workshops. Today we call this *combined heat and power* (CHP). Because the heat is not wasted, combined heat and power allows extremely efficient operation. On the other hand, it gives away one of the advantages of the grid, which is to separate the air pollution from the customers. In addition, CHP is less flexible than generators that only produce electricity and furnaces that only produce heat, because the generators need to run when either heat or electricity is required.

7.1.2 Electricity Generation History

Figure 7.4 shows the history of world electricity generation. Because of Edison, the US got off to a fast start, with an annualized growth rate of 15%/y from 1902 to 1912. In 1912, on the eve of World War I, 60% of the world's electricity

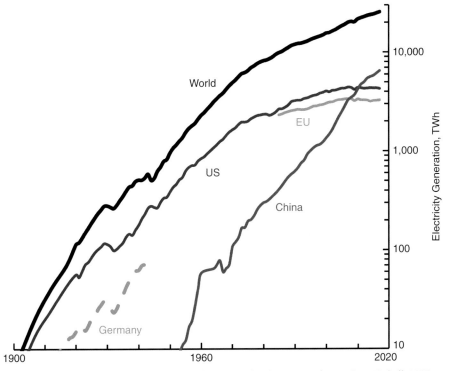

Figure 7.4 Annual electricity generation. Before 1985, the data comes from Brian Mitchell, 2007, *International Historical Statistics*, Palgrave Macmillan. The data from 1985 through 2017 are from the BP *Statistical Review.*

generation was American. Germany was also an early leader, particularly in the AC technology that eventually replaced Edison's DC system. DC distribution survives today in cars, boats, recreational vehicles (RVs), and perhaps surprisingly, in some long-distance high-voltage transmission lines. In recent years, Chinese generation, based primarily on coal, has passed the US and the EU. China's annualized growth from 2007 to 2017 was 7.1%/y, while the US and the EU have declined slightly, at 0.3%/y. World growth was 2.5%/y.

7.1.3 Daily Demand Curves

Modern grids use alternating current (AC) and they cover millions of square kilometers. The electricity consumed is called the *demand* or *load*. The demand varies by the hour during the day and with the season. Some hydroelectric plants can act either as load or supply. These plants store energy by pumping water uphill and then supply electricity when they let water fall back through turbines. Similarly, rechargeable batteries can store energy and then supply electricity later. Figure 7.5 shows the hourly load for the California Independent Systems Operator (CAISO) for a winter day, Tuesday, January 3, 2017. CAISO is one of the largest electricity markets in the US, covering 80% of the electricity consumed in the state. In the

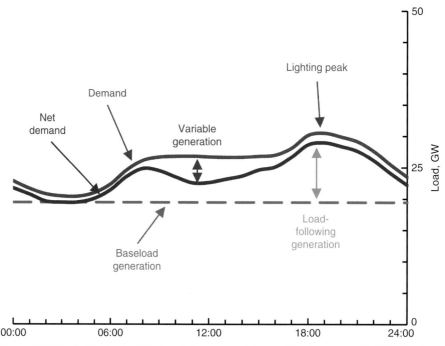

Figure 7.5 The California electricity hourly load on a winter day, Tuesday, January 3, 2017. The data come from the California Independent System Operator (CAISO). The points plotted are the average powers consumed during the previous hour. Data are available at www.caiso.com/green/renewableswatch.html

figure, the demand is lowest in the morning before dawn, 20 GW at 04:00. From 05:00 to 09:00, the demand rises as people wake up and turn on lights and heaters. By 09:00 the load has risen to 27 GW, where it stays until 16:00. At 17:00 it starts to rise again as people go home and turn on their lights and heaters. The peak demand for the day is 31 GW at 19:00. After 21:00, the load falls as people turn off their lights and heaters and go to bed.

California has commercial solar farms and wind turbines that sell their output to utilities. These generators accounted for 27% of the electricity produced in California for the CAISO market in 2018.[1] The commercial wind and solar generation are classified as *variable generation*. Variable generation is not under the control of the utility, except to the extent that it can be curtailed if the grid is not able to accept the power. Utilities are required to take the output of wind and solar generators preferentially, and they do so under long-term purchase agreements at pre-negotiated prices. From the utility perspective, the wind and solar farm generation functions like negative demand. We define the *net demand N* to be the difference between the demand D and the variable generation V. We write

$$N = D - V. \qquad 7.2$$

The utilities must secure the net demand by buying electricity in the market from generators that can control their output. These are called *dispatchable* generators. The minimum net demand is called the *baseload*. Depending on the amount of solar generation, this minimum can occur in the early morning before sunrise or during the middle of the day. We write the baseload B as

$$B = \text{minimum}(N). \qquad 7.3$$

In Figure 7.5, the minimum of 19 GW occurs at 03:00. The difference between the net demand and the baseload is the *load-following generation*. We write

$$L = N - B \qquad 7.4$$

where L is the load-following generation. The process of adjusting the supply to follow the load is called *balancing*.

---

[1] For a country or an individual state, the solar and wind shares of production and consumption will be somewhat different because of imports and exports. For the CAISO market, imports met 27% of consumption in 2018. Thus the in-state contribution of solar and wind to demand is 20%, less than the 27% production share. In interpreting statistics in publications, one must be aware of whether the writer is calculating production shares or consumption shares. Production shares are more common, but the State of California uses consumption shares in its legislated goals. In addition, the shares will be somewhat different depending on whether one is talking about the CAISO market or the entire state of California. A further complication is that some solar generating plants also use natural gas as a secondary fuel source, and some of the electricity that is classified as solar in the CAISO market is actually produced by natural gas.

Figure 7.6 shows the pattern for a summer day, Friday, September 1, 2017. The demand rises gradually as customers turn on air conditioners. The peak is at 17:00 at 50 GW. This is close to sunset, which was 17:22 that day. This is considerably higher than the winter lighting peak in Figure 7.5 of 31 GW. In fact, this was a record demand for CAISO. In California, the highest loads during the year are on hot summer days, particularly if the previous day was also hot. Houses can take more than a day to heat up fully. At northern latitudes the winter lighting peak is the largest load of the year. The baseload of 27 GW occurs at 04:00. The variable generation is largest at 13:00, reaching 10 GW, including 9 GW of solar generation. This is much larger than the winter variable peak in Figure 7.5 of 4 GW. The net demand peak was 47 GW at 20:00. This was also a net demand record for CAISO. At this time, the majority of the net demand was met by in-state natural gas, at 26 GW.

Figure 7.7 shows the generation on a spring day, Monday, March 6, 2017. The variable generation reached 10 GW from 12:00 to 14:00. The solar production share at 12:00 reached 43%. This pushed the net demand down to 13 GW from 12:00 to 13:00. The lowest net demand is now in the middle of the day rather than before dawn, and the 12-GW baseload is much lower than baseloads we saw for the winter and summer generation, which were around 20 GW. One problem for utilities is that the net demand grows very rapidly late in the day as solar

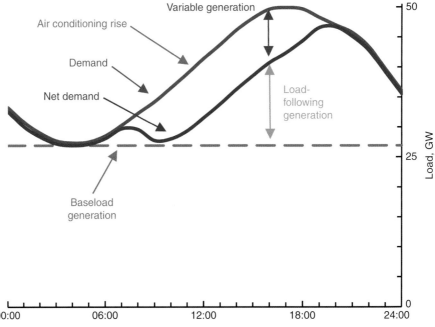

Figure 7.6 California electricity hourly load on a summer day, Friday, September 1, 2017. The data come from the California Independent System Operator (CAISO).

generation falls. People call this the *duck curve*, visualizing a duck between the demand and the net demand curves with the head later in the day.

The large solar production in the middle of the day can collapse the wholesale prices. These are shown in Figure 7.8. The wholesale prices were negative from 07:00 to 15:00. When prices are negative, the generators without a pre-arranged price agreement have to pay a customer to take their generation if they cannot shut generation down quickly enough. However, even though the wholesale prices are often negative during the day, California residential electricity prices have been increasing. The residential prices are set to recover the average costs to the utilities.

The baseload plants and the load-following generators are different. The generators with large steam boilers like nuclear and coal plants run best at a level power output. It is difficult to ramp them up and down quickly, and trying to do this increases air pollution and maintenance costs. This makes them appropriate for baseload. They can be shut them down seasonally when the baseload is reduced, like spring in California. On the other hand, natural gas plants are ideally suited for dispatchable load-following generation. Modern natural gas plants are combined-cycle designs, where the major component is a gas turbine. In a *gas* turbine it is the combustion gas that turns the blades. It is also called a *combustion* turbine. The power can be adjusted relatively quickly by throttling the fuel. Jet airplane engines are also gas turbines. In a combined-cycle generator, a steam turbine is

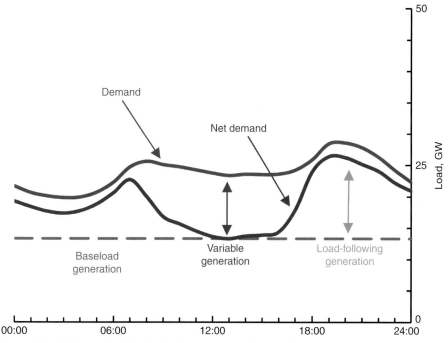

Figure 7.7 California electricity hourly load on a spring day, Monday, March 6, 2017. The data come from the California Independent System Operator (CAISO).

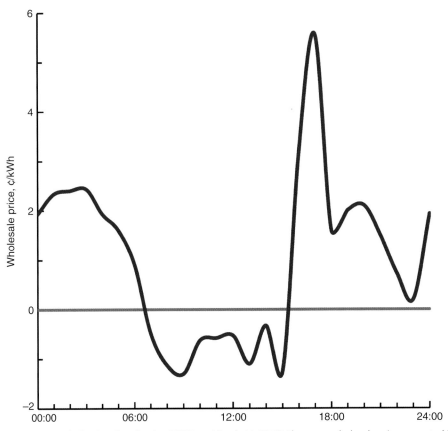

Figure 7.8 Wholesale prices for the CAISO on March 11, 2017. These are wholesale prices reported by the EIA, April 7, 2017, "Today in Energy."

driven by the hot exhaust from the gas turbine. This can improve the efficiency of the electricity generation to 60%. We have only shown hourly data here, but there are also stand-alone natural-gas combustion turbine generators called *peaking plants* that respond on time frames of a few minutes.

7.1.4   Capital Investments in Electricity Grids

Capital investments in the world electrical grids in recent years have been dominated by renewables and by network integration. Figure 7.9 shows the history, distinguishing between renewables, network, and fossil-fuel generation. The renewables and network investments have both increased substantially during this time, and were both close to 300 G\$ in 2017. These investments have been associated with larger and larger annual increments of renewable power. These are shown in Figure 7.10. Note that in 2001 the power increment was negative. Renewables include hydroelectric power, which can have wet and dry years.

Leaving aside a relatively small amount of generation from wood pellets and biogas, there is no fuel to buy for most renewables generation. This suggests an

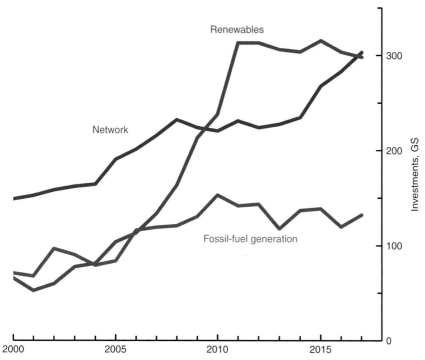

Figure 7.9 World capital investment history for electricity from 2000 to 2017. The investments are in 2017 dollars calculated at market exchange rates (MER). Nuclear power investments are relatively small, 17 G\$ in 2017, and are not included. The data come from various editions of the IEA *World Energy Investment*.

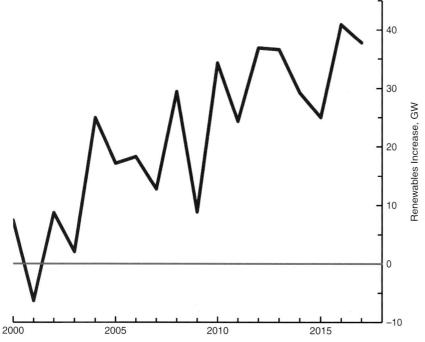

Figure 7.10 Annual increase in world renewables electricity generation. Average power is plotted rather than capacity. Nuclear power is not included. The points are calculated from data in the BP *Statistical Review*.

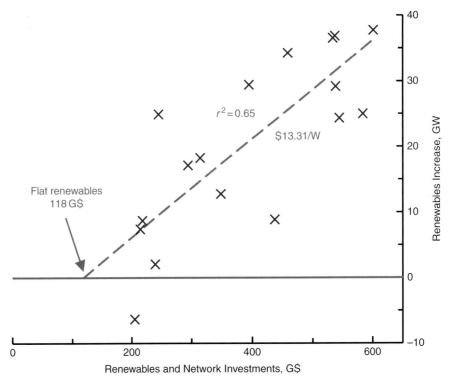

Figure 7.11 Renewables power increase from Figure 7.10 plotted against the renewables and network investments from Figure 7.9.

interesting calculation. We plot the renewable power increase against the renewables and network investments in Figure 7.11. The regression line is shown on the figure. The r^2 value is 0.65. We interpret the slope of the regression line as the increase in renewables power, 75 mW, that is associated with one dollar of capital investment. Usually people think in terms of the inverse, $13.31/W. Then we can calculate the component of the electricity sales price that would be needed to repay the capital investment with a 10-year payback time. A one-watt source supplies 88 kWh in ten years, so this would be $13.31/88 kWh, or 15¢/kWh.

7.1.5 European Residential Electricity

Capital costs are only one component of retail electricity prices. Figure 7.12 compares national residential electricity prices in Europe with per-person wind and solar capacity. The countries with higher wind and solar capacity tend to have higher residential electricity prices. The highest price, 33.7¢/kWh, is in Denmark, and the lowest is in Norway at 9.5¢/kWh. The demand-weighted average is 21.8 ¢kWh. A regression line is shown on the figure. The r^2 value is 0.71. The y-intercept of the regression line, 13.5¢/kWh, can be interpreted as the price in a mythical European

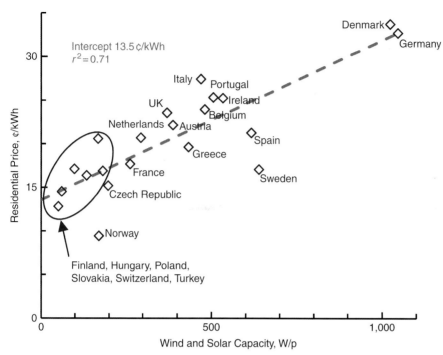

Figure 7.12 Residential electricity prices versus per-person wind and solar capacity in 2015 for the European OECD countries that are tracked by the BP *Statistical Review*. Residential prices (MER) for 2015 are taken from the IEA, *Electricity Information 2017*. The population data come from the UN Population Division.

country that has no wind or solar capacity. This is 8.2¢/kWh lower than the average. We associate this difference with the wind and solar capacity and calculate an effective wind and solar surcharge *s* by dividing by the 2015 wind and solar generation share in these countries, 11.6%, and write

$$s = \frac{8.2\,¢\,/\,\text{kWh}}{11.6\%} = 71\,¢\,/\,\text{kWh}. \qquad 7.5$$

This effective surcharge is five times as large as the no-sun, no-wind price.

No government publicizes the correlation between wind and solar capacity and residential electricity prices. On the contrary, governments typically calculate the cost of generating electricity from a solar panel or a wind turbine that is isolated from the grid. However, home owners are not buying electricity from an isolated wind turbine or solar panel. They are buying electricity at the flip of a switch, any time of the day, any day of the week, any week of the year. Backup must be supplied for the times that the wind does not blow and the sun does not shine, and the backup generators must shut down when the wind and solar generation pick up.

7.1.6   The Electricity Alternatives

Figure 7.13 shows the history of the alternative electricity generation, expressed as shares of world primary energy. Hydroelectricity has the largest share and it has been increasing, from 5.9% in 2007 to 6.5% in 2017. Hydroelectric generation goes back to the very earliest days of the electrical grid. when power plants were built at Niagara Falls starting in 1882. A major advantage of hydro is that it can be load following. A major disadvantage is that it can vary greatly from year to year. Moreover, water supply and flood control are often higher priorities for a dam than generating electricity. The electricity may simply be a by-product. Nuclear power rose rapidly in the 1970s and the early 1980s. However, after the Chernobyl explosion in 1986, the expansion slowed and then reversed. The prospects for nuclear power were further damaged by the Fukushima disaster. The advantage of nuclear power has been that the plants have provided baseload power reliably and in the past, inexpensively. However, somewhere along the line, the recipe was lost in the OECD countries – their newer plants are extremely expensive. The share of the category of geothermal, biomass, and biogas electricity has risen from 0.6% in 2007 to 0.9% in 2017. Geothermal can make a significant contribution in seismically active areas like Iceland and California. The geothermal plants use steam turbines and are typically run as baseload generators. Geothermal energy is also used for residential heating, making hot water, and heating sidewalks in

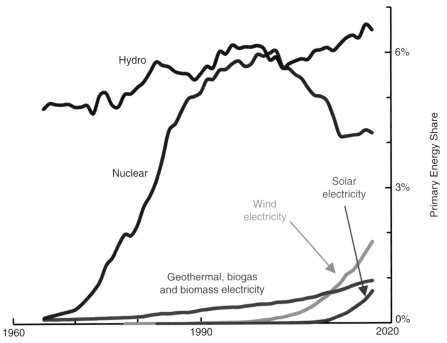

Figure 7.13 Shares of world primary energy for the alternative electricity sources. The data come from the BP *Statistical Review*.

the winter. Biomass in the form of wood chips can be stored at the plant like coal, and it can even be burned in coal plants. The principal component of biogas is methane, like natural gas. Biomethane has the potential to substitute for natural gas in electricity generation.

The wind and solar shares are modest, with wind at 1.8% and solar at 0.7% in 2017. However, their growth rates are spectacular. The 10-year annualized growth rate for wind generation is 21%/y. For solar, the growth is an incredible 50%/y. This reflects the fact that wind turbines and solar panels are standardized and made in factories. There are many suitable locations for them and they can be set up quickly. The major disadvantage is variability in output power. Wind generation can be extremely low over areas of hundreds of thousands of square kilometers for days on end. For solar, there is no generation at all at night, and the generation in the high latitudes is lowest in winter when the demand may be largest. Natural gas generators are often used to balance the grid against these fluctuations today. An outstanding question is the extent to which this role could be filled by batteries in the future.

In this chapter, we will introduce the electrical components of the grid from the supply side: the transmission lines, transformers and generators. We follow with a survey of the different alternative sources. It is interesting to consider what an electricity grid that does not use fossil fuels might look like, and we conclude the chapter by evaluating a model grid for California that is based on solar panels, batteries, and biomass and biogas.

7.2   Transmission Lines and Transformers

A fundamental problem that Edison faced with his grid was that the transmission-line resistance R_t is proportional to the length of the line, and it is in the denominator of the formula for efficiency, Equation 7.1. The longer the line is, the larger R_t is, and the lower the efficiency. This limited his grids to a few kilometers. However, a technical solution was developed that solved the problem. This was the transformer. Transformers allow the transmission-line current to be reduced and this reduces the line losses. The price for this improvement was that the grids had to be AC. Transformers do not work in DC circuits. In addition, while transformers can reduce the current, the voltage goes up by the same factor that the current goes down. The high voltages were a concern for Edison, and he did not accept an AC grid.

A transformer is based on magnetic coupling between two circuits. Michael Faraday demonstrated the first transformer in 1831, fifty years before Edison lit J. P. Morgan's home. Faraday was not attempting practical electrical engineering. Rather he was trying to understand the relationship between electricity and

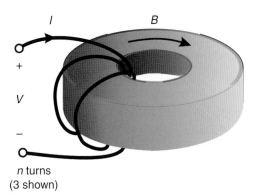

Figure 7.14 An inductor constructed as a coil on a torus of magnetic material.

I   *B*

+

V

−

n turns
(3 shown)

magnetism. A transformer is a complicated device, and we will start our discussion with a simpler component called an *inductor*. Figure 7.14 shows a coil of wire wrapped on a donut shaped core. Mathematicians call this shape a *torus*. The core material is magnetic, usually a steel alloy. The current in the wire produces a magnetic field that circulates in the torus. Each loop contributes to the field. We describe this by Ampere's law, writing

$$B \propto nI \qquad 7.6$$

where $\propto$ indicates a proportional relation. In the formula B is the magnetic field[2] in the torus, I is the current in the wire, and n is the number of turns in the coil. If the current is DC, the B-field is DC. If the current is AC, then the B-field will be AC. A B-field that is time varying will produce a voltage across the wire terminals. This is described by Faraday's law as

$$V \propto n\frac{dB}{dt}. \qquad 7.7$$

The way we visualize this is that each loop contributes to the voltage that we see across the terminals. We can combine these two formulas and write

$$V = L\frac{dI}{dt} \qquad 7.8$$

where L is a proportionality constant called the *inductance*. In words, we say that the voltage across the terminals is proportional to the time derivative of the current. This effect is called self-inductance, and it was discovered by Princeton professor Joseph Henry. The unit of inductance is called *henry* in his honor. The effect

---

[2] This is physicist's language. In an engineering class, the B-field is called the *magnetic flux density*, and a different field called H is the magnetic field. The B- and H-fields are related by the formula $B = \mu H$, where μ is called the *permeability*. Underlying the difference in terminology in physics and engineering are different units and a different emphasis that are not easy to sort out. It is safe and quite proper to call them the B-field and the H-field. Engineers and physicists will both understand these terms.

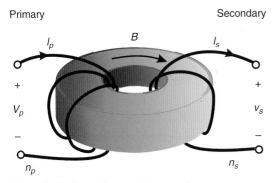

Primary

Secondary

Figure 7.15 A transformer with two coils.

is quite surprising. The wire is continuous and ordinarily we would expect no voltage across the two ends of a continuous wire because it looks like a short circuit.

In a transformer, more than one coil is wrapped on the core. Figure 7.15 shows a transformer with two coils. We call one the *primary* and the other the *secondary*. Usually we think of the primary coil as the input and the secondary coil as the output. In the figure the current directions on the coils are reversed to reflect this. We can repeat Equation 7.7 for each coil. This gives us

$$V_p \propto n_p \frac{dB}{dt} \tag{7.9}$$

for the primary coil and

$$V_s \propto n_s \frac{dB}{dt} \tag{7.10}$$

for the secondary coil. From the symmetry of the torus, the proportionality constant for each coil is the same. This means that when we divide the equations, we get

$$\frac{V_s}{V_p} = \frac{n_s}{n_p}. \tag{7.11}$$

In words, the voltage ratio is the same as the turns ratio. If we want to step the voltage up, the secondary should have a larger number of turns than the primary. If we want to step the voltage down, the secondary should have a smaller number of turns. Note that this is an AC expression. In a DC circuit, the time derivatives are zero, so the induced voltages are zero.

For the current relation we return to Equation 7.6. There are two currents to include now. We write

$$B \propto n_p I_p - n_s I_s. \tag{7.12}$$

The term for the secondary current has a minus sign because the current directions of the coils are reversed in Figure 7.15. We divide by n_s and rearrange to get

$$I_s + \frac{\alpha B}{n_s} = \frac{n_p}{n_s} I_p \qquad\qquad 7.13$$

where α is a proportionality constant. The term $\alpha B / n_s$ is called the *magnetizing current*. It can be made small by using a large number of turns. In an initial analysis, the magnetizing current is often taken to be smaller than the other current terms and neglected. This is called the *ideal transformer* approximation. The term cannot be strictly zero, because the B-field is needed to produce the voltages. With the ideal transformer approximation, we get

$$\frac{I_s}{I_p} = \frac{n_p}{n_s} \qquad\qquad 7.14$$

Note that this is the reverse of the voltage relation, Equation 7.11. When the voltage steps up, the current steps down by the same factor. This means that

$$V_s I_s = V_p I_p. \qquad\qquad 7.15$$

In words, the output power is equal to the input power.[3] In a practical transformer, there will be losses of a few percent because the coils have resistance that absorbs power, and the magnetic core material has some loss.

Figure 7.16 shows how transformers are used in a grid. In the figure, a voltage step-up transformer connects a generator to a transmission line. We return to our efficiency expression for the grid, Equation 7.1 and write

$$\eta = \frac{I_l^2 R_l}{I_g^2 R_g + I_t^2 R_t + I_l^2 R_l} \qquad\qquad 7.16$$

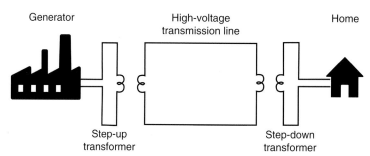

Figure 7.16 Schematic of a generator, step-up transformer, transmission line, step-down transformer, and a home.

---

[3] The power expression $P = VI$ is a DC expression. In an AC circuit there are different conventions for writing the sinusoidal voltages and currents. The most common one in AC power circuits is the rms (root-mean-square) value. For rms values the expression $P = VI$ still holds, but the peak voltages and currents are $\sqrt{2}$ times the rms values. For example, in Germany, the AC rms voltage is 230 V and the peak voltage is 325 V.

where the generator, transmission line, and the home have different currents. The step-up transformer increases the transmission-line voltage, but reduces the current I_t. This increases the efficiency by lowering the $I_t^2 R_t$ loss term. The total distribution loss in practice might be 5%. The voltages on the lines can be hundreds of kilovolts, and the lines can be a thousand kilometers long. The step-down transformer reduces the voltage to a level that is safe to use in a home. In the United States the residential voltage is 120 V. In Europe it is 230 V. The European voltage is high enough that a shock is quite painful, and for this reason, receptacles in many European countries are recessed for safety.

There are also long-distance, high-voltage DC lines. These lines use transistor switching circuits to step voltages up and down. The DC lines can bridge between two unsynchronized AC grids. In addition, DC lines have lower loss than AC lines and this can outweigh the expense of the switching circuits. This is the result of a magnetic phenomenon called the skin effect. In an AC line, the skin effect reduces the current in the center of the conductor. DC lines use the entire conductor and this gives them lower resistance than comparable AC lines. DC lines also have an advantage under water. Underwater cables have a large capacitance that shunts current away from the load in AC grids. DC lines do not have this problem. The switching circuits are more expensive and they require much more space than transformers.

7.3 Electricity Generators

Most of the electricity in the world is generated by rotating machinery. The great exception is solar panels, where light produces electrical currents in semiconductor material (Section 7.8.3). The development of electrical machines was the great triumph of nineteenth-century electrical engineering. When we write *machines*, we mean both generators and motors, which we will discuss in Section 8.3. Figure 7.17 shows an AC generator. In the figure, a C-shaped electromagnet produces a

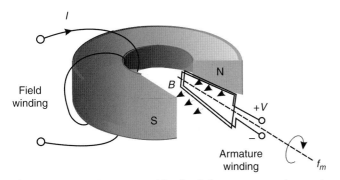

Figure 7.17 An AC generator with a fixed electromagnet and a rotating coil.

B-field in the gap between its faces. The direction of the *B*-field is determined by the direction of the current in the magnet coil. By convention, for consistency with a magnetic compass, the face with the *B*-field leaving is the *north* pole, and the end with the *B*-field entering is the *south* pole. The coil that establishes the *B*-field is called the *field* winding. Permanent magnets can be also used in place of electromagnets. Between the faces of the magnet there is another coil that rotates around the axis indicated by the dotted line through an applied torque. The torque could come from an engine or even a hand crank. We will let the frequency of the mechanical rotation be f_m. In machines, it is common to use frequency units of revolutions per minute, abbreviated rpm. The rotation causes the magnetic field through the loop to change with time and by Faraday's law, a voltage *V* is generated across the terminals of the loop. If we attached an appropriate load, like an incandescent lamp, the lamp would light up. The coil that is associated with power in an electrical machine is called the *armature* winding. There are also terms for the mechanical parts of a machine. The rotating piece is the *rotor* and the fixed piece is the *stator*. In this generator, the field coil is on the stator and the armature coil is on the rotor, but it can also be the other way around.

A more complex generator is shown in schematic form in Figure 7.18. This time the magnets are on the rotor and the armature coils are on the stator. There are two magnetic pole pairs on the rotor. On the stator there are six armature windings denoted A_1, B_1, C_1, and A_2, B_2, C_2. Each winding is indicated in cross section by a circled dot and a circled cross. The dot represents currents in a wire coming out of the page and the cross represents the return current on a wire back going

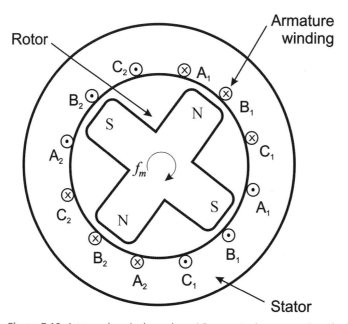

Figure 7.18 A two-pole-pair, three-phase AC generator in cross section. The field windings are omitted.

back into the page. As the rotor turns, each pair of poles produces an AC voltage in the armature windings. In this generator, there are two pole pairs, so the rotor generates two electrical cycles for each mechanical cycle. In general, if there are p pole pairs the electrical frequency f_e is given by

$$f_e = pf_m. \qquad\qquad 7.17$$

We will see the same relationship in Section 8.3 for motors. To produce the electrical power frequency f_e of 60 Hz in the United States with our generator, we would need a mechanical rotation frequency f_m given by

$$f_m = f_e \,/\, p = 60 \text{ Hz} \,/\, 2 \text{ pole pairs} \times 60 \text{ s/minute} = 1800 \text{ rpm}. \qquad 7.18$$

For European generators producing 50 Hz, the rotation frequency would be 1,500 rpm. Adjusting the number of pole pairs allows the mechanical rotation frequency to be set at a convenient value. For example, the Hoover Dam has 130-MW hydroelectric generators with 20 pole pairs. The mechanical frequency f_m is 180 rpm, or 3 revolutions per second.

7.3.1 Three-Phase Systems

In Figure 7.18 the A_1 and A_2 coils have the same relation to the corresponding magnet poles, so they will have the same voltage outputs and they can be connected in series. Similarly, B_1 and B_2 can be connected in series, and C_1 and C_2. This gives us three combined outputs, A, B, and C. Because of their layout in the stator, the B voltage lags 120° from A and C lags 120° from B. Figure 7.19 shows the voltages. This is called a

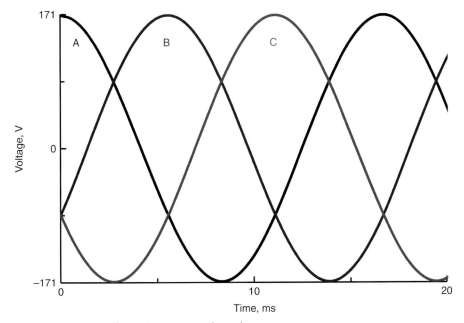

Figure 7.19 Voltages from a 120-V, 60-Hz, three-phase generator.

three-phase system. Three-phase systems have important advantages. One is that the sum of the voltages is zero at all times. This means in a three-phase system with similar loads on each line, the return wire can be smaller with a reduced current capacity than the other lines. It also allows a three-phase transformer to be relatively compact. Another advantage is that a stator wound like Figure 7.18 can produce a rotating magnetic field inside a motor to apply torque to the rotor. We will return to this in Section 8.3. Three-phase systems dominate commercial and industrial electricity.

7.3.2 Fuel Costs in Fossil-Fuel Generation

An electrical generator needs a source of mechanical power. This could be a diesel engine or a turbine. Figure 7.20 shows a diesel generator rated at 200 kW, or 270 hp. This generator could be used to supply off-grid power to a fishing lodge or backup power to a building. It produces three-phase output. The mechanical speed is 1800 rpm, the same as we calculated for the 2-pole-pair generator in Figure 7.18. For comparison, the power is comparable to the engine power rating of a pickup truck for towing a trailer or a powerboat. The diesel engine is in the center of the figure, the cooling radiator on the right, and the generator is tucked behind the control panel on the left. The assembly is mounted on a sled. Its weight is 2 tonnes, light enough to be delivered by a pickup truck. The cost was $40,000, or 20¢/W. This is dramatically lower than other generators. However, the fuel costs for a diesel generator are high. Diesel fuel costs vary widely because of different tax rates. We will use $1/l for comparisons. Taking the energy density to be 35 MJ/l and the efficiency to be 35%, we find the fuel cost c as

$$c = \$1/l \, / \, 35 \text{ MJ/l} \times 3.6 \text{ MJ/kWh} \, / \, 35\% = 29¢/\text{kWh}. \qquad 7.19$$

Diesel engine

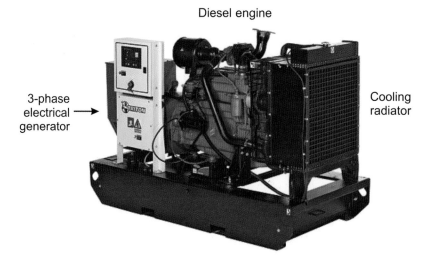

3-phase electrical generator →

Cooling radiator

Figure 7.20 A 200-kW three-phase diesel electric generator. This is a figure with labels added from a product that was at one time advertised on Amazon.com.

Figure 7.21 shows a natural gas plant that operates on an altogether different scale. This is the 600-MW Siemens natural-gas generator in the city of Dusseldorf, Germany, commissioned in 2016. It replaced an earlier coal plant at the same site. The cost was 600 M$, or $1/W. Its best measured efficiency is 61.5%. The design is a combined-cycle gas turbine (CCGT). The primary turbine is a 400-MW gas turbine. The exhaust from the gas turbine heats steam for a second 200-MW turbine. The plant can ramp up and down at 55 MW/m for load-following operation. The system is also a combined heat and power (CHP) system. The exhaust from the steam turbine provides 300 MW of heat to produce hot water for customers in the city. In CHP operation, the electrical output is reduced to 500 MW.

We can calculate a fuel cost for comparison with the diesel generator. The BP *Statistical Review* gives the price of imported natural gas for Germany in 2016 as $5.62 per million Btu, or 0.53¢/MJ. We will take the average efficiency of the electricity generation to be 60%, and write the fuel cost c as

$$c = 0.53¢ \, / \, MJ \times 3.6 \, MJ/kWh \, / \, 60\% = 3.2¢/kWh. \qquad 7.20$$

The fuel cost for gas is one-ninth of the diesel generator's fuel cost. This is because natural gas is cheaper on an energy basis than diesel fuel and because a CCGT power plant is more efficient than a diesel engine.

Figure 7.21 The Lausward power plant in Dusseldorf, Germany. Credit: Jochen Tack/Alamy Stock Photo.

In Figure 7.12, we saw that German residential electricity was one of the most expensive in Europe. If Siemens has this outstanding technology for natural gas plants, why isn't German electricity cheaper? Part of the answer lies in the access priority of the different sources that is set by national policy. Wind and solar generation are given the highest priority, then nuclear, then coal, and finally natural gas. Natural gas plants in Germany have closed because their priority is so low that they have few customers.

7.4   Hydroelectric

Hydroelectric power is the most important alternative energy source, accounting for 6.5% of the world's primary energy in 2017. Hydroelectric power is often the cheapest source of electricity. In addition, the plants can act as load-following generators, coming on during the time of the day that the price is the highest. Hydroelectric power does have disadvantages. Rain and snow can be extremely variable on all time scales: seasonal, annual, decadal, and centennial. Enormous areas are required for reservoirs. The reservoirs may also be useful for flood control, irrigation, and for recreation, but they do permanently alter the landscape by submerging the land and changing the river flow.

7.4.1   Hydroelectric Generation History

Figure 7.22 shows the generation history. The annualized growth rate from 2007 to 2017 for the world was 2.8%/y, comparable to world electricity demand as a whole, 2.5%/y. China is the largest producer by far, with 28% of world production and a high 10-y annualized growth, 9%/y. China's hydroelectric production is 18% of its overall electricity production. The general pattern for hydroelectric development has been high growth in the early years and then a gradual saturation as the good sites are filled. The continent of South America shows this pattern, with an annualized growth rate of only 0.5%/y. Hydroelectric power dominates electricity production for the continent, with a 58% share in 2017. However, the share is dropping because the growth in hydroelectric power has not kept up with overall electricity demand. Ten years earlier, the hydroelectric share for South America was 72%. The profiles for Norway and Canada are similar. Hydroelectric power accounts for over half of their electricity production. Norway has by far the lowest prices among the OECD EU countries in Figure 7.12 in 2015 at 9¢/kWh. It exported 16% of the electricity it produced that year. Likewise, Canada's residential electricity price in 2015 was 9¢/kWh, and it exported 18% of its production. Selling hydroelectric power to other countries when the market price is high helps Canada and Norway keep their domestic prices down.

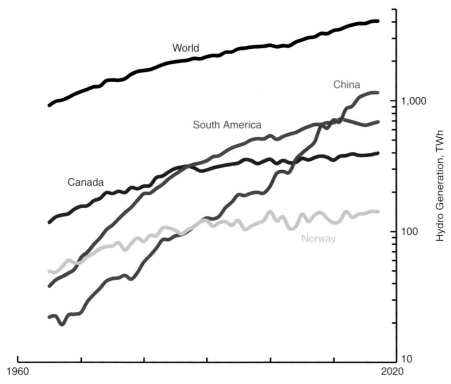

Figure 7.22 Hydroelectric power generation history. The data come from the BP *Statistical Review*.

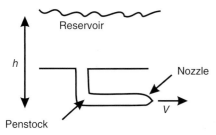

Figure 7.23 Converting the gravitational energy of water to kinetic energy.

7.4.2 The Pelton Wheel and Hydroelectric Power in California

Next we introduce the equations we need for analyzing hydroelectric power. Figure 7.23 shows a pipe draining a reservoir. The pipe is called a *penstock*. It has a nozzle at the end that produces a water jet to turn a turbine. We let v be the velocity of the water at the nozzle and h be the height of the surface of the reservoir above the nozzle. In fluid dynamics h is called the *head*. It is used as a measure of energy. Head is a useful concept and it appears in other areas of engineering. For example, railroad engineers refer to the *velocity head* of a train, which is the height that a train would coast uphill.

We let ψ be the mass flow of water out the nozzle in kg/s. We equate the change in gravitational energy to the kinetic energy of the water leaving the nozzle and write

$$P_h = \psi gh = \psi v^2 / 2 \qquad\qquad 7.21$$

where P_h is a hydraulic power and g is the acceleration of gravity, $9.8\,\text{m/s}^2$. This equation gives us the power that would be available to turn a turbine, expressed both in terms of the head and in terms of the nozzle velocity. In practice the frictional losses in the penstock might reduce the head by 10% from the value we would calculate using the change in elevation alone. Engineers specify a *net head* that takes these losses into account. We can solve for the velocity as

$$v = \sqrt{2gh}. \qquad\qquad 7.22$$

Hydroelectric power was critical for the development of electricity in California because the state has little coal. Figure 7.24 shows a turbine for a power station in California's Sierras. The wheel itself is called a *runner*. This type of turbine was invented by American Lester Pelton (1829–1908). A Pelton turbine was used for California's first hydroelectric power station in 1887. Each bucket has a ridge across the middle. The nozzle is aimed at the ridge. The water jet divides at the ridge and reverses direction. In physics a change of momentum is called an *impulse*, and Pelton's design is called an *impulse turbine*. Initially Pelton had trouble selling his ridge-bucket runners. However, he won a turbine contest, demonstrating an

Figure 7.24 A Pelton wheel for the Moccasin Power Station in California. William Dies is polishing the runner. In the actual power station there are six nozzles aimed at the bucket splitters, which are the ridges across the buckets. Courtesy of Matthew Gass.

efficiency of 90%, 14% higher than his closest competitor. After the contest, sales took off.

The goal in operating the Pelton turbine is to extract the energy completely from the water jet. The kinetic energy of the reversed water will be zero when its velocity is zero. We let v be the velocity of the input jet and u be the velocity of the bucket. We can write the velocity of the jet relative to the bucket as $v - u$. When the jet hits the bucket, the relative velocity of the water changes sign, so we have $u - v$ for the relative velocity of the reversed spray. To return to the velocity of the water in the original reference frame of the nozzle jet, we add the bucket velocity u back, and set the result equal to zero for maximum efficiency. We write

$$0 = u - v + u \qquad\qquad 7.23$$

which has the solution $u = v/2$. When the buckets are moving half as fast as the jet, the hydraulic power of the water, P_h in Equation 7.21, is completely converted to rotational energy in the runner. In practice there are frictional losses, and 90% is a typical efficiency for a Pelton wheel. Alas, we will see in Section 7.7.3 that this is not the case with wind turbines.

Figure 7.25 shows the Moccasin Power Station in California. The four pipes coming down the hill are the penstocks that feed two 50-MW Pelton turbines that were installed in 1968. The source of the water is the Hetch Hetchy Reservoir, shown in Figure 7.26. This reservoir was created by the O'Shaughnessy Dam. Figure 7.27 shows how the valley looked before the dam was built. The reservoir is part of a tunnel and aqueduct system that has supplied most of San Francisco's water since 1934. The valley walls are granite, and the water is so clean that it is not filtered. The water passes through two hydroelectric power stations. The higher station is Kirkwood, with three 40-MW Pelton turbines. The lower one is the Moccasin Power Station. One thing that makes the reservoir unusual is that it is entirely inside Yosemite National Park. It is inconceivable that a reservoir like this would be built in an American national park today, and over the years there have been proposals to remove the dam to restore the Hetch Hetchy Valley to its original condition. In 2012 the city of San Francisco voted on a proposition to do this. The proposition was defeated.

The head provided by the Moccasin penstocks, net of frictional losses, is 350 m. We can calculate the jet velocity from Equation 7.22 as

$$v = \sqrt{2gh} = 83 \text{ m/s} = 298 \text{ km/h}. \qquad\qquad 7.24$$

This is an extremely high velocity, and the water must be quite pure. Otherwise, sand in the water would destroy the buckets. In a test of one of the turbines, a mass flow of 18.5 t/s gave an electrical power P_e of 56.7 MW. We can write the hydraulic power input from Equation 7.21 as

$$P_h = \psi gh = 18.5 \text{ t/s} \cdot 9.8 \text{ m/s}^2 \cdot 350 \text{ m} = 63.5 \text{ MW}. \qquad\qquad 7.25$$

Figure 7.25 The Moccasin Power Station and penstocks in California. The first power station here was built in 1925. Credit: B. Christopher/Alamy Stock Photo.

We can write the electrical efficiency η as

$$\eta = P_e / P_h = 89\% .$$

7.26

Figure 7.26 The Hetch Hetchy Reservoir in the California Sierras in 2010. The white strip is a high water indicator. Credit: Neil Cronin/CC BY 2.0, https://creativecommons.org/licenses/by/2.0/

Figure 7.27 The Hetch-Hetch valley with the Tuolomne River flowing through it before the valley was flooded in 1923. Credit: Isiah West Taber.

Pelton wheels work best when the head is relatively high, and the mass flow is relatively low. For lower head and higher mass flow, the Francis turbine is more suitable.

7.4.3   The Francis Turbine and the Hoover Dam

The Francis turbine was developed by engineer James Francis (1815–1892). Francis was born in the UK, and he came to the United States as a teenager. In the Francis turbine, there is a gradual transfer of angular momentum to the runner blades rather than a sharp impulse. As with the Pelton turbine the efficiency can be very high, around 90%. Figure 7.28 shows a Francis Turbine. In the figure, water enters from the input pipe at the lower left. It circulates around the outside of the turbine. The white arrows show the path of the water. The guide vanes can be rotated to control how fast the water gets to the runner blades inside. After the water has lost its angular momentum to the runner blades, it is dumped into the output pipe at lower right.

Figure 7.29 shows the Hoover Dam, built in 1936 on the Colorado River along the border of the states of Nevada and Arizona. At the time it was completed, it was the largest hydroelectric plant in the world. However, the main purpose of the dam is not power generation, but rather water – urban water supply, flood control, and irrigation water for Arizona, Nevada, California, and Mexico. This means that the releases are set for the water customers rather than the power customers. The Colorado River empties into the Sea of Cortez. For many years the withdrawals from the river were so large that the river actually disappeared into its delta, but more recently enough water has been left in the river to provide water for the delta

Figure 7.28 A Francis turbine. This is an annotated still from an animation by Andritz Hydro, 2015, "Animation of a Francis Turbine." A link to the animation is given in the references at the end of the chapter.

Figure 7.29 Hoover Dam, on the Colorado River, between the states of Nevada and Arizona. The reservoir behind the dam is Lake Mead. The white stripe above the reservoir gives an indication of the high water level. Credit: Graham Thomson/CC BY 3.0, https://creativecommons.org/licenses/by/3.0/

and to reach the sea. The dam reservoir, Lake Mead, has an area of $350\,\mathrm{km}^2$ and the average head is $160\,\mathrm{m}$. The watershed of Lake Mead has an area of $440{,}000\,\mathrm{km}^2$, about the size of Sweden.

The dam has 17 Francis turbines. A row of these turbines is shown in Figure 7.30. The total capacity is 2.1 GW. The generation in 2017 was 3.4 TWh, corresponding to a capacity factor of 19%. We can express this power as an average power density of 1.1 W per square meter of reservoir area. This is an order of magnitude less than for a solar photovoltaic farm.

The water supply for the Hoover Dam actually involves two lakes, Lake Mead and Lake Powell. Lake Powell was created by the Glen Canyon Dam 300 km up the Colorado River from Lake Mead. It also supplies cooling water for the Navajo

Figure 7.30 A row of 130-MW Francis turbines on the Nevada side of the Hoover Dam in 2009. Photograph by Dale Yee.

Power Plant (Figure 4.7). Figure 7.31 shows the history of the annual inflow for Lake Powell, the outflow for Lake Mead, and the combined storage of the two lakes. At the beginning of the time period of the graph, the inflow into Lake Powell was considerably larger than the outflow from Lake Mead. The average inflow for the 10-year period from 1963 to 1972 was 13.6 km³/y, compared with the outflow of 10.0 km³/y. Consequently, the reservoir water volume rose, from 25 km³ in 1963 to 38 km³ in 1972. The rise is not the same as the difference in flows. Lake Mead has some small creek sources in addition to the Colorado River, but the most important reason for the difference is evaporation. These lakes are in the desert, and the water loss is 2 km³/y. This is a lot of water. For comparison, the Los Angeles Department of Water and Power supplies 0.6 km³/y to its four million customers.

The lake storage volume continued to grow until 1983, when it reached 60 km³. The following year, 1984, had the greatest inflow, 27 km³ and the greatest outflow, 26 km³. That year the electricity generation was a record 10 TWh. In recent years the inflow has dropped and the outflow has risen, compared with 1963 to 1972. The average inflow for 2008 to 2016 is 12.2 km³/y and the outflow is 11.5 km³/y. Evaporation has continued, of course, and the storage volume has fallen to 29 km³

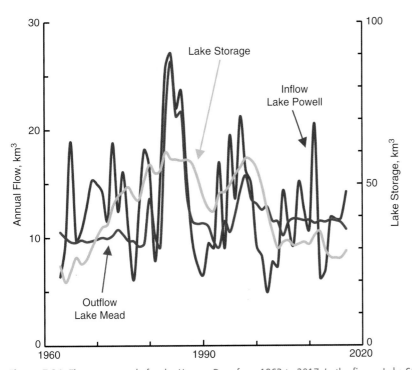

Figure 7.31 The water supply for the Hoover Dam from 1963 to 2017. In the figure, Lake Storage is the combined volume of Lake Mead and Lake Powell. The water year is calculated from October 1 of the previous year to September 30 of the current year, the same as the fiscal year in North America. The data come from water-data.com.

during this time, half of the peak in 1983. The main source of the Colorado River water is melting snow and it is extremely variable. For example, in 1981, only three years before the maximum of $27\,km^3$ in 1984, the inflow was only $8\,km^3$.

Hydroelectric power is the most important alternative today. It is not perfect. It has the largest footprint of any of the alternatives. Generation varies seasonally and from year to year. However, on a daily basis it is dispatchable and in many areas it provides the least expensive power. It is likely to remain significant for the foreseeable future.

7.5   Nuclear

By 1945, World War II had been raging for eight years in Asia and six years in Europe. It was a time of inconceivable violence. Fifty million people were dead. Germany surrendered on May 8, but Japan fought on. Harry Truman had recently become the American president after the death of Franklin Roosevelt. Roosevelt had started a project in 1942 to develop a bomb based on nuclear fission. The impetus for the project was a demonstration in 1938 by the German chemist Otto Hahn at the Kaiser Wilhelm Institute in Berlin. He and Fritz Strassmann bombarded uranium with neutrons and detected barium. Hahn and Strassman's key contribution was to recognize that barium had been produced in the reaction. The reaction was identified by a former collaborator, Austrian physicist Lise Meitner, and her nephew Otto Robert Frisch, as the splitting of a uranium atom into a barium atom and a krypton atom. Frisch gave the reaction the name *fission*, after the biological process where cells divide. The number of protons is conserved in the reaction. Uranium has atomic number 92, barium 56, and krypton 36.

Frisch then demonstrated in a laboratory in Denmark that the atoms had very high velocities, giving them enormous kinetic energy. The reason for this is that once the two nuclei are out of the range of the strong interaction force, they are accelerated by intense electric forces because they carry large positive charges. The energy released was millions of times that of a chemical reaction. The process also produced three neutrons that could induce additional fission reactions.

7.5.1   The Atomic Bomb

Physicists around the world realized that with a sufficient mass of uranium there would be a runaway chain reaction. The minimum mass that achieves this is called the *critical mass*. This would make it possible to construct an enormously powerful bomb simply by firing a uranium ring out of a gun at a cylinder that was shaped to fill the hole in the ring. Germany, Japan, and the Soviet Union all had bomb projects during the war. However, they were hampered in their efforts because they were under attack. In addition, they lacked a heavy bomber to deliver the bomb.

The Americans were able to bring the required resources to its bomb project free of interference, and they had a suitable bomber, the B29. Frisch played a role in the development, as did Richard Feynman. In addition, they had British help and it was agreed that dropping the bomb would require the consent of both governments. The first test, code named Trinity, was on July 16, 1945 near Socorro, New Mexico. The energy released by the explosion was 2 ktoe. On July 26, the United States, the UK, and China issued a declaration that set surrender terms for Japan. It concluded with, "We call upon the government of Japan to proclaim now the unconditional surrender of all Japanese armed forces, and to provide proper and adequate assurances of their good faith in such action. The alternative for Japan is prompt and utter destruction." The message was broadcast by shortwave radio. However, the Japanese government did not surrender. After the agreement of the British government was secured, B29s dropped bombs on Hiroshima, a city with an army headquarters, on August 6 and Nagasaki, a port with a naval shipyard, on August 9. British Prime Minister Winston Churchill later wrote that he believed the shock of the bombs might persuade the Japanese armed forces to surrender. Both Truman and Churchill were themselves former soldiers who had been in combat and who appreciated what might happen if the war continued.

The energy released by each bomb was comparable to the Trinity test. More than a hundred thousand people died as a result of the atomic bomb attacks. A significant number who survived the blasts died later from radiation sickness. Later in the day after the Nagasaki explosion, the Japanese Supreme Council met and discussed what to do, but they could not agree. At ten minutes before midnight, they met the emperor and asked him to make the decision. The emperor said that he wished to surrender. Telegrams were sent out early the next morning, August 10, to Switzerland and Sweden to let the world know that World War II was finally over.

Did the atomic bombs end the war? Some historians prefer to emphasize the fact that the Soviet Union declared war on Japan on August 9, and the Red Army attacked Manchuria the same day. The emperor was also influenced by American air raids that used incendiary bombs to set fires. At the time, Japanese buildings were primarily made of wood. The attacks had killed hundreds of thousands of people and had burned millions of homes, and the attacks were intensifying. Moreover, the Japanese people were starving. The country is utterly dependent on imports by freighters and on shipping in its Inland Sea. This shipping was at a complete standstill from submarine attacks, mines dropped in the Inland Sea from B29s, and from attacks by carrier planes. We leave the last word to the emperor. On August 15, he announced the surrender to the Japanese people by radio. He said,

> But now the war [against the United States] has lasted for nearly four years. Despite the best that has been done by everyone – the gallant fighting of the military and naval forces, the diligence and assiduity of our servants of the state, and the devoted service

of our one hundred million people – the war situation has developed not necessarily to Japan's advantage, while the general trends of the world have all turned against her interest.

Moreover, the enemy has begun to employ a new and most cruel bomb, the power of which to do damage is, indeed, incalculable, taking the toll of many innocent lives. Should we continue to fight, not only would it result in an ultimate collapse and obliteration of the Japanese nation, but also it would lead to the total extinction of human civilization.

So far, Hiroshima and Nagasaki have been the only nuclear attacks. However, the atomic bomb has cast a long shadow over the development of nuclear power. Uranium fuel intended for power plants has been diverted and processed to make bombs in Pakistan. Reactors intended to generate electricity have made plutonium for bombs in North Korea. A contributing factor to the Chernobyl catastrophe in the Soviet Union was that the reactor was designed both to generate electricity and to make fuel for bombs. The containment vessel had been omitted to make it easy to access the bomb fuel. Historically nuclear power has increased the chance that countries will get nuclear weapons, but there has not been another Hiroshima. Yet.

In the United States nuclear waste disposal proposals became entangled in the legacy of earlier bomb testing. More than one thousand nuclear weapons were exploded in the state of Nevada, mostly underground.[4] People and livestock were exposed to radiation and the mushroom clouds were visible from Las Vegas 120 km away. The testing area is heavily contaminated. In later years, the federal government decided that nearby Yucca Mountain would be the best place to store the dangerous high-level nuclear waste. Currently this waste is kept in ponds at the power plants until they cool off enough to put them in specially constructed casks. Yucca Mountain does have much to recommend it. Water is the scourge of storage schemes and Yucca Mountain is so dry that the waste could have several hundred meters of dry rock overhead and still be several hundred meters above the water table. However, because of its early experience with atomic weapons, Nevada has fought the federal government. The state was helped in the struggle by a judge who required that the Yucca Mountain facility be proven to be safe for one million years. People who are not geologists often have a difficult time understanding what one million years means in practical terms. During the last million years, there have been ten ice ages that put northern states under hundreds of meters of ice. The Yellowstone Supervolcano last erupted 600,000 years ago and covered

---

[4] The damage can be seen on Google Maps by entering these latitude and longitude coordinates: 37.116667, −116.05.

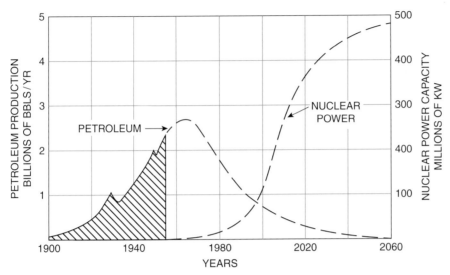

Figure 7.32 King Hubbert's idea of a transition from oil to nuclear power in the United States. While oil production did peak in 1970, that production level was surpassed in 2014. The graph is from King Hubbert, 1956, "Nuclear Energy and the Fossil Fuels," Shell Development Company, Houston, Texas, publication 95. Courtesy of the Shell Oil Company.

the western states with ash. The continuing failure of the Yucca Mountain storage plans in turn hinders new plants, because many states have legal requirements that there be a national repository for nuclear waste before a construction permit can be issued.

7.5.2   Failed Dreams

After World War II, there was speculation that nuclear energy might become the dominant power source. Isaac Asimov, often considered the greatest science fiction writer ever, was also a professor of biochemistry. In his classic 1951 novel *Foundation*, there is nuclear power at all scales from personal reactors to global power plants. King Hubbert foresaw a transition from oil to nuclear power. He sketched a curve indicating a nuclear capacity for the United States of 370 GW in 2020 (Figure 7.32). This transition is not happening. US nuclear capacity has been stuck at 100 GW since the Three Mile Island accident in 1979. A more fundamental criticism of Hubbert's curves is that nuclear power has not become a substitute for oil in transportation, although nuclear power has advantages for submarines and aircraft carriers. Nuclear generators do not require oxygen, and this allows submarines to operate underwater for long periods. The advantage for a carrier is that the reactors are quite compact compared to the space required for fuel in an oil-powered carrier and this leaves space for aviation fuel.

7.5.3   Nuclear Generation History

At 4.2% of world primary energy in 2017, nuclear power is the second most important alternative energy source. Figure 7.33 shows the production history. In recent years the largest expansion has been in China, with an annualized growth from 2007 to 2017 of 15%/y. In the early years the world growth rate for nuclear was spectacular, like wind and solar today. From 1965 to 1975, generation increased at an annualized rate of 30%/y. For the largest producer, the United States, generation grew by a factor of 25 from 1965 to 1979, the year of the Three Mile Island accident. The fundamental safety problem in nuclear reactors is that heat generation in a reactor core cannot be stopped. The fission reaction can be blocked by inserting materials into the core that absorb neutrons. However, the core is still radioactive, and the decay process generates heat, tens of megawatts in a power reactor. Without a flow of cooling water, the temperature rises and it does not stop rising. At high enough temperatures, any metal reacts with steam to produce an oxide. Often it is the zirconium cladding on fuel rods that oxidizes. This releases hydrogen that accumulates until it explodes. In the Three Mile Island reactor in Pennsylvania, the cooling water was lost because a valve stuck open. There were no instruments that directly showed the water level in the reactor, and the operators did not figure out the problem until it was too late. There was a containment vessel, so no one was killed. But the reactor was destroyed. After the accident, no new power plants were built in the

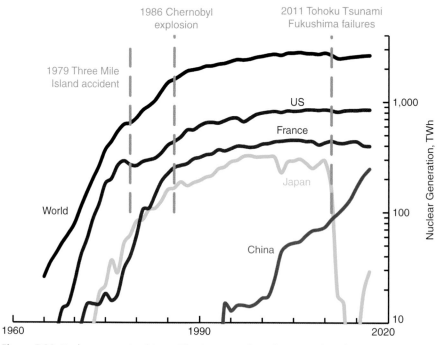

Figure 7.33 Nuclear generation history. The data come from the BP *Statistical Review*.

United States for many years. Some plants reached the end of their forty-year certification and were shut down. Others have been recertified for an additional twenty years. Then in 2012 and 2013, four new nuclear plants were started in South Carolina and Georgia. These plants used the new Westinghouse AP1000 design. However, Westinghouse declared bankruptcy in 2017 and the two South Carolina plants were abandoned. The Georgia plants are ten billion dollars over budget and five years late, but as of 2018, George Power still planned to finish them. However, these may be the last nuclear plants built in the United States. If no new plants are built, the end of nuclear power there is inevitable. It is just a matter of time as the certifications expire.

France, the number two producer, made an exceptional commitment to nuclear power. Figure 7.34 shows the history of nuclear power generation in France, compared with coal and hydrocarbon consumption. France was running out of coal during this time (Table 4.2). During the 1960s consumption shifted away from coal to oil. Hydrocarbons reached an all-time high share of 77% of French primary energy in 1973, the year of the Yom Kippur War and the Arab Oil Embargo. After the Iranian Revolution in 1979, hydrocarbons fell and the nuclear share rose. However, nuclear power mainly replaced oil in electricity generation. In 2017, nuclear power was 38% of primary energy, while hydrocarbons were still 50%. Nuclear power provides 72% of France's electricity generation. Oil still dominates

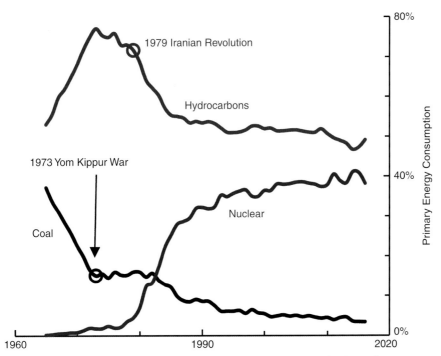

Figure 7.34 France's shift from coal and hydrocarbons to nuclear power. The share of primary energy consumption is shown. The data come from the BP *Statistical Review*.

in transportation, although the country has developed an excellent high-speed train network that uses electricity. France does not have its own uranium supply. It imports uranium primarily from Canada and Niger. Nuclear power plants are not load following, and France needs other countries to take its excess electricity. The country exported 11% of the electricity it generated in 2016. France has proposed banning diesel and gasoline cars by 2030, so it can be anticipated that demand for electricity would increase for charging electric vehicles. However, the future of French nuclear power is unclear. Plans to replace the current generators with a new design have bogged down.

7.5.4 The Chernobyl Disaster

The worst disaster in the history of nuclear power occurred in 1986 in the Soviet Union. It was at reactor number 4 in the Chernobyl complex on the Pripyat River in the Ukrainian Republic. Shortly after midnight on April 26, the operators were conducting a poorly designed test to verify operation with *the electrical supply for the cooling pumps turned off* [author emphasis]. The reactor overheated and there was a hydrogen gas explosion (Figure 7.35). Graphite fires burned for more than a week, causing a massive release of radiation into the air, hundreds of times that of the nuclear weapons dropped on Japan. Firemen eventually put the fires out, but there was no way to protect them from exposure to radiation. Thirty-one people were killed directly by the radiation and epidemiologists predict that many

Figure 7.35 Reactor number 4, Chernobyl. From www.globalsecurity.org/wmd/world/russia/rbmk.htm

more will die of cancers induced by it. The reactor is now covered by an enormous concrete structure.[5]

The Chernobyl explosion has been the most consequential accident in the history of energy. From a world energy perspective, Chernobyl was a shock that caused a transition in nuclear generation from rapid growth to level generation. The local impacts were also severe. An exclusion zone 60 km across was established around the site and hundreds of thousands of people were resettled. Figure 7.36 shows abandoned boats in the area. The political effects were also important. Four years later in 1990, Ukraine declared independence from the Soviet Union. Chernobyl was a contributing factor in the declaration. The following year the entire Soviet Union was dissolved.

7.5.5 The Great Tohoku Tsunami

After Chernobyl, world nuclear generation was relatively steady for many years. However, on March 11, 2011, there was an earthquake off the coast of the Tohoku region in northeastern Japan, 120 km east of the city of Sendai. This was a magnitude 9 quake, the most powerful ever recorded in Japan. There was severe shaking, with accelerations of $g/4$ for more than a minute. For people who do not live in earthquake country, a minute in an earthquake is an eternity. A strong tsunami

Figure 7.36 Abandoned boats near Chernobyl. Credit: Iryna Rasko/Shutterstock.

---

[5] To see the cover in Google Maps, enter the latitude and longitude: 51.38911, 30.09403.

was generated by the quake. Figure 7.37 is a horrifying photograph of the tsunami arriving at the city of Miyako. The earthquake with its tsunami was the most expensive natural disaster in history, with damages of hundreds of billions of dollars. Twenty thousand people were killed and over a hundred thousand people were displaced.

Two hundred kilometers to the south of Miyako, at the Fukushima nuclear complex, three reactors were generating electricity and three reactors were shut down for maintenance. All six reactors came through the shaking of the earthquake without problems and the three reactors that were operating shut down safely. However, the tsunami arrived ten minutes later. The wave overtopped the seawall and flooded the plants. Thirteen backup generators for the cooling pumps were knocked out. Four of the six reactors were left without working cooling pumps. These reactors were of the boiling water type that had systems to cool the reactor temporarily without electricity, but over the next few days the cooling systems were overwhelmed. Temperatures began to rise in the reactors and in the cooling ponds that held spent fuel rods. Eventually the temperatures reached the level where the zirconium cladding on the fuel rods reacted with steam to form zirconium oxide and hydrogen, which exploded. No one was directly killed by radiation, but radioactive gas was released to the atmosphere and the area around the plant was evacuated to a distance of 30 km.

Figure 7.37 The tsunami generated by the Great Tohoku earthquake at Miyako, Iwate Prefecture in 2011. Photographer unknown.

After the Fukushima disaster, Japan shut down all of its nuclear reactors. These had been providing 25% of Japanese electricity. The restart has been slow. By 2017, nuclear power had only recovered to a 3% share of Japanese electricity.

Operations safety is deeply embedded in human psychology and culture. We can be humbled by the willingness of Ukrainian firemen to die to protect their citizens from radiation exposure, but still criticize Soviet reactor design and operator training. However, the Fukushima disaster shows that the problem did not go away with the Soviet Union. In hindsight, the Tokyo Electric Power Company (TEPCO) had made a catastrophic design decision at the very beginning. Originally the Fukushima site was a bluff thirty meters above the ocean, safe from any tsunami. However, the elevation was reduced by excavation to make it easier to bring construction materials in from the sea.

Where does nuclear power go from here? It has often been argued that underground coal mining has killed more people than nuclear power. However, coal mining accidents do not depopulate vast areas. Coal has enabled many countries to escape poverty. We were all poor once. Underground coal mining develops within communities where the risks, if not accepted, are at least understood. As countries become richer, they turn from coal to natural gas. A major problem for nuclear power now is that large amounts of solar and wind capacity are being added to electrical grids. This variable generation requires that the other generators be flexible in their output. It is possible that future nuclear power plants will be designed so that they can respond quickly, as in a nuclear submarine. Otherwise, in many countries we will just be watching the existing plants marking time until their permits expire. Then we will be left with decommissioning for decades and radioactive wastes for thousands of years.

7.6 Geothermal

Geothermal energy comes from heat inside the earth. This heat can produce hot water for buildings and steam for generating electricity. The sources of the heat are primordial. The earth was formed by the accretion of debris from the solar nebula. In the process the kinetic energy of the incoming fragments was dissipated as heat. This heat has been working its way up to the surface for billions of years. Another heat source, probably comparable in magnitude, is radioactivity, mostly from uranium and thorium. In most places the geothermal power density at the surface is small. The exception is near tectonic plate boundaries, where the hot molten magma comes near the surface. This happens in the Western United States, Iceland, and Japan. In these places we see steam vents, geysers, and hot springs. Figure 7.38

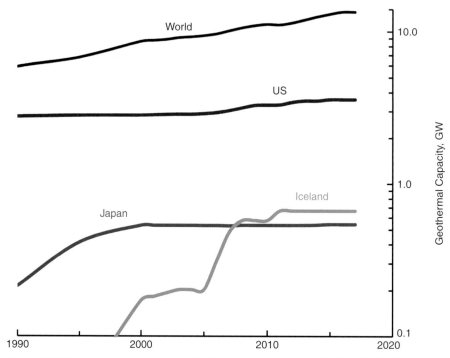

Figure 7.38 Geothermal generating capacity. The data come from the BP *Statistical Review.*

shows the geothermal electrical capacity for these countries and for the world. The world capacity was 14 GW in 2017. The 10-y annualized growth in capacity was 3.3%/y, larger than the growth of electricity as a whole, 2.5%/y. Geothermal plants typically operate at a capacity factor of 80%, so we can calculate that annual generation was 100 TWh. This is 0.2% of the world primary energy consumption.

7.6.1   The Geysers Field in California

The largest geothermal power system in the world is the Geysers Field in northern California (Figure 7.39). This field has many steam vents that make the geothermal potential obvious. The country is rugged and there are limited economic opportunities otherwise. Despite the name, there are no geysers in the field. The first steam wells were drilled in 1921 to drive a 35-kW generator. The Pacific Gas and Electric Company built the first utility-scale generator in 1960. It had a capacity of 11 MW. By 2016 there were 20 plants operating with a total capacity of 1.6 GW.

Four hundred steam wells have been drilled in the Geysers Field. The history of the steam production is shown in Figure 7.40. There is a characteristic pattern that is seen in many geothermal fields. At first the wells produced steam without any help and the amount of steam rose steadily. However, there was a peak in production in 1987 because the water does not replace itself quickly enough. Because of this, 50 additional wells were drilled to inject replacement water. Waste water from sewage plants and condensed steam from the power plants are used. Over time, the

Figure 7.39 The Geysers Geothermal Field, California in 2009. Credit: Julie Donnelly-Nolan, United States Geological Survey.

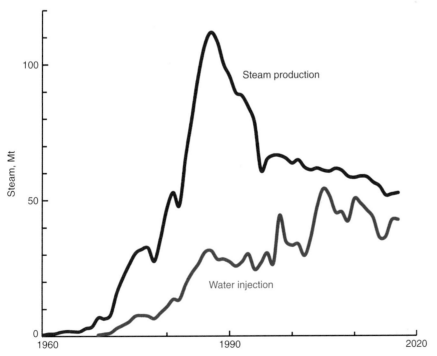

Figure 7.40 Steam production and water injection in the Geysers Field. The data come from the California Department of Conservation.

mass of replacement water has been approaching the mass of the steam. In 2016, the ratio was 81%. However, the injection wells do not restore the steam production to its peak, but rather they slow the decline. Steam production in 2016 was only 46% of the production at the peak in 1987. Earthquakes have been associated with the water injection. The largest was magnitude 4.6 (Table 5.2). At this level people feel the shaking and they hear rattling noises, but there is little damage.

7.6.2   Geothermal Heating in Iceland

The country of Iceland is on the boundary between the North American and the Eurasian tectonic plates. The country extends 500 km east to west and 300 km north to south, but the population is only 300,000. This reflects the fact that before modern times it would have been difficult to feed a large population there. The main industry is fishing. Figure 7.41 shows a geyser in Iceland. Geysers are hot springs that boil over periodically. The geothermal resource for the island is 7 GW, 23 kW per person. This is enormous, equivalent to 18 toe/p/year. In comparison, the world hydrocarbon consumption is 1 toe/p/y (Figure 1.6).

Geothermal energy provides heating for the country's homes, office buildings, and greenhouses. At this latitude, the winters are dark and cold and it is important for safety to keep sidewalks clear of ice. In most northern countries, the sidewalks are sanded and snow shoveled, but in Iceland sidewalks are kept clear by running hot water in tubes under the surface.

Figure 7.41 A geyser erupting in Iceland in 2008. The word "geyser" comes from the Icelandic language. Photograph by the author.

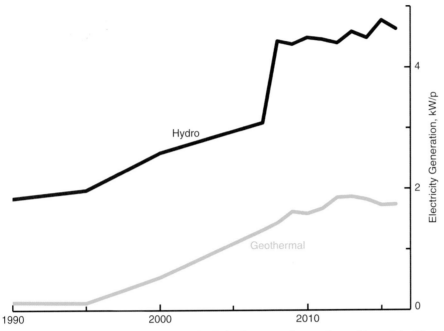

Figure 7.42 Per-person electricity use in Iceland. The data come from various editions of the IEA, *Renewable Energy Information.*

Virtually all of the electricity generation in Iceland comes from renewable sources. The resources are so large that it has been difficult to figure out what to do with it all. Figure 7.42 shows the electricity production on a per-person basis. Geothermal electricity alone in 2016 was 1.7 kW/p, which is higher than the US consumption, 1.4 kW/p, but the geothermal electricity is dwarfed by hydroelectric power at 4.6 kWh/p. It was decided that the best thing to do with the electricity was to make aluminum. Making aluminum takes a lot of electricity, 14 MWh/t. In 2015, aluminum production used 71% of the electricity generated in the country. A new, higher risk market for the electricity is mining bitcoin (Section 8.5.2).

7.6.3   The Japanese *Onsen*
Japan also has large geothermal resources, but it has not emphasized electricity production. Its geothermal electricity capacity is less than Iceland's. Japan uses its hot springs as resorts for communal bathing (Figure 7.43). These are called *onsen* in Japanese. There are thousands of *onsen*, and the *onsen meguri*, or hot springs tour, is a popular vacation. Nobel Prize winning author Yasunari Kawabata portrayed the *onsen* life in his 1947 novel, *Snow Country*. To the Japanese people, the *onsen* have some of the qualities of national parks, and there is opposition to building geothermal electrical plants in these areas. This is in contrast to the Geysers Field in California, where hot springs resorts have not been successful.

Figure 7.43 A Japanese hot springs resort in Akiu, near Sendai. Credit: Hotel Iwanumaya.

After nuclear, geothermal is the most important energy source that does not ultimately depend on radiation from the sun. While it is classified as renewable, geothermal energy is associated with wells. As in oil wells, the production is managed by water injection. Geothermal energy can provide baseload electricity, but it suffers from the fact that the resources are near tectonic plate boundaries. Next we turn to wind, where the resources are much greater and are more widely distributed.

7.7 Wind

Wind is the first variable source that we consider; solar is the other. Wind speeds vary dramatically hourly, daily, seasonally, and annually. The fundamental way to accommodate a variable output is storage. Wind turbines have been used to grind grain and pump water for more than a thousand years. Grain can be stored in silos and water in ponds and tanks. A machine that grinds grain is a *mill*, and a wind turbine combined with a mill is called a *windmill*. This name has carried over to water, and a wind turbine combined with a water pump is usually called a windmill also. Figure 7.44 shows a windmill on a ranch in Oklahoma. There are 18 blades. Having many blades helps start the pump when the wind speed is low. The pump fills the stock tank with water for the cattle to drink. In an electrical grid, wind generation may be balanced by natural gas and coal and the storage is indirect. Natural gas can be stored in depleted oil and gas fields and coal can be stored in a mound at the power plant. There is great interest now in using batteries

Figure 7.44 A windmill pumping water to a stock tank on a ranch in Oklahoma. Courtesy of Dori Troutman.

for storage. From the point of view of grid control, the flexibility of batteries is attractive. However, it is still unknown whether they can be made at reasonable cost on the scale required. We will return to grid storage in Section 7.9.

7.7.1   Wind Resources

Figure 7.45 shows a map of the world wind resources on land at a height of 80 m. This is a typical height for wind turbines that generate electricity. Generally, the higher latitudes have higher wind speeds, for example, in northern Europe, Greenland and Tierra del Fuego. Sailors speak of the roaring forties, the furious fifties, and the screaming sixties. Some of these places have few people and the challenge would be to find customers for the electricity. Good winds are also associated with high plains, like Tibet and the Great Plains of North America. The map does not show wind speeds over the ocean, but generally they are higher than over the land because there are no obstacles. In some places where the water is shallow, like the North Sea, it can be practical to erect wind turbines.

There also latitudes where the winds are weak. One is a band around the equator. Sailors call this the *doldrums*. The doldrums are characterized by low winds and violent thunderstorms. Brazil, the Congo and Indonesia are in this range. The thirties is another latitude band of poor wind resources. These are the sailor's *horse latitudes*. Examples are the southeastern United States and much of eastern China. However, even in areas which are otherwise unfavorable, winds may be locally strong on ridges.

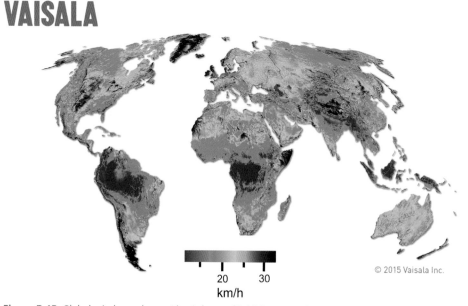

Figure 7.45 Global wind speed map. Blue is low, red is high. Copyright © 2017 Vaisala Company, Vantaa, Finland.

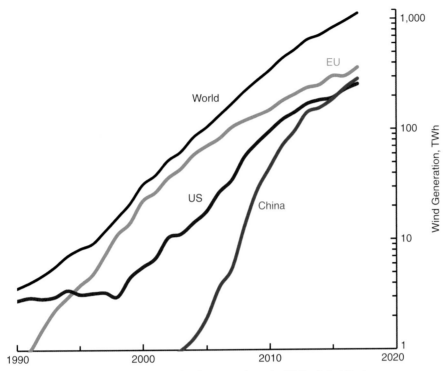

Figure 7.46 Wind generation history. The data come from the BP *Statistical Review*.

7.7.2 Wind Generation History

Figure 7.46 shows the history of wind electricity generation. World generation has grown rapidly, with 21%/y annualized growth from 2007 to 2017. Worldwide, wind accounted for 1.9% of world primary energy in 2017. The EU, China, and the US all now have comparable generation.

7.7.3 The Betz Limit

When we analyzed the Pelton wheel in Section 7.4.2, we found that the conversion efficiency from hydraulic power to turbine power could be 100% in principle. The corresponding limit for wind was developed by Albert Betz in 1920. Betz was a professor at the University of Gottingen in Germany. The Betz limit is given by

$$\eta = 16/27 \approx 59\% \qquad\qquad 7.27$$

where η is the maximum efficiency. The approach we will take to derive the Betz limit is similar to the one we used for the Pelton wheel, but it is much more complicated. Here are the steps so that you can see where we are headed.

1. Calculate the flux ψ of molecules on a surface from the wind.
2. Calculate the available wind power P_w.
3. Calculate the maximum force against a surface F_m.
4. Calculate the power P gained by the machine.

5. Calculate the speed of the surface that gives maximum power P_m.
6. Calculate the maximum efficiency as the ratio $\eta = P_m / P_w$.

We start by finding an expression for the molecular flux ψ through a surface with area A, shown in Figure 7.47:

$$\psi = nAv \qquad\qquad 7.28$$

where n is the molecular density, that is, the number of air molecules per unit volume, and v is the velocity. The fact that the flux is proportional to the area is a critical part of the derivation. In the Pelton wheel, the ridged bucket was larger than the jet of water, so it does not determine the flux.

The wind power P_w is written in terms of the kinetic energy of the molecules and the molecular flux ψ as

$$P_w = \psi m v^2 / 2 \qquad\qquad 7.29$$

where m is the average mass of the air molecules. We write the atmospheric volume mass density ρ (the Greek letter rho) as

$$\rho = mn. \qquad\qquad 7.30$$

At room temperature and sea level ρ is $1.2\,\mathrm{kg/m^3}$. We rewrite the wind power as

$$P_w = \rho A v^3 / 2. \qquad\qquad 7.31$$

The power varies as the cube of the wind speed, which means we really, really want to look for windy places for our turbines. It also helps to make the turbines tall because the wind speeds are higher away from the ground, where friction slows the wind down. Finally, we want the propellers to have a large diameter, because the area increases as the square of the length of the propeller.

Here we take a detour to introduce the idea of a drag coefficient. We can rewrite the expression for wind power (Equation 7.31) as

$$P_w = \rho v^2 / 2 \cdot Av. \qquad\qquad 7.32$$

The expression $\rho v^2 / 2$ has units of $\mathrm{N/m^2}$, force per unit area. These are the units of pressure. It is not a pressure like the water pressure that one feels as one submerges

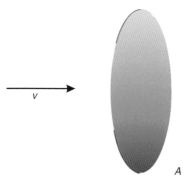

Figure 7.47 Surface for deriving the Betz limit. The surface has area A and faces the wind. The wind speed is given by v.

v

A

in the ocean. Rather, it is an expression associated with wind that has the same units. It is called the *dynamic pressure* p_d, defined as

$$p_d \equiv \rho v^2 / 2. \qquad 7.33$$

It is conventional to write the drag force F_d of the wind on an object as

$$F_d \equiv p_d A C_d \qquad 7.34$$

where C_d is called the *drag coefficient*. The drag coefficient does not have units. The drag coefficient does vary somewhat with velocity, so the range of velocities should be made clear. The force is proportional to the square of the velocity.

Sometimes people want a large drag coefficient. When a sailboat is going downwind, the crew may set a large baggy sail called a spinnaker (Figure 7.48). The drag force on the spinnaker helps propel the boat, so the drag coefficient should be large. A spinnaker can have a drag coefficient as large as 1.7. Parachutes have similar drag coefficients. A drag coefficient greater than 1 indicates that the surface is causing some reversal of the momentum of the wind, like the reversal of water momentum in the Pelton wheel. When the boat is moving,

Figure 7.48 A sailboat in a race. The blue sail is the spinnaker. Credit Johann-Nikolaus Andreae/CC BY-SA 2.0, https://creativecommons.org/licenses/by-sa/2.0/

the force would be calculated with the relative velocity $v - u$, where u is the boat velocity. Sailors call this relative velocity the *apparent wind velocity*, while v is called the *true wind velocity*. One disadvantage of sailing directly downwind is that it reduces the apparent velocity, which in turn reduces the force. For this reason, sailboats typically go faster at an angle to the wind than they do directly downwind.

On the other hand, cars are designed to have low drag coefficients. For example, the Tesla Model S has a drag coefficient of 0.24. If the speed of the car u is much larger than the wind speed, we could write the power lost to drag as

$$P_d = F_d u = \rho A C_d u^3 / 2 \qquad\qquad 7.35$$

where u is the car velocity. This drag loss increases as the cube of the speed and it is important for cars and trucks at highway speeds.

To return to our derivation for the Betz limit, recall that Newton's Second Law states that the force is equal to the rate of change of momentum. First we consider an individual molecule incident on a surface. The largest possible change in momentum occurs if the molecule reflects back in the direction it came from without losing speed. In this case the change of momentum would be $2mv$. The maximum force F_m would occur if all of the air molecules did the same thing. This force would be given by

$$F_m = 2mv \cdot \psi = 2\rho A v^2. \qquad\qquad 7.36$$

This is for a stationary surface, if the surface is moving with a velocity u, we rewrite the maximum force in terms of the relative velocity $v - u$ as

$$F_m = 2\rho A \left(v - u \right)^2 . \qquad\qquad 7.37$$

The force goes to zero as the surface velocity becomes close to the wind speed. We can calculate the power P transferred to the surface by multiplying by the velocity u,

$$P = F_m u = 2\rho A \left(v - u \right)^2 u. \qquad\qquad 7.38$$

This function rises from zero at $u = 0$, reaches a maximum, and then falls to zero at $u = v$.

To find the maximum power, we need to set the derivative of P with respect to u equal to zero. There is a standard approach to finding the maxima of power-law expressions like this. We take the natural logarithm of both sides of the equation

$$\ln(P) = \ln(2\rho A) + 2\ln(v - u) + \ln(u). \qquad\qquad 7.39$$

The chain rule implies that the logarithm of P will have a maximum at the same place that P does. To find it, we set the derivative equal to zero

$$0 = \frac{d\ln(P)}{du} = -\frac{2}{v - u} + \frac{1}{u} \qquad\qquad 7.40$$

with the solution $u = v/3$. In contrast, for the Pelton wheel the optimum velocity of the bucket was half the velocity of the water. The maximum power P_m is given by

$$P_m = 2\rho A \left(2v/3\right)^2 \left(v/3\right) = \rho A \left(2v/3\right)^3.$$ 7.41

Finally, we calculate the efficiency η as the ratio of the maximum power to the wind power P_w from Equation 7.32 as

$$\eta = \frac{P_m}{P_w} = \frac{\rho A \left(2v/3\right)^3}{\rho A v^3 / 2} = \frac{16}{27}$$ 7.42

which establishes the Betz limit.

Figure 7.49 shows the power output for the GE 2.5-103 turbine as a function of the wind speed. It has a capacity of 2.5 MW and a rotor diameter of 103 m. The hub is 100 m above the ground. The Betz limit is plotted for comparison. The wind speed must reach 20 km/h before there is significant electricity generation. At speeds above 90 km/h, the turbine will feather the blades, orienting them so that they present little surface area to the wind. This prevents damage to the turbine.

From a distance, it often appears that the blades of a wind turbine are moving slowly. This is an illusion caused by the fact that they are so large. At the maximum power output, the tips of the rotor are moving at 270 km/h. This will kill a bird if it collides with a blade. Many of the birds killed are hawks and eagles that do not avoid the blades. In many countries these birds are strictly protected in other contexts and killing one would result in a large fine.

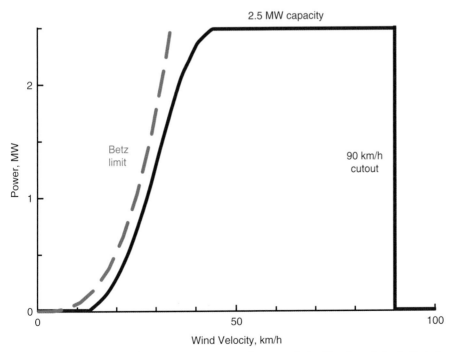

Figure 7.49 Power output of the GE 2.5-MW, 103-m diameter wind turbine. The wind velocity is the speed at the hub height. For reference 1 m/s = 3.6 km/h. The data come from the GE product sheet.

7.7.4   Wind Variability

A major problem with wind power is its extreme variability, from day to day and from year to year. Figure 7.50 and Figure 7.51 illustrate the problem for California's CAISO market. On December 25, 2015, the average generation was 1.6 GW, above the 2015 annual average of 1.4 GW. The next day, the average power dropped to 300 MW. The day after that, it was back up to 1.7 GW. In the previous year, there was an 11-day period starting January 1, 2015 where the average was 70 MW. Wind power was useless over the entire state for more than a week.

Wind turbines must be spaced out to avoid downwind wake interference. The wake patterns can be seen in Figure 7.52. This is the Horns Rev wind farm off the coast of Denmark. The facility has 80 turbines with a total capacity of 160 MW. It was completed in 2002. These turbines are a half kilometer apart. The depth of the sea is 10 m. Offshore wind farms are more expensive than land farms, but they have high capacity factors. For the Horns Rev wind farm, the capacity factor is 40%.

The spacing between wind turbines allows other activities to continue, fishing, for example, on Horns Rev. Wind turbines on the Great Plains are compatible with livestock, and they provide well-paying jobs and royalties to people who appreciate them. The next source, solar power, has a widely distributed resource, like wind.

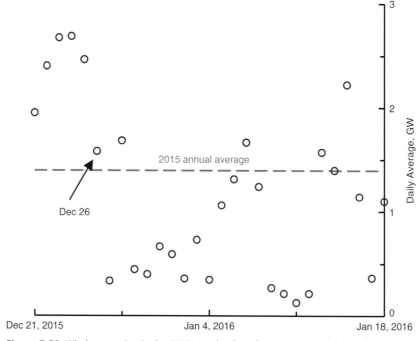

Figure 7.50 Wind generation in the CAISO market for a four-week period during the winter of 2015–2016. The data points are the power averaged over the day. The dashed line shows the annual average for 2015. The data come from CAISO.

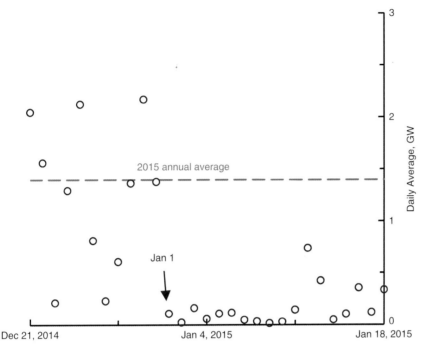

Figure 7.51 Wind generation in the CAISO market for a four-week period during the winter of 2014–2015. The weeks of the year are the same as in Figure 7.50. The data points are the power averaged over the day. The dashed line shows the annual average for 2015. The data come from CAISO.

Figure 7.52 Wind turbines at Horns Rev in the North Sea. *Rev* means "reef" in Danish. This picture was taken on February 12, 2008. When the air is humid the turbulence can create condensation, making it possible to see the wake pattern behind the wind turbines. Credit: Christian Steiness.

The factory learning curve for solar panels has been so spectacular that costs are dominated by the balance of the system rather than the solar panels themselves.

7.8 Solar

7.8.1 Solar Resources

The solar resources are larger than any conceivable need for energy, now or in the future. To get a perspective on this, consider the Kayenta Solar Facility shown on the cover of this book. The plant occupies 900,000 m², including the inverters and transformers. The average electrical output is 8.7 MW, or 9.7 W/m². This is an order of magnitude higher than for hydroelectric power and the land does not need to be submerged. At this power density, matching the primary energy consumption of the world, 14 Gtoe in 2017, would require 700,000 km². This is comparable to the area planted for biofuels. As the cover photograph shows, solar panels do not need agricultural land and compared with steam turbines, the water consumption is minimal, an occasional wash. The panels do not kill hawks and eagles as wind turbines do. Figure 7.53 shows a map of world solar resources. *Insolation* is the solar power density incident on the surface. In some ways this map is complementary to the one for wind resources (Figure 7.45). The doldrums and horse latitudes have weak winds but they are sunny. The high latitudes have strong winds but little sun.

7.8.2 The Ivanpah Solar Thermal Plant

Solar energy has long been used for heating water. With focusing mirrors, the temperature of the water can be raised to make steam and drive a turbine to

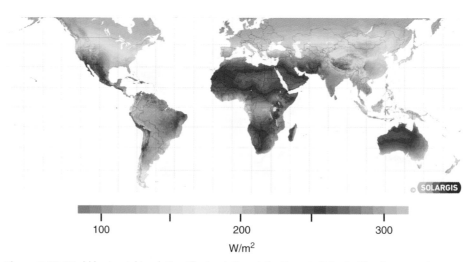

Figure 7.53 World horizontal insolation. The insolation at the Kayenta Solar Facility shown on the cover is 260 W/m². Courtesy of Solargis.com.

generate electricity. These are called *solar thermal* plants. An example is the Ivanpah Solar Electricity Generating System in the Mojave Desert in California, shown in Figure 7.54. In this plant, computer-controlled mirrors focus the light on three receivers on towers. Inside the receiver, water is heated to make steam. The receivers become so hot that they incandesce and the gleaming towers are a spectacular sight for passengers flying to Los Angeles. The steam drives turbines

Figure 7.54 The Ivanpah solar thermal power plant in California near the Nevada border town of Primm. This plant came on line at the end of 2013. Courtesy Bright Source.

on the ground by each tower. The total capacity of the turbines is 377 MW. Ivanpah's challenges with desert tortoises were discussed in Section 6.5. Another problem is that insects are attracted to the light and birds follow the insects into the beam.

One advantage of solar thermal plants is that it is possible to store thermal energy by adding a heat exchanger to heat up a vat of molten salt. The plants were once thought to be promising, but they have been completely overwhelmed by solar photovoltaics (PV). The cost of the Ivanpah plant was $6 per watt of capacity, while the cost for a solar PV farm is $1/W. The solar thermal plants are complicated to run and maintain. They need a skilled staff of engineers, and in the case of Ivanpah, biologists to manage the tortoises. In contrast. solar panels can operate unattended for years, and their AC inverter electronics give great flexibility in controlling the output.

7.8.3   Solar Cell Efficiency

The first practical solar cells were developed at the Bell Telephone Laboratories in 1954 by Daryl Chapin, Calvin Fuller, and Gerald Pearson. Their cell was a silicon diode. It converted 6% of the power in sunlight to electrical power. The invention of the silicon solar cell was based on a revolution in the understanding of electrical conduction in silicon crystals. The mechanism is quite different from metals. In metals, electricity conduction occurs readily at all temperatures whether the metal is pure or impure, crystalline or not. However, in pure crystalline silicon, conductivity is poor at room temperature, and it drops further as the temperature is reduced. The electrons in the material are either attached to the atomic nuclei or they participate in covalent bonds with other silicon atoms. However, light shining on the material gives the covalent electrons enough energy to break loose and move through the crystal. In the process, the electron gains 1.1 V of electrical potential. By applying a voltage across the silicon chip and measuring the photocurrent, one can make a sensitive optical detector. However, the energy that the electrons absorb is lost at the connections to the crystal, so a piece of silicon by itself does not make a solar cell. However, silicon diodes have a built-in internal voltage, and if the light is absorbed within the diode, it can deliver the photocurrent with a voltage of 0.5 V. In a solar panel a number of cells are connected in series to provide a larger voltage.

One source of loss in the solar cell is that the voltage delivered to an outside circuit is less than the voltage the electrons gain when they absorb light energy. There is another loss related to the energy in the light itself. The light is absorbed discretely, with one photon giving energy to one electron. The energy that the light can deliver to the electron is proportional to the frequency of the light. This relationship is called the photoelectric effect, and it was proposed by Albert Einstein

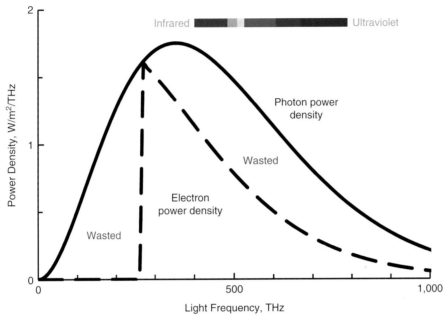

Figure 7.55 The photoelectric effect in silicon. The color bar at the top of the figure shows the relationship between the color and frequency. The photon power density is calculated from the Planck formula for black-body radiation with a temperature of 6,000 K to represent solar radiation.

in 1905. Einstein won a Nobel Prize for his paper. Figure 7.55 shows the relationships. The solid curve gives the solar power density in W/m^2/THz. At 266 THz, the photon energy is just sufficient to produce a conduction electron. This is called the *gap frequency*. Below the gap frequency the photon energy is not sufficient to create conduction electrons. Above the gap frequency, a covalent electron is turned into a conduction electron with a larger energy initially, but it rapidly loses energy to the crystal in the form of heat, so the excess photon energy is wasted. The number of photons is dropping at the higher frequencies, so the electron power drops with it. The overall ratio of the electron power to the photon power, calculated by integration of the curves in the graph, is 44%. If we multiply this by the ratio of output voltage to the voltage gained by a conduction electron, we get an estimate of the efficiency

$$\eta = 44\% \cdot 0.5\,\text{V} \,/\, 1.1\,\text{V} = 20\%. \tag{7.43}$$

This is a typical efficiency for commercial solar cells.

7.8.4 Solar Generation History

Figure 7.56 shows the solar production history. In 2017, solar electricity was 0.7% of the world primary energy consumption. The growth rate for solar power is the highest of any of the alternatives. The ten-year annualized growth rate for the world is 50%/y. The generation for the European Union, China, and the United

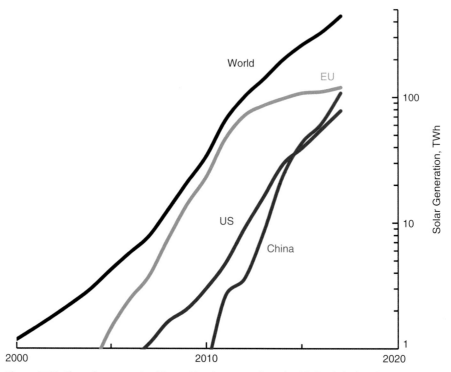

Figure 7.56 The solar generation history. The data come from the BP *Statistical Review.*

States have been comparable. Moreover, there does not appear to be any obstacle to further increases in the production of the solar panels, which are called *modules* in the industry. The most important component in a module is silicon, which is abundant. The sand at many beaches is silicon dioxide, which is the feedstock for the production process.

The main factor that has enabled the growth of solar power is the spectacular drop in the price of solar power plants. This reflects the learning curves for the module manufacturing process and the plant construction process. Figure 7.57 shows the price history for utility scale power plants in the United States. The price has dropped by a factor of four from 2010 to 2017, to $1/W of capacity. The price of modules has fallen even more rapidly, and is now less than half the cost of the plant.

7.8.5   Solar Variability

The biggest challenges for solar power are its daily variation and the changes with weather and the seasons. There is no generation at night and the output is lower on cloudy days. Generation is reduced in the winter because the days are short and the elevation of the sun drops. Winter is a particular problem at higher latitudes. Figure 7.58 shows a comparison between the summer and winter insolation at the

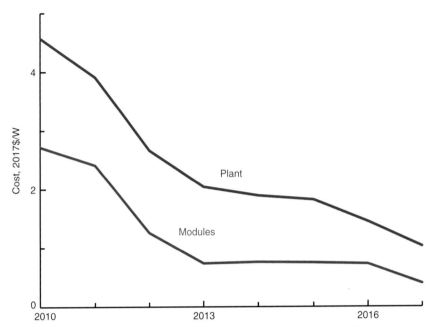

Figure 7.57 Cost of a 100-MW capacity, fixed-tilt solar power plant in the United States. The cost is what a developer would pay, excluding government subsidies and profits. From NREL (National Renewable Energy Laboratory), 2017, "Solar Cost Benchmark."

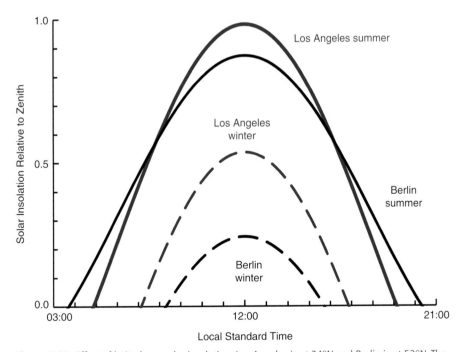

Figure 7.58 Effect of latitude on solar insolation. Los Angeles is at 34°N and Berlin is at 53°N. The curves are plotted from formulas given in Robert Jaffe and Washington Taylor, 2018, *The Physics of Energy*, Cambridge University Press.

latitudes of Los Angeles and Berlin. The insolation is normalized to the zenith, that is, to the insolation when the sun is directly overhead. The curves are drawn for standard time (not daylight savings time) at the solstices for cloud-free days. It should be appreciated that since Berlin is cloudier than Los Angeles these calculations are biased in favor of Germany. In addition, the electricity loads in Los Angeles are a better match to solar seasonal generation. The highest loads in Los Angeles are in the summer, while the highest loads in Berlin are in the winter. The total daily summer insolation is the same in Los Angeles and Berlin to a tenth of a percent. The elevation angle at noon is higher in Los Angeles, 79°, compared with 61° in Berlin. However, the day is longer in Berlin than in Los Angeles, and the effects cancel out at the solstice. However, Los Angeles has 2.8 times the winter insolation that Berlin has.

In comparison with the other alternatives, solar power has many advantages. The resources are ample, the environmental impacts are modest, and the costs of generation are manageable. The land requirements, while larger than fossil-fuel facilities, do not appear to be limiting. It can be convenient to put solar farms on abandoned agricultural land because it makes it easier to get the environmental permits. However, something must be done during the time that the sun is not shining. In principle, this could be addressed by battery storage, to which we turn next.

7.9 Batteries

The most important alternative energy storage approach so far has been pumped water storage, where pumps move water uphill to a reservoir. Later the water flows back down through turbines to generate power. The energy is stored as the gravitational potential of the reservoir water. The plant can buy electricity when the price is low and sell it back when the price is high. Biogas and wood chips can also be stored. Wood chips can be piled up at the power plant the way coal is. The principal component of biogas is methane, just as it is in natural gas, and it could use the same kinds of storage facilities that natural gas does.

However, these storage approaches are limited in scale. Batteries do not have fundamental scale limits, because they are made as a manufacturing process, like solar panels. Batteries are used on a small scale for storage in recreational vehicles (RVs), where it is common to combine solar panels with battery storage. When it is sunny, the solar panels provide the power to operate lights, fans, and pumps, and to charge the batteries. When it is cloudy or dark, the batteries take over.

7.9.1 Battery Grid Storage

Battery storage facilities for the electrical grid are just beginning to be developed. Figure 7.59 shows a battery plant that was built in 2016 by the Mitsubishi Electric Company. It has a power capacity of 50 MW and an energy capacity of 300 MWh, 6 hours at the rated power capacity. The area of the facility is 14,000 m². This is small compared to a solar farm with a comparable power capacity. For comparison, the Kayenta Solar Facility has a capacity of 37.6 MW and an area of 900,000 m². The plant uses sodium–sulfur batteries developed by NGK Insulators in Nagoya. Sodium–sulfur batteries operate at 300 °C, so they are not suitable for cars. One advantage of the sodium–sulfur battery is that sulfur and sodium supplies are abundant, sufficient to scale up production to any conceivable level. Sulfur is recovered from oil refining and sodium is produced through the electrolysis of table salt, which is sodium chloride. In 2017, a lithium-ion battery storage facility with an energy capacity of 129 MWh was built at Hornsdale, in the state of South Australia. The plant was built by the Tesla Company, which is also a leading electric vehicle manufacturer. The CEO, Elon Musk, said that the facility would be free if it was not completed in 100 days. Tesla met the deadline.

The price of these projects is not readily available, although it has been speculated that Mr. Musk would have lost 50 M$ if the facility had not been delivered on time. That would be $400/kWh for the plant. The most important cost for the

Figure 7.59 Battery storage facility with an energy capacity of 300 MWh in Buzen, on the island of Kyushu, in Japan. Credit: Mitsubishi Electric.

plant is the lithium-ion batteries themselves. Figure 7.60 shows the history of the price of lithium batteries. Prices fell by almost a factor of four from 2010 to 2016, to $1,000/kWh to $273/kWh.

It appears that lithium resources are adequate to scale up battery production for grid storage and for electric vehicles (EVs). Chile and Australia are the largest producers. The majority of lithium production comes from brines pumped up from wells. Figure 7.61 shows lithium evaporation ponds in the Atacama Desert of Chile. When the water is gone, lithium salts are left behind. To get a sense of the scale of the operation, each pool in the foreground is 270 m long by 150 m wide, or 4 hectares in area.

Cobalt is a different story. Cobalt is a component of the cathode in the majority of the lithium-ion batteries manufactured today. The Democratic Republic of the Congo dominates production, and it has been plagued by violence for decades. While acknowledging this problem, we will assume for discussion purposes that the challenges of scaling up the battery production for grid storage and EVs could be met one way or the other.

7.9.2   A Model Renewable Grid for California

To give a sense of how a battery grid storage system could work on a larger scale, we will develop a scenario for a grid based on the California Independent System

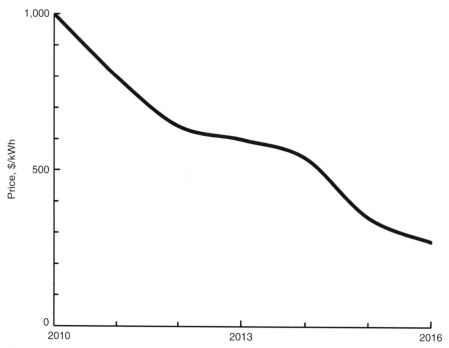

Figure 7.60 The price of lithium-ion batteries over time. The data come from a survey reported in Claire Curry, 2017, "Lithium-ion battery costs and market," *Bloomberg New Energy.*

Figure 7.61 Aerial view of the lithium brine pools of the Sociedad Quimica Mineral de Chile in 2014. These are in the Atacama Salar. *Salar* means "salt flat" in Spanish. Credit: Hemis/Alamy Stock.

Operator (CAISO) market that we discussed in the introduction to this chapter. California was early to adopt electricity in cities and farms compared with the rest of the United States. Most of this generation was hydroelectric power, primarily from Pelton wheels in the Sierras (Section 7.4.2). At the beginning of World War II, hydroelectric power supplied 80% of California's electricity. The state has limited coal resources, and when the Sierra hydroelectric plants could not meet the demand, it began building natural gas plants. In addition, two nuclear power stations were constructed on the coast, San Onofre 90 km south of Los Angeles, and Diablo Canyon 260 km north. The San Onofre plant was closed in 2012 and it is planned that Diablo Canyon will shut down in 2025. Figure 7.62 shows the different contributions to the supply over time. The state deregulated its electricity market in the late 1990s. The deregulation was botched and there were blackouts in 2000 and 2001. It did not help that there was a drought in 2001 that reduced the Sierra hydroelectric generation. The largest utility in the state, Pacific Gas and Electric Company, declared bankruptcy. Governor Gray Davis was recalled by the voters and Arnold Schwarzenegger was elected governor. Schwarzenegger is an actor best known for his role as the implacable Terminator. In California lore, the villains of the crisis were the traders at the Enron Corporation who manipulated the electricity market. However, the state was naïve to enter the pit assuming that trader mores would not apply. California also started the deregulation process with an inadequate in-state power infrastructure. Finally, the state adopted a market

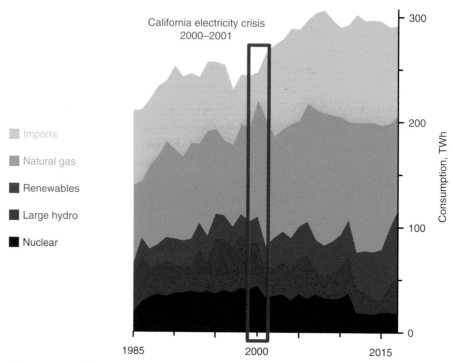

Figure 7.62 California electricity consumption. The source is the online California Energy Almanac.

design that was easy to manipulate. To give one example, out-of-state generators were allowed higher prices than in-state generators. This allowed swaps where an in-state-generator sold its electricity outside the state at the same time an out-of-state generator sold inside.

After the crisis, the major change was to import more power from outside the state. During the three years before the crisis, imports averaged 21% of demand. In the three years after the crisis, imports averaged 31% of demand. This dependence on outside power has continued to the present day. In 2017, 29% of the power was imported. The most important import is hydroelectric power from dams on the Columbia River that forms most of the border between Oregon and Washington. In recent years, there has been an emphasis on alternative sources, particularly in-state solar power. Table 7.1 shows the in-state generation shares in 2011 and six years later in 2017. The most important increase is the solar share, which rose from 1% in 2011 to 12% in 2017. This share was higher than for any country in the world in 2017. The closest international competitor that year was Italy at 8.5%. For one hour on Sunday, February 11, 2018, the solar share of California generation hit 60%. In spite of these large increases the natural gas share was little changed because the solar increase was mostly offset by the closure of the San Onofre nuclear plant, which reduced the nuclear share by 9%. The same offset will occur if the remaining nuclear plant, Diablo Canyon, closes in 2025 as planned. One thing

Table 7.1 California electricity generation shares.

	2011	2017
Natural gas	47%	44%
Nuclear	18%	9%
Large hydro	18%	18%
Renewables	17%	30%
Geothermal	6%	6%
Small hydro	4%	3%
Biomass and biogas	3%	3%
Wind	4%	6%
Solar	1%	12%

This table does not include imports. A small amount of production from oil and coal is included with natural gas. The data come from the online California Energy Almanac.

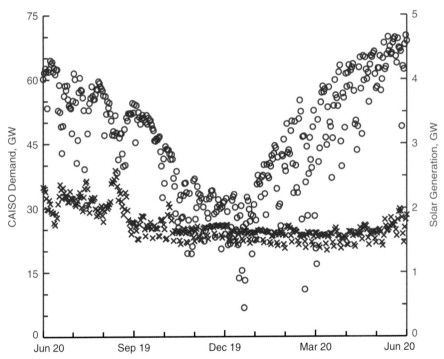

Figure 7.63 California daily average demand in the CAISO market and in-state solar generation for one year starting on June 20, 2017. The demand scale is 15 times the solar scale. The data come from CAISO.

that the table does not convey is that in-state hydro drops in a drought year. In 2015 the hydro share of in-state generation was 7% compared with 21% in 2011 and 2017.

Our scenario model grid includes solar power plants, batteries to supply electricity at night, and biomass and biogas generation for the cloudy periods. Figure 7.63

shows the daily average demand and the daily average solar generation in the CAISO market for one year starting in 2017 at the summer solstice, June 20. The demand plot shows a two-level structure. The higher level is weekdays, while the lower level is weekends. There are several sharp spikes in demand in the summer. These are from air conditioning on hot days. The solar and demand scales differ by a factor of 15 for comparison with scaled-up solar generation. There would be a great deal of surplus electric power available for most of the year. It is easy to discard this power at the inverters in the solar plants. This process is called *curtailment*. It is possible that uses could be found for this power. For California, one application could be desalting sea water, which requires $3\,kWh/m^3$. There are a few cloudy days in the late fall, winter, and early spring where the solar generation would be too small. In our model, this hole would be filled by biomass and biogas generation.

Our goal is to develop a cost model based on construction expenses of the solar and battery plants. We will neglect the costs for the biomass and biogas generators and the transmission-line costs. The construction expenses for the solar plants S in $/y can be written as

$$S = \frac{sdm}{ct_s} = \frac{sdm}{4y} \qquad\qquad 7.44$$

where s is the price of the solar plant capacity in $/W, c is the capacity factor, estimated to be 20%, d is the average annual demand in W, t_s is a plant lifetime, assumed to be 20 years, and m is a *solar multiple* that takes into account the fact some of the solar generation will be curtailed. We can write a similar expression for the construction costs of the battery plants B given by

$$B = \frac{bdT}{t_b} = \frac{bdT}{8y} \qquad\qquad 7.45$$

where b is the price of a battery plant in ¢/(Wh), t_b is the battery replacement time in years, assumed to be 8 years, and T is the number of hours of storage at the average demand. We add the solar and the battery costs and normalize the sum to the cost of a system with a solar multiple of 1 and no battery storage. This gives us a construction cost multiple M, given by

$$M = m + (b\,/\,s)\,T\,/\,2 = m + 0.2\,T \qquad\qquad 7.46$$

where we have taken the ratio $b\,/\,s$ to be 0.4/h, based on costs of 40¢/(Wh) for battery farms and $1/W for solar farms. In the model, solar generation is compared to the demand on an hourly basis. If solar generation exceeds demand, the batteries charge. If the demand exceeds the solar generation, the batteries discharge. Figure 7.64 shows examples of the daily variation in the battery charge levels. In (a), the charge is shown for a week during the spring when the solar generation exceeds the demand. In this case the batteries are charged to the maximum level

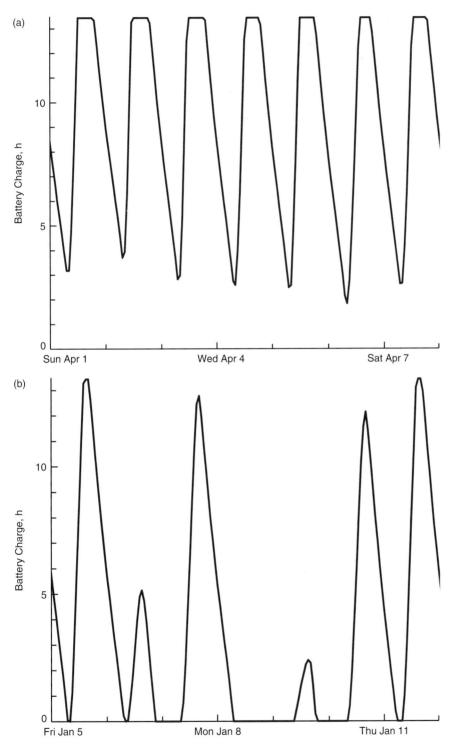

Figure 7.64 Modeled hourly battery charge during a week in the spring (a), and during a cloudy week in the winter (b). The solar multiple m is 1.5 and the battery storage time T is 13.4 h.

each day, and then are partially discharged during the night. In (b), the charge is shown for a cloudy week in the winter. On some days, the solar generation is not sufficient to fully charge the batteries and the batteries cannot meet the demand after dark. In these cases the excess demand is assumed to be met by biomass and biogas generation.

For each solar multiple, the battery storage time T is chosen so that the biomass and biogas share is 3%, which was the biomass and biogas share of California generation in 2017.

Figure 7.65 shows the results. The construction cost multiple has a broad minimum of 4.2 at a solar multiple of 1.5, where the required battery storage is 13.4 h. At lower solar multiples, the battery storage requirements increase quickly and this pushes the overall cost up. Taking into account the biomass and biogas generation, a solar multiple of 1.5 implies an average curtailment equal to 35% of generation.

Several issues are not addressed in the model. The costs for both battery farms and solar farms are declining, and the ratio could change in the future. Battery lifetime and solar panel lifetime could change. Battery losses are not included. There is no safety margin and there are no imports or exports. Other renewables

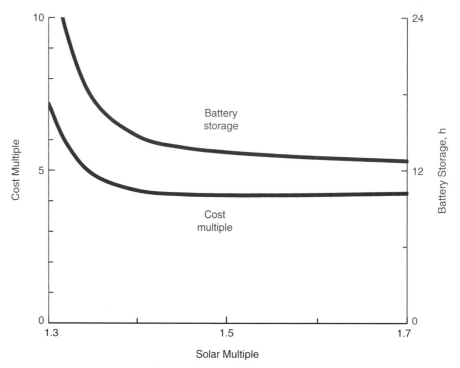

Figure 7.65 Construction cost for a California grid with solar plants, battery plants with a 3% biomass and biogas generation share. The cost is compared to a solar system with a solar multiple of 1 and no battery storage. The details for the calculations are available in an Excel workbook at energybk.caltech.edu.

like wind, hydroelectric and geothermal power are not included. The model does not consider how to establish the frequency of a grid based on solar power. Today the grid frequency is stabilized through the combined angular momentum of the turbines. In a grid without rotating machinery, the frequency would need to be established in a different way. One possibility would be to synchronize the clocks of the electronic inverters that convert the DC output of solar modules and batteries to AC. This could be done with GPS receivers, which include precise clocks.

The grid is technically feasible, although it would be expensive. Achieving a solar multiple of 1.5 would require 13 times the solar capacity that was in the CAISO market in 2017. The batteries would cost more than the solar panels. In addition, the model does not make an allowance for the possibility that people might increase demand by charging electric cars. Moreover, because electricity has been relatively expensive for decades in California, Californians often use natural gas for appliances that are electrical elsewhere, like space heaters, water heaters, stoves, and clothes driers. If they switch to electrical appliances to reduce the use of natural gas, electricity demand would increase. Finally, we have not included the additional costs associated with biomass and biogas generation. This capacity would be significant because these generators would need to handle the entire night-time load when the batteries are at the minimum charge level.

7.10 Prospects

Because the fossil fuel supply is finite, we can be confident that there will eventually be a transition back to the alternatives, but we do not have enough information yet to project when alternatives will provide the majority of the world's primary energy. There is a long way to go; the alternatives share of primary energy was 17% in 2017. There are important questions that must be resolved. Most of the alternative sources generate electricity, but electricity itself was only 41% of the primary energy in 2017. While wind and solar capacity have been expanding rapidly, we have seen that high wind and solar capacity are associated with high prices for European residential customers. Electrification of transportation has just begun, and battery grid storage is at the demonstration stage. The success of battery grid storage and electric vehicles (EVs) depends critically on the evolution of the price of batteries. It is foreseeable that the manufacturing learning curve will make batteries cheap enough for EVs and for balancing wind and solar generation. But it is not guaranteed. Finally, the scenario model grid in the last section applies to California only. Many countries have little sun in the winter, many countries do not have California's supply of land for solar farms and many countries do not have California's financial capabilities. Alternative energy solutions are local rather than global.

Concepts to Review

- Evolution of the electricity share of primary energy
- Thomas Edison and the electricity grid
- Transformers in transmission lines
- AC generators
- Fossil-fuel generators
- Hydroelectric generation history
- Pelton wheels
- Francis turbines
- Nuclear weapons
- Nuclear generation history
- The Chernobyl disaster
- The Fukushima disaster
- The California Geysers Field
- Geothermal energy in Iceland
- The *onsen* culture in Japan
- Windmills
- Wind generation history
- The Betz limit
- Solar cell efficiency limit
- Solar generation history
- Latitude effects on generation
- Sodium–sulfur grid storage batteries
- Lithium-ion grid storage batteries
- Construction cost model for a solar grid with battery storage

Problems

Problem 7.1   Transmission line losses

Most long-distance transmission lines operate with three wires that carry currents that differ in phase by 120°. We will calculate the losses in a three-wire line transmitting a total of 400 MW that uses 3M's 795-kcmil composite conductors. Locate the 3M product information for this conductor online and find the resistance in Ω/km for the 795-kcmil conductor at 50 °C.

a. For a voltage supply of 345 kV AC, calculate the AC current in amperes on one wire.
b. What percentage of the input power would be absorbed in a 1,000-km line? Note that this fraction could be cut in half by running half as much power on the line.

Table 7.2 Parameters for the Kirkwood Pelton wheel.

Net head h	382 m
Volume flow Q	13.68 m³/s
Turbine speed f	327.7 rpm
Turbine diameter D	2.321 m
Mechanical output power P_m	46.75 MW

Matthew Gass, the engineer who led the development for the Pelton wheels in the Hetch Hetchy system, kindly provided this data.

Problem 7.2 The Kirkwood Pelton wheel

The Kirkwood power station is supplied by the Hetch Hetchy Reservoir in the California Sierras. A Pelton wheel was installed with the measured parameters in Table 7.2.

a. Calculate the jet velocity v in m/s.
b. Calculate the ratio φ of the bucket velocity u to the jet velocity.
c. Calculate the hydraulic power P_h in MW.
d. Calculate the turbine mechanical efficiency, η_m.

Problem 7.3 Rail storage

Rail storage is a storage system based on the gravitational potential energy of loaded rail cars proposed by the Ares North America Company. We will calculate the expected performance of a rail storage system based on a train with four cars, where each car carries a container of gravel with internal dimensions 10 m long by 2 m wide by 5 m high. The gravel density is 1.6 t/m³.

The city of Needles is on the California side of the Colorado River that forms the border with Arizona. It was originally a railroad town, but today it is on Interstate 40. For many people, Needles is the first city they see when they come to California. Often they do not forget Needles, because Needles is hot, very hot. The average daily high in the middle of the summer is 43 °C. West of Needles, I-40 heads into the Mojave Desert towards Barstow. The road climbs gradually for 50 km, with an average gradient of 1.4%. We will assume that a rail storage line has been built along this ascending section of the highway.

a. Calculate the gravitational energy in GJ stored in the gravel by moving a train from the bottom to the top of the rail line.
b. In the original Ares proposal, there was a system for leaving gravel containers off the cars at the top and the bottom of the rail lines. Allowing for area at both the top and the bottom, calculate the area storage density in GWh/km² for the gravel in the containers, neglecting conversion losses and the thickness of the container walls.

Problem 7.4   An eclipse in California

Solar generation by homes and businesses reduces the demand that electricity markets see. A solar eclipse on August 21, 2017 gives a chance to estimate this self-generation. Locate the CAISO website and download the generation statistics for that day. The demand can be calculated by summing all of the generation components.

a. Identify the hour of the eclipse in the morning. At that hour, the commercial solar PV generation dips. However, the demand rises because the self-generation drops, presumably by the same factor that the commercial generation does.

b. Use the dip in commercial solar PV generation and the rise in demand at the time of the eclipse to estimate the self-generation at 2pm that afternoon.

Further Reading

- Aldo da Rosa, 2012, *Fundamentals of Renewable Energy Processes*, 3rd Edition, Academic Press. This is an excellent resource for the fundamental physics and chemistry behind different renewable energy sources. Our derivation of the Betz limit is in the style of da Rosa's.
- Andritz Hydro, 2015, "Animation of a Francis Turbine." Andritz is a large engineering company with its headquarters in Graz, Austria. Available online at www.youtube.com/watch?v=S3MQJSDoTuw
- Andritz Hydro 2015 "Animation of a Pelton turbine." Available online at www.youtube.com/watch?v=Qwh6N_PSZ_Q
- Lee Buchsbaum, 2016, "Dusseldorf's Lausward Power Plant Fortuna Unit wins POWER's highest award," *Power* magazine.
- Ronald DePippo, 2012, *Geothermal Power Plants*, 3rd Edition, Butterworth-Heinemann. This book has a detailed discussion of California's Geysers Field.
- Matthew Josephson, 1959, *Edison, A Biography*, History Book Club. Thomas Edison contributed much more than the electrical grid. He developed the first industrial research laboratory. He received over 1,000 American patents. His inventions included audio recording, movies, a rechargeable electric battery, and the carbon microphone that was standard in telephones for a hundred years.
- MIT, 2015, *The Future of Solar Energy*. This gives much more technical information than we do here.
- Ruth Sime, 1996, *Lise Meitner, A Life in Physics*, University of California Press. Lise Meitner's career is a recapitulation of the early history of nuclear physics. She was a pupil of Ludwig Boltzmann, and a colleague and friend of Max Planck and Albert Einstein. Meitner was born in 1878 in Vienna but came to Germany to study. She worked her way up the academic ladder to become professor of physics in the Kaiser Wilhelm Institute. Meitner had initially led Otto Hahn's project, but she had to flee Germany because her parents were Jewish. She was living in Sweden at the time of Hahn and

Strassman's experiment. Hahn received the Nobel Prize in chemistry in 1945. The Nobel committees do not divide a single prize into more than three parts. With the benefit of hindsight, it would have been appropriate to award Hahn and Strassmann the Nobel Prize in chemistry and Meitner and Frisch the Nobel Prize in physics. Historians give credit to all four scientists.

- James Williams, 1997, *Energy and the Making of Modern California*, University of Akron Press. A detailed discussion of the California energy supply. It ends before the California Electricity Crisis of 2000 and 2001.

8 Stationary Demand

Thanks to central air-conditioning
many in these parts will remember the brutal
heat wave as nothing more than a minor inconvenience.
It is a life-and-death matter only for those who
can't afford the luxury of an air-conditioned existence.

Jerry Heaster, journalist on the 1980 Missouri
heat wave that killed 1,265 people [1]

8.1   Introduction

Stationary demand is the energy that is not used for transportation. It includes homes, offices, factories, mines, and farms. Stationary demand was 77% of total primary energy demand in 2017, so it is much larger than transportation demand. The share has been stable; it was also 77% in 2000. Figure 8.1 shows the history of world stationary demand. The character of demand differs from supply. Much of demand comes from families and small companies, while energy suppliers are typically large companies. For this reason, we will be plotting demand on a per-person basis. Demand data are not as good as supply data. The demand data series do not go back far in time and there are gaps. Lights, heaters, refrigerators, and air conditioners have had significant efficiency improvements in recent years, but changes in stationary demand are not dramatic. There is nothing in the history of stationary demand comparable in scale to the fracking revolution or the explosive growth of solar PV. The International Energy Agency classifies stationary demand as *buildings, industry,* and *other,* a category that includes agriculture and hydrocarbons used as petrochemical feedstock for plastics and fertilizer. In Figure 8.1, the IEA industry and other categories are combined into one category, industrial. Petrochemical feedstock was 10% of industrial demand in 2017. Building demand on a per-person basis has been almost flat, with an annualized decline of 0.2%/y from 2007 to 2017. It was 621 kgoe/p in 2017. On the other hand, per-person industrial demand has been growing at 1.3%/y, reaching 682 kgoe/p in 2017.

---

[1] Quotation from Ackermann, Marsha. *Cool Comfort: America's Romance with Air-Conditioning,* 2002, Smithsonian Institution.

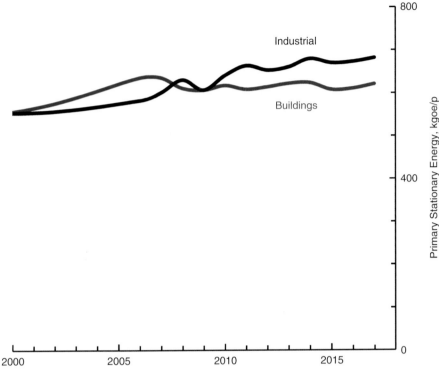

Figure 8.1 World stationary energy demand on a per-person basis from 2000 to 2017. The energy data come from various editions of the IEA *World Energy Outlook*. The energy calculations are done on a primary energy basis, with the electricity generation efficiency taken to be the BP *Statistical Review* value, 4.4 TWh/Mtoe, close to 38%. The population data come from the UN Population Division.

8.1.1 Passive Cooling

One factor that has restrained building demand is the development of compact fluorescent lights and LED lamps that are much more efficient than incandescent bulbs. An offsetting factor is the growing use of air conditioning. Buildings have been heated with wood fires for thousands of years, but buildings have traditionally been cooled passively by architectural design. For example, in the American Southwest, where it is dry and the daily temperature swings are very large, buildings are built with thick walls. The *adobe* bricks are made of mud and straw. An example is the church shown in Figure 8.2. This church was built at the end of the eighteenth century when the region was part of the Spanish Empire. The town of Taos can be hot during the day, but usually the humidity is low and the sky is clear, so the temperature drops rapidly at night. The average temperature is comfortable. The thick walls increase the thermal time constant of the building so that the temperature inside stays close to the average outside temperature.

Passive cooling is limited, however. If the climate is humid and cloudy, the outside temperature drops little at night. The humidity is itself a problem because it

Figure 8.2 The San Francisco de Assisi Church in Ranchos de Taos, New Mexico entrance (a) and rear (b). Architecturally, this is an exquisite building. Photographs by Dale Yee.

leads to mold, which causes allergies and breathing problems. Historically people avoided moving to places that are hot and humid like the American South. It is also difficult to cool multi-story buildings passively. The top floors of the beautiful Haussmann six-story walk-up apartments in Paris were death traps for the elderly in the 2003 European Heat Wave. Air conditioning solves these problems. In addition, the systems have filters that block allergens. The IEA reckoned that the world investments in air conditioning in 2016 amounted to 80 G$. Air conditioning does increase the electricity demand in the summer. From a broader perspective, when people move from colder areas to warmer areas, the increase in energy use for cooling in the summer is offset by a reduction in heating energy in the winter. In addition, there is a shift from direct fossil-fuel burning to electricity.

8.1.2 Temperature and Mortality

Figure 8.3 shows how mortality varies with the daily temperature for several countries. In the figure, mortality is normalized to the temperature with the minimum mortality. The most important conclusion from the graph is that people can adapt to a wide range of temperatures. The minimum mortality for Canada is at 15 °C, compared with 29 °C for Thailand, a difference of 14 °C. The temperature ranges in Italy and the United States are similar, but the increase in mortality away from the optimum temperature is much sharper in Italy than in the United States. In the US,

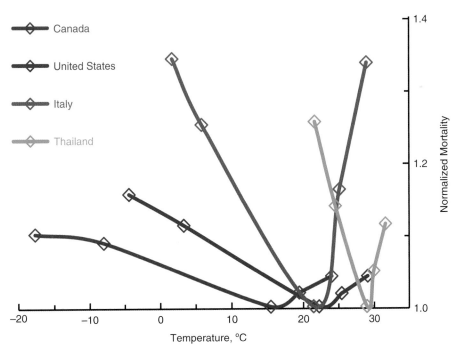

Figure 8.3 Normalized mortality versus temperature for selected countries. The data come from Yuming Guo et al., "Global variation in the effects of ambient temperature on mortality," *Epidemiology*, vol. 25, 2014, pp. 781–789. The study followed 10 million deaths in the United, States, 3 million in Canada, 2 million in Thailand, and 1 million in Italy. The points at the end of each curve are the extreme daily average temperatures that occur 1% of the time, averaged over a group of cities in each country. The adjacent points are the 10th percentile points.

where 87% of the houses had air conditioning in 2011, there is little excess mortality associated with heat. This does require electricity – 18% of residential electricity consumption in the United States goes to air conditioning. More generally, we can associate high electricity use with a flattened mortality curve. In 2016, the per-person electricity generation for Canada was 2.1 kW/p, for the US 1.5 kW/p, Italy 0.5 kW/p, and Thailand 0.3 kW/p. Italian residential electricity prices are more than twice as high as American and Canadian prices and high prices discourage electricity use.

Even in the United States, where heating and cooling is available in the great majority of homes, there is still a strong seasonal component to mortality. Figure 8.4 is a plot of the normalized mortality against the daily average heating degrees in °C for the United States. The normalized mortality here is the ratio of the mortality for a month compared to the August mortality, which is the lowest. Heating degrees are calculated as the number of degrees the average daily temperature in a state is below the reference temperature of 18.3 °C (65 °F). The heating degree-days for the entire country are calculated from the state heating degree-days by weighting with the state populations. The mortality rates are calculated on a daily basis to allow for differences in the number of days in a month and offset

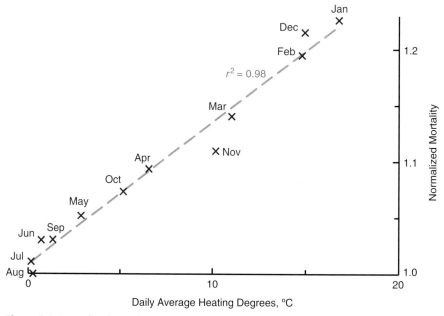

Figure 8.4 Normalized mortality ratio versus daily average heating from 1979 to 2013 for the United States. The data for daily deaths from all disease for 49 million people was kindly provided by Professor David Phillips at the University of California at San Diego. The heating degree days are taken from the *EIA Monthly Energy Review*.

one week to allow for a delay in the effect of the weather. The relationship between heating degrees and the mortality is linear, and statistically it explains almost all of the variation in the mortality ratio. The value of $r^2 = 0.98$. The January normalized mortality was 1.23.

8.1.3   OECD vs. Non-OECD Stationary Demand

Figure 8.5 is a comparison of the OECD and the non-OECD stationary demand. In 2017, people in the OECD countries used 3.9 times as much energy for buildings as people in the non-OECD countries. OECD buildings are better lit, better heated, and better cooled. Homes and office buildings in the OECD countries are more comfortable to live in and work in than in the non-OECD countries. This ratio has been dropping; it was 5.1:1 in 2000. Industrial demand has evolved differently because manufacturing has shifted from OECD countries to non-OECD countries where wages are lower and environmental restrictions are fewer. OECD industrial demand has been declining at a 10-y annualized rate of 1.1%/y, while non-OECD demand has been rising at 3.3%/y. OECD national policies often encourage the shift to the non-OECD countries through tariff reductions and pollution policies. The negative part of these policies is that industrial workers in the OECD countries

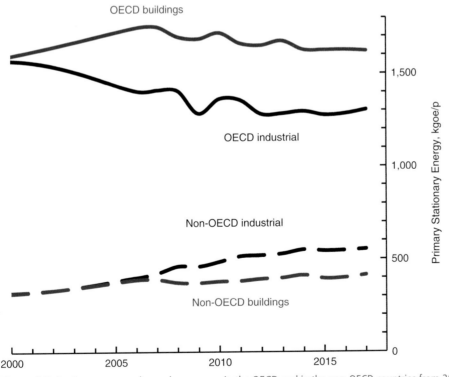

Figure 8.5 Stationary energy demand per person in the OECD and in the non-OECD countries from 2000 to 2017. The data come from various editions of the IEA *World Energy Outlook*.

lose their jobs, and the losses are concentrated in factory towns. In addition, the pollution does not disappear. It is simply transferred to the non-OECD countries. On a per-person basis industrial demand is still 2.4 times larger in the OECD countries as in the non-OECD countries. The OECD countries remain major manufacturing powers and they dominate the design of manufactured products.

We argued in the last chapter that because the alternatives are primarily electrical, a shift in demand to electricity is a prerequisite for a transition from fossil fuels to alternatives. Figure 8.6 shows the electricity shares of building demand and of industrial demand over time. The electricity share for buildings has been rising, from 50% in 2007 to 56% in 2016. For buildings it is often attractive to get electricity from the grid rather than burning fossil fuels on site, particularly for oil and coal, which need on-site storage. In addition, in many places there are pollution restrictions on burning oil and coal. However, natural gas can be competitive with electricity because it is distributed by pipeline and because it burns cleanly. In addition, heating a building with electricity presents a risk in the winter because a storm can take down the grid. Compared with buildings, the industrial electricity share has been smaller and more stable, 39.3% in 2007 and 39.8% in 2017. In a factory, it often cheaper to burn hydrocarbons or coal to heat water in a

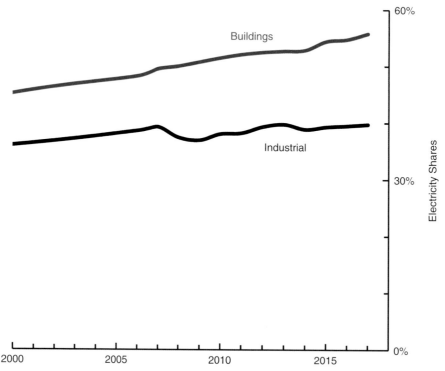

Figure 8.6 Electricity shares of building demand and of industrial demand from 2000 to 2017, calculated on a primary energy basis. The data are from various editions of the IEA *World Energy Outlook*.

local boiler than to use electricity. The heat can be distributed either as hot water or as steam in pipes.

In this chapter, we consider the different components of stationary demand. Lighting has had remarkable efficiency improvements, starting with candles, and moving on to kerosene lanterns, incandescent light bulbs, fluorescent light bulbs, and more recently light-emitting diodes (LEDs). An important component of industrial demand is for rotating machinery. The dominant machine is the induction motor, invented by Nikola Tesla. Heating and air conditioning are essential for many manufacturing processes and are important for productivity in offices and for comfort in homes. Finally, we consider information technology, a new component of stationary demand that has had large efficiency improvements, but nevertheless has seen increasing energy demand.

8.2 Lighting

Modern lighting allows us to work and play inside and outside, day and night. Artificial lighting is particularly important at high latitudes, where it is mostly dark during the winter. It can be difficult to get a feeling for how different life was before modern lighting. Sometimes a period movie can do it. Films based on Jane Austen novels show dinners lit by candles. The dining rooms are quite dark. Recall from Section 2.2.4 that by definition, the standard sperm candle had a luminous flux of 4π lumens. Large modern candles have a comparable output. The unit of luminous flux density is the lux, equal to $1 \, lm/m^2$ and abbreviated lx. The standard candle would have a flux density of $1 \, lx$ at a distance of $1 \, m$. A guest at a candle-light dinner would see a lighting level of $10 \, lx$. The lighting level in a dining room today might be $100 \, lx$, and it can be controlled by a switch. An office might have a lighting level of $800 \, lx$.[2]

Lighting is embedded in virtually all buildings and industrial facilities, and this makes it difficult to separate lighting demand from other stationary demand. The lighting data that exist are quite limited compared to energy production statistics. What is available is a hybrid of survey data and modeling. This makes it somewhat speculative. The best-known series is one compiled by Roger Fouquet, a professor at the London School of Economics. Figure 8.7 shows lighting production in lumens per person for the UK. There is an enormous increase in luminous flux per person between 1850 and 2000, 940:1. There is evidence of a shock in the data starting around 1973, the year of the Yom Kippur War and the Arab Oil Embargo.

---

[2] The natural outdoor illumination range vastly exceeds the artificial indoor lighting range. The flux density from sunlight is $100 \, klx$, while the light level on an overcast night is a billion times lower.

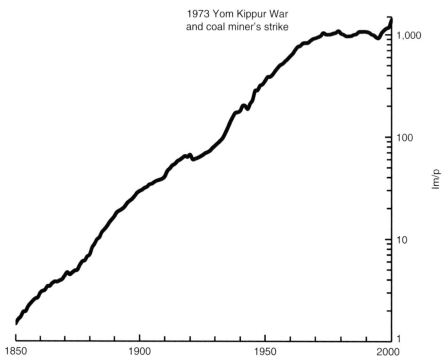

Figure 8.7 Lighting in the UK. The scale is logarithmic. The luminous flux data are from Roger Fouquet, 2008, *Heat, Power, and Light*, Edward Elgar Publishing. The population data come from the UK Office of National Statistics.

In the UK, the coal miners' union sensed an opportunity and declared a strike. The strike brought down the Conservative government of Edward Heath in the following year. In numbers, from 1850 to 1973 the annualized growth rate was 5%/y, while from 1973 to 2000 the growth was only 1%/y.

Lighting technology has had large improvements in efficiency over the years and this makes it an excellent subject to consider the Jevons Paradox. This was Stanley Jevons' observation in his 1865 book, *The Coal Question*, that efficiency improvements are not necessarily associated with reductions in energy use (Section 1.3.2). Figure 8.8 shows the efficacy plotted against lighting primary energy consumption for the UK from 1850 to 2000. From 1850 to the end of World War I in 1918, the efficacy improved by a factor of five, from 0.13 lm/W in 1850 to 0.69 lm/W in 1918. This was a result of the technology changing from candles to kerosene lanterns and coal gas lighting. However, this efficiency improvement did not result in less energy being used for lighting. On the contrary, lighting primary energy per person increased 10:1. There is evidence of a shock around 1918, the end of World War I. This is when electric lighting began to be significant. From 1918 to 2000, efficacy improved by a factor of 40:1. However, lighting primary energy per person in 2000 was 33% larger than it was in 1918.

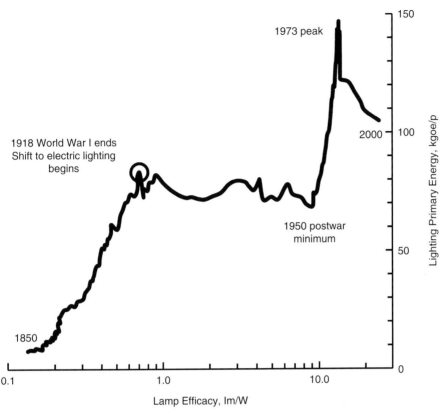

Figure 8.8 Lamp efficacy plotted against per-person lighting primary energy for the UK from 1850 to 2000. Efficacy and lighting technology shares from Roger Fouquet, 2008, *Heat, Power, and Light*, Edward Elgar Publishing. Fouquet's data show an increase in per-person energy in the 1950s and 1960s. This was a time of a post-war building boom in the UK. After 1973 the primary energy falls, but not back to the 1950 minimum.

8.2.1   Nick Holonyak and His Light-Emitting Diode (LED)

In the 1990s, another lighting technology emerged, light-emitting diodes (LEDs). Nick Holonyak is credited with the invention of the first practical light-emitting diode in 1962. Holonyak was working at the GE laboratories in New York. He made his LED on a gallium-arsenide-phosphide crystal. The light was red. Red LEDs became popular for indicator lamps in electronics but they were not suitable for general lighting. For this, a blue LED was needed. Making a blue LED turned out to be quite difficult and many researchers failed in their attempts over the next thirty years. Finally, in 1993, Shuji Nakamura, working at Nichia Chemical Company in Japan, succeeded. The crystal material was gallium nitride. Nakamura won the Nobel prize for his invention in 2014. White light can be produced from a blue LED by covering it with a yellow phosphor (Figure 8.9). The phosphor is designed so that some of the blue light is absorbed by the phosphor and some makes its way

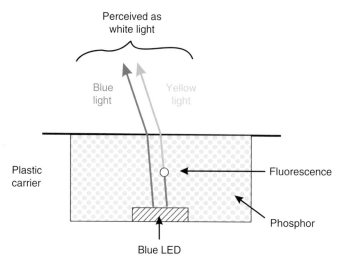

Figure 8.9 Cross section of a white LED lamp with a blue LED and a yellow phosphor.

through the phosphor unaltered. The light that is absorbed causes the phosphor to fluoresce at a longer wavelength, yellow. The eye perceives the combination of the unaltered blue light and the yellow fluorescence to be white.

Blue LEDs can also be used directly. For example, some companies are developing indoor farms that grow vegetables in shipping containers (Figure 8.10). Red and blue light are good for photosynthesis, and these colors can be efficiently provided by LEDs. Plants actually do not need green light. The leaves reflect green light well and that is why they appear green. The plants are grown hydroponically, without soil, and the CO_2 concentration can be increased to the level that provides the best growth rate. No pesticides are needed because there are no insects. Water losses are eliminated. No ordinary agricultural land is needed at all; the shipping containers can be stacked near city markets.

Lamps are characterized by a color temperature that is based on its correlation with a black-body radiator. The color temperature does not refer to the operating temperature of the LED, which might be 50 °C, depending on how good the engineers are at getting the heat out. Rather it is the temperature of an incandescent light whose color spectrum correlates best with the LED lamp output. A lower temperature lamp, *warm* lighting in the language of interior designers, has a yellowish tinge. These lamps provide pleasant lighting for homes. A typical lamp for a residence might have a color temperature of 2,700 K. *Cool* lighting, with a color temperature of 4,000 K, is closer to natural light.

8.2.2 LED Efficacy Limits

We can calculate the maximum possible efficacy of a white lamp with a given color temperature. For this we need the two curves shown in Figure 8.11. The standard luminosity curve shows the response of our eyes to different colors of

Figure 8.10 Container farming under LED lights. Courtesy of Freight Farms.

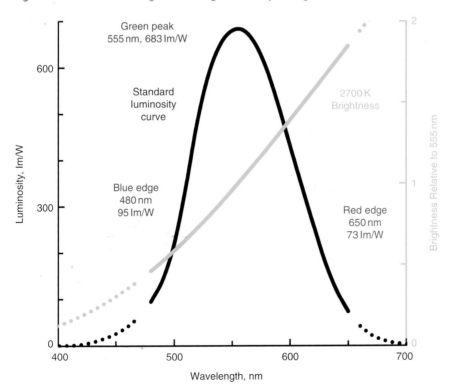

Figure 8.11 Functions for calculating luminous efficacy limits for lamps. The standard luminosity data come from the National Physics Laboratory in the UK.

light. Our eyes are most sensitive to green light. Possibly this was adaptive for our tree-climbing ancestors in a world of green leaves. The peak response is defined to be 683 lm/W at a wavelength of 555 nm. For this definition the power is the power in the light itself, rather than the power supplied to a practical lamp. This definition for the lumen replaced the sperm candle definition that we discussed in Section 2.2.4. As the wavelength gets longer or shorter, our sensitivity becomes poorer. For an ideal source at 650 nm, red light, the response is 73 lm/W, while at 480 nm, blue light, the response is 95 lm/W.

The other curve is the brightness as a function of wavelength for a 2,700-K lamp. *Brightness* here is light intensity, W/m^2, on a per-unit wavelength basis calculated from Planck's radiation formula. In the figure, we have normalized the brightness to the value at 555 nm, the peak of the standard luminosity curve. To find the maximum efficacy, we calculate the average of the standard luminosity curve weighted by the brightness curve. The 2,700 K brightness curve is weighted heavily toward to the red, where our eyes are less sensitive and this reduces the maximum efficacy. The band edges for the calculation are 480 nm and 650 nm, where the standard luminosity has fallen to just above 10% of the maximum.

The result of the calculation is 407 lm/W for the maximum efficacy of a 2,700-K lamp. The STS-DA1-4854 R70 from the Nichia Company is rated at 35.8 lm with a voltage of 2.73 V and a current of 65 mA. The efficacy is 202 lm/W. This is half the maximum possible efficacy. The development of the LED means that the long march of efficacy improvements that started at 0.1 lm/W in 1850 is near its end.

8.2.3 Prospects for LEDs

It is likely that the transition to LEDs will be as revolutionary as the shift to electric lighting a century ago. LEDs seem to be following the same kind of manufacturing learning curve as solar cells and we can anticipate that they will continue to get cheaper. Lighting designers use the word *luminaire* for the assembly that includes the fixture and the lamps. LED luminaires can be thin and flat compared with incandescent and fluorescent luminaires. The LED chips can be mounted efficiently by pick-and-place robots on circuit boards inside the luminaires so that the luminaires become permanent fixtures with no replaceable parts. This encourages interior designers to use large numbers of LED luminaires in new buildings. In addition, LEDs are the light sources in most computer displays and televisions. LEDs have much better shock resistance than incandescent and fluorescent bulbs and they are much more reliable. Incandescent lamps typically last a thousand hours and fluorescent lamps last a few thousand hours. With LEDs, 50,000-hour lifetimes are feasible. This makes them suitable for traffic lights and railroad signals.

The LED technology makes new integrated lighting products possible. An example is the solar sensor light shown in Figure 8.12 that combines a motion detector,

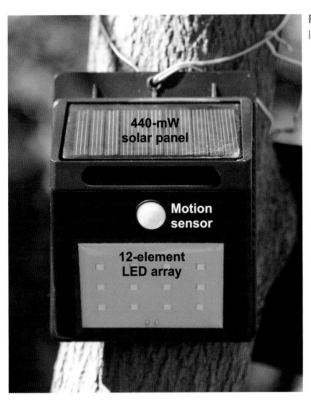

Figure 8.12 A solar sensor LED lamp. Photograph by Dale Yee.

a solar panel, a lithium-ion battery, and a flat luminaire with 12 LEDs. The solar panel charges the battery during the day and allows the device to sense night. The lamp lights up a walkway at night when the motion sensor detects a person approaching. Several vendors sell competing versions for $5. Previously, outdoor light fixtures required expensive underground wiring.

Integrated devices like the solar sensor lamp present challenges for energy accounting and it is appropriate to maintain a skepticism about lighting demand statistics. The lamp includes an electricity source, the solar panel, storage in the form of the lithium-ion battery, and an LED load. However, the electricity supply, storage, and demand are not accounted for anywhere. An earlier lighting system for a walkway based on underground wiring would appear as a load to the local electricity grid and would be included in residential demand. The solar sensor lamp, on the other hand, is isolated from the grid. The energy consumption that could be measured is the energy that is consumed in manufacturing the device. This would be reckoned as industrial demand, possibly in a different country from where the lamp is set up.

Modern lighting has revolutionized our lives. The most remarkable aspect of lighting technology has been the spectacular increase in efficacy over time. LEDs are a thousand times as efficient as a candle, and they are approaching physical limits. In energy terms, the cost of lighting is modest. The lighting energy consumption shown in Figure 8.8 represented only 3% of British primary energy consumption in 2000. The British lighting data give perhaps the best example of the Jevons Paradox. In spite of the large improvements in efficacy, lighting primary energy per-person is larger than it was a hundred years ago.

8.3   Electric Motors

Lighting became more much useful when the technology changed from lamps based on combustion to lamps based on electricity. Something similar happened with motors. A photograph can give us a sense of what early industrial work was like. Figure 8.13 shows a student machine shop in the 1890s. Twenty students are at work at their benches. The power for the lathes comes from a single steam engine outside the room. The power is distributed to each bench by a system of belts. The speed of each lathe is controlled by selecting wheels for the belts. In a modern machine shop, there are no belts. At each bench there would be a variety of different machine tools, each with its own variable-speed electric motor. Compared to the modern shop, the steam-powered shop is expensive, inefficient, inflexible, unreliable, and hazardous.

Figure 8.13 Student machine shop in the 1890s at the California Institute of Technology in Pasadena, California, then called the Throop Polytechnic Institute after its founder, businessman Amos Throop. Credit: California Institute of Technology Archives.

The goal of this section is to explain how electric motors work. Recall that the electric forces between charges are repulsive if the charges have the same sign, and attractive if one charge is positive and one charge is negative. It is difficult to use electrostatic forces to make rotating machinery, and this has been achieved only for small motors. Most electric motors are based on magnetic forces. The mathematical description of magnetic forces is more complicated than electric forces. The magnetic force is given by the Lorentz force law, which can be written as

$$\mathbf{F} = I\ \mathbf{L} \times \mathbf{B} \qquad\qquad 8.1$$

where **F** is the force in newtons, abbreviated N.[3] In this formula, force is a vector and it is expressed as a cross-product. We will use bold-face non-italic font for vectors. Figure 8.14 shows the geometry. There is a straight wire carrying a current I in amperes, abbreviated A. In the formula **L** is a vector whose direction is the direction of the wire, and whose length is the length of the wire. The magnetic flux density is given by the vector **B**. The units of magnetic flux density are teslas, abbreviated T. The tesla unit is in honor of Nikola Tesla, the Serbian-American electrical engineer who invented the induction motor. In more fundamental units, $1\,\mathrm{T} = 1\,\mathrm{Vs/m^2}$. Strong magnets have B-fields of around $1\,\mathrm{T}$. Only the component of the wire direction that is perpendicular to the B-field contributes to the force. The force is at right angles to both the magnetic field and the current carrying wire.

8.3.1   The Universal Motor

Figure 8.15 is a drawing of a *universal* motor, called that because it can use either an AC or a DC supply. Many power tools use universal motors. In the figure, there is a loop of wire with a circulating current I that is provided by a power supply.

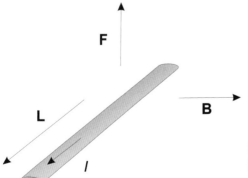

Figure 8.14 Geometry for the Lorentz force law. The direction for **L** and the sign for I are chosen to be consistent.

---

[3] We have not discussed the origin of the magnetic force. At a fundamental level the magnetic field is best thought of as a relativistic correction to the electric field from moving charges. This is beautifully described in Edward Purcell and David Morin, 2013, *Electricity and Magnetism*, 3rd edition, Cambridge University Press. This may be the best physics textbook ever written.

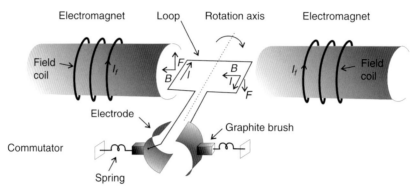

Figure 8.15 The universal motor, a popular design for power tools.

At the end of the loop, there is a pair of electrodes that make electrical contact to the supply through a pair of graphite blocks called *brushes* that are held against the electrodes by springs. As the loop rotates, the electrodes swap the brushes they make contact with, reversing the current. This switching process is called *commutation* and the switching assembly is called a *commutator*.

The loop lies between the two poles of an electromagnet with a field coil with a current I_f that establishes a B-field across the loop. In a DC motor, permanent magnets can be substituted for the electromagnet. There are forces associated with the right and left sides of the loop. The force is downward on the right and upward on the left. The two forces result in a torque that causes the loop to rotate. The rotation continues until the left side rises to the top and right side descends to the bottom. At this point the current in the loop switches direction and the forces change sign so the rotation continues. One advantage of this motor is that the torque can be controlled by the loop current I. Turn I up and the speed goes up. One disadvantage of the design is that the brushes grind down and have to be replaced. In addition, the commutation process makes sparks that are a source of radio static. Usually there will be magnetic material inside the loop to enhance the B-field. This is not shown in the figure. The figure shows a single loop. However, we can use a coil, with the torque proportional to the number of turns. This ability to create a large torque by adding more turns is a fundamental reason why magnetic forces are so useful in rotating machinery.

8.3.2   Nikola Tesla and His Induction Motor

The workhorse motor in buildings and industry today is the three-phase AC induction motor. The invention of the induction motor was one of the great intellectual achievements of the nineteenth century. The advantage of the induction motor is that it has no commutator, and this means it needs little maintenance. Early induction motors had a nearly fixed rotation speed, but modern switching electronics allows them to operate over a wide range of speeds, so that they are suitable for electric vehicles and locomotives. The induction motor was invented in 1877 by Nikola Tesla, a Serbian immigrant engineer working in New York City (Figure 8.16). Tesla was born in Smiljan, a town in the Croatian Military Frontier, a territory of the Austro-Hungarian Empire. Tesla's ancestors had fled from Serbia, then a part of the Ottoman Empire, 150 years earlier. He attended the Joanneum (now the Graz University of Technology), in Graz, Austria. He studied mathematics and physics there but he spent so much time gambling that he did not finish his degree. In 1881, he got a job as a draftsman in the Central Telegraph Office in Budapest. In later life, Tesla recalled that he began thinking about AC motors while he was in Budapest. In 1882 he moved to Paris to take a job at the Société Electrique Edison. In 1884, he immigrated to New York and began working at the Edison Machine Works. He continued to work on AC motors in his spare time, but was not able to interest anyone at the Edison companies in them. In 1885 he left Edison to go out on his own. By the fall of 1887, he had a working induction motor to show his investors and he filed for the patent at that time.[4] He sold

Figure 8.16 Nikola Tesla, inventer of the induction motor (1856–1943). Tesla became an American citizen in 1891. He was a friend of the noted author Mark Twain. In addition to motors, Tesla was also interested in radio communications. His work in radio communications was supported by J. P. Morgan.

---

[4] Primacy in the invention of the induction motor is contentious. In Europe, it is common to credit physicist Galileo Ferraris, who was working in Turin, Italy, with the invention. The two men developed their motors independently over extended periods of time that overlapped. By 1885, Ferraris had constructed a two-phase motor that he showed visitors to his lab. This motor used two coils of wire to produce a rotating magnetic field. The rotor was a copper cylinder. The torque to turn the rotor came from eddy currents in the cylinder. Ferraris wrote a paper on his motor that was published by the Royal Academy of Sciences of Turin in 1888. Bernard Carlson's book in the references at the end of the chapter gives the details.

the patent rights to George Westinghouse for $316,000 the next year. Although Westinghouse was quick to recognize the potential of the device, the German company AEG (Allgemeine Elektricitäts-Gesellschaft), led by Russian refugee Mikhail Dolivo-Dobrovolsky, won the race to demonstrate a practical three-phase induction motor in 1889.

Understanding Tesla's motor has challenged generations of electrical engineering students. There were two different inventions in the motor. The first was a multi-phase stator to produce a rotating magnetic field. The second was a rotor that supported an induced electric current. The combination of the rotating magnetic field and the current produce a Lorentz force that provides the torque on the rotor. We consider the rotating B-field first. Figure 8.17 is a figure from Tesla's patent. It shows a circle of magnetic material with six coils on stubs arranged at

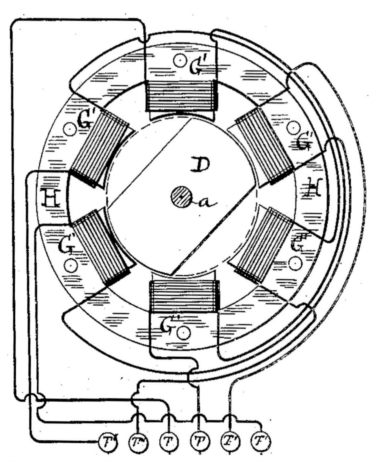

Figure 8.17 A stator that produces a rotating magnetic field. The diagram is from Nikola Tesla's US Patent 381968, "Electro Magnetic Motor," granted May, 1888. One of the things that discouraged people from using Tesla's system was that there were six supply wires. However, it was discovered that it could be done with three wires by shorting three of the terminals together.

60° intervals around the circle. Let us start with the coils at the top and bottom of the figure. If you follow the path of the wires connected to the two terminals labeled T in the middle, you will see that that it is actually a single wire that connects the top and bottom coils in series. An AC current through this wire will produce a vertical field component B_y that we can write as

$$B_y = B\sin(\omega t) \qquad 8.2$$

where B is the B-field magnitude, ω is the frequency in radians/s, and t is the time.

We can also write expressions for the B-field components produced by the two coils connected to the two terminals on the left labeled T''. We will take the B-field magnitude to be the same, but will include a phase shift of 1/3 cycle (120°), or in radians, $2\pi/3$. This time we will have both B_x and B_y components. We can write B_y as

$$B_y = B(-1/2)\sin(\omega t + 2\pi/3) = B(1/2)\sin(\omega t - \pi/3) \qquad 8.3$$

and B_x as

$$B_x = B(\sqrt{3}/2)\sin(\omega t + 2\pi/3) = B(\sqrt{3}/2)\cos(\omega t + \pi/6). \qquad 8.4$$

Finally, we write down the B-field components for the two coils connected to the terminals on the right labeled T', including a phase shift of $2\pi/3$ radians in the other direction. We get

$$B_y = B(-1/2)\sin(\omega t - 2\pi/3) = B(1/2)\sin(\omega t + \pi/3) \qquad 8.5$$

and B_x as

$$B_x = B(-\sqrt{3}/2)\sin(\omega t - 2\pi/3) = B(\sqrt{3}/2)\cos(\omega t - \pi/6). \qquad 8.6$$

To get the total B-field, we need to sum the x components and the y components of the fields separately. We are interested in the orientation angle θ of the B-field, which can be written in terms of the ratio of the sum of all the y components to the sum of all the x components as

$$\theta = \tan^{-1}\left(\frac{B\sin(\omega t) + B(1/2)\sin(\omega t + \pi/3) + B(1/2)\sin(\omega t - \pi/3)}{B(\sqrt{3}/2)\cos(\omega t + \pi/6) + B(\sqrt{3}/2)\cos(\omega t - \pi/6)}\right). \qquad 8.7$$

We can simplify the numerator by noting that $\sin(\omega t + \pi/3) + \sin(\omega t - \pi/3) = \sin(\omega t)$ and the denominator by seeing that $\cos(\omega t + \pi/6) + \cos(\omega t - \pi/6) = \sqrt{3}\cos(\omega t)$. We can rewrite Equation 8.7 as

$$\theta = \tan^{-1}\left(\frac{B(3/2)\sin(\omega t)}{B(3/2)\cos(\omega t)}\right) = \tan^{-1}(\tan(\omega t)) = \omega t. \qquad 8.8$$

This describes a B-field of constant magnitude $3B/2$ rotating counterclockwise at the radian frequency ω. In this example, the rotation speed is the AC line frequency. However, it is typical to use additional pole pairs to slow the motor down as with the electric generators in Section 7.3.

Figure 8.18 shows a rotor for an induction motor. This rotor is different from the universal motor in Figure 8.15 in that there is no commutator. This makes the motor cheaper and improves its reliability. Currents are induced in the rotor conductors in the same way they are induced in transformer coils, through a time-varying B-field. These currents cause a Lorentz force on the conductors and the rotor turns. However, this means that the rotor must rotate at a different speed from the B-field. If they rotate at the same speed, the B-field that the rotor sees will be constant and there will be no induced current and no Lorentz force.

The difference between the rotation speed of the B-field and the speed of the rotor is called the *slip* speed s. The slip is usually normalized to the B-field rotating speed. A slip of zero means that the rotor is rotating at the same speed as the B-field. If the rotor is stopped the slip is 1. When the slip is between 0 and 1, the machine acts as a motor. When the rotor speed exceeds the B-field rotation speed, $s < 0$, and the machine acts as a generator. This is the basis of the regenerative braking system in electric vehicles and in trains. The advantage of a regenerative braking system is that the electricity generated can charge the battery in a car, or a train can put electricity back into the grid. In addition, there is no mechanical wear on the brake pads.

Electric motors can be made in an enormous range of powers, from 1-W appliance motors to 100-MW water pumps for reservoirs. They have significant torque even at low speeds, and this means they do not need the gears and clutches that gasoline engines require. Electric motors run compressors in refrigeration systems

Figure 8.18 A drawing of a rotor from an induction motor. Drawing from Rankin Kennedy, 1909, *Electrical Installations*, Vol. 2, Caxton Publishing Company.

that keep our food from spoiling and in air conditioners that allow us to live and work comfortably in a hot climate. Many air conditioning systems can now be reversed to act as heat pumps. In this case one electrical system provides cooling and heating, eliminating the need for a furnace.

8.4 Heating and Cooling

Being able to control the temperature and humidity in a room contributes greatly to our productivity and our quality of life. People who grew up in hot climates before air conditioning was available in schools are well aware that heat and humidity affect cognitive responses. After working outdoors in the sun, it is refreshing to come into an air-conditioned building to cool off. The continuing American migration from North to South is inconceivable without air conditioning. Similarly, after exercising outdoors in the winter, it is invigorating to come into a heated room. Finns relax in a sauna, where the temperature is typically 70 °C.

8.4.1 The Comfort Zone

Both temperature and humidity are important for comfort. The most common measure of the water content of the air is the relative humidity. *Relative humidity* is the ratio of the partial water vapor pressure in the air to the equilibrium water vapor pressure that is found above a surface of liquid water. The equilibrium water vapor pressure increases strongly with temperature, from 1.2 kPa at 10 °C to 4.2 kPa at 30 °C. This means that the relative humidity changes when the temperature changes even if the mass fraction of water in the air does not change. The temperature and relative humidity ranges for comfort depend on many factors: age, gender, clothing, season, the air circulation, and whether the room is an office, a bedroom, a factory floor, or a gym. In a home, people can adjust to different conditions by changing clothing. However, in an office, people often must follow a dress code, and the rooms are often set to a single temperature and relative humidity, typically 23 °C and 50%. If the relative humidity is too low, people get chapped lips and bloody noses. If the relative humidity is too high, mold may grow on the walls and cause respiratory problems. It is common to specify a target comfort zone as an area on a graph of relative humidity and temperature. Figure 8.19 shows a representative comfort zone.

For analyzing heating systems we define a *humidity ratio*, conventionally written as w and given by

$$w = m_v / m_a \qquad\qquad 8.9$$

where m_v is the mass of the water vapor and m_a is the mass of the dry air in the same volume. The importance of the humidity ratio is that it does not change as

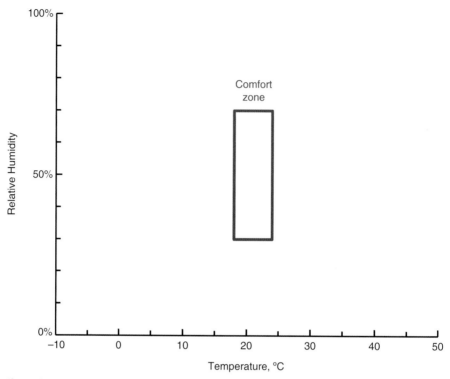

Figure 8.19 A representative indoor comfort zone with a minimum temperature of 18 °C and a maximum of 24 °C, and a minimum relative humidity of 30% and a maximum of 70%.

we heat air, whether the heater is a hot water radiator, an electrical resistive heater, a furnace, or a stove.[5] Electrical resistance heaters are inexpensive and are useful in heating small areas without generating smoke. Figure 8.20 shows curves of relative humidity and temperature at fixed humidity ratios. As we move along one of these curves to a higher temperature, the relative humidity drops.

Data loggers are available that measure temperature and relative humidity conveniently. Figure 8.21 is a photograph of a data logger than can store 16,000 readings. This allows a full year of hourly measurements. At the end of a measurement run, the data are uploaded to a computer through the USB connector. The manufacturer specifies the typical accuracy as ±0.5 K and ±2% relative humidity.

8.4.2 Heating with a Stove

In Figure 8.22, the gray line shows the locus of temperature and relative humidity inside an unheated building during the winter. The building was the log cabin shown in Figure 1.3. The logger was the Lascar EL-USB-2+, the type shown in

---

[5] Infrared lamps heat differently. The infrared radiation propagates until it hits an absorbing object. The advantage of infrared lamps is that something specific can be heated, like food, without heating up the rest of the room.

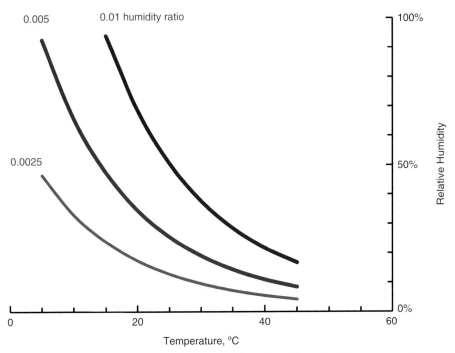

Figure 8.20 Relative humidity vs. temperature at different humidity ratios. When we heat air we move down one of these curves. The plots are based on empirical formulas given in Donald Gatley, 2005, *Understanding Psychrometrics*, American Society of Heating.

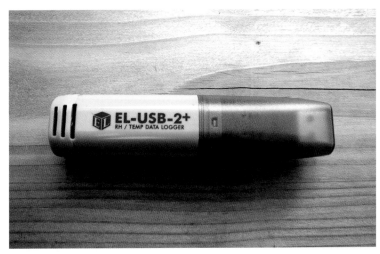

Figure 8.21 The Lascar EL-USB-2+ data logger that measures temperature and relative humidity. The logger is 10 cm long and the cost is $100. Photograph by the author.

Figure 8.21. The data were taken at hourly intervals during a three-month period from December through February. The minimum indoor temperature during this time was −13 °C on January 15. It was considerably colder outside that day, −28 °C. The typical inside conditions during the winter were −5 °C at 50% relative humidity.

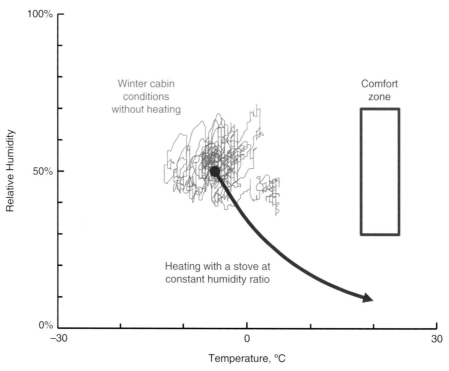

Figure 8.22 Heating a cabin in the winter. The data for winter conditions were taken by a data logger during the three months of December 2012 through February 2013 at hourly intervals.

Starting at that point, if we heat the air with a stove, we move down a line of constant humidity ratio. This is the red curve shown in the figure. By the time the air has heated up to 20 °C, the relative humidity has dropped to 9%. This is a comfortable temperature, but the room will feel dry. The traditional approach to this problem is to heat up an open pot of water on the stove to increase the humidity. Today one can buy an ultrasonic mist humidifier.

8.4.3 Evaporative Coolers

We turn next to cooling. Buildings are usually cooled by evaporating a refrigerant. When a liquid evaporates the gas molecules have higher energy than they did as a liquid. This energy is called the *heat of vaporization*. When air provides this energy, the air cools. Figure 8.23 shows an *evaporative cooler* that works through the evaporation of water. In an evaporative cooler there is a pump that drips water down absorbent pads. A fan forces outside air through the pads into the building. The water drops on the pads evaporate in the flow of air, and this cools the air, typically by 10 K. The water that does not evaporate collects at the bottom of the cooler, where it is drawn into the pump again. To get a sense of the size of the effect, the heat of vaporization of water is 2.3 MJ/kg. The heat capacity of dry air is 1 kJ/kg/K. This means that a kilogram of water can cool 230 kg of air by 10 K.

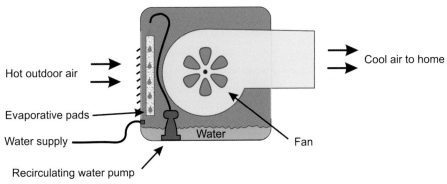

Figure 8.23 Schematic of an evaporative cooler. Not shown are a float valve to set the level of the water and an overflow tube.

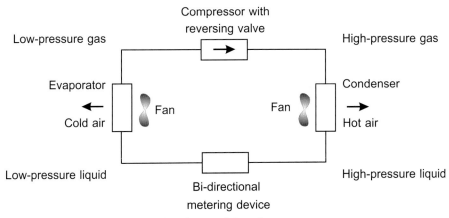

Figure 8.24 A schematic for a combined air conditioner/heat pump system.

The cooler must have a water supply to make up for the water that is lost through evaporation. A major advantage of evaporative coolers is that they use little electricity. The only draws are for the fan and a small water pump. In contrast, air conditioners have a powerful compressor to condense the refrigerant. Our bodies also cool themselves by evaporation. A breeze increases the cooling effect by increasing the evaporation rate. When the humidity is high, we are less comfortable because the sweat does not evaporate as easily. Similarly, evaporative coolers are not as effective when the relative humidity is high.

8.4.4 Air Conditioners and Heat Pumps

An air conditioner does not have the limitations that an evaporative cooler does, but it is more expensive and it uses more electricity. In an air conditioner, the refrigerant moves in a closed cycle, so it does not need to be continually replenished. Figure 8.24 shows a schematic of the process. Starting at the lower left the liquid refrigerant moves into an evaporator coil. The process of evaporating the liquid to a gas absorbs the heat of vaporization, which cools the pipe coil. The coil is embedded in an array of metal fins. A fan blows air from the room across the fins and this cools the inside air. The refrigerant then goes to a compressor which increases the pressure of the gas. The pressure increase causes the temperature of the gas to rise. The gas moves into the condenser coil. The process of condensing the gas into a liquid releases the heat of vaporization, which warms the pipe coil. The condenser coil is embedded in metal fins. A fan blows outside air across the fins and this heats the outside air. The liquid refrigerant moves to the metering device which regulates the flow of liquid into the evaporator pipe. The metering device absorbs the pressure drop between the high-pressure condenser line and the low-pressure evaporator line. Then the cycle begins again. The compressor consumes most of the electrical power, more than 90%, with the rest going to the fans. The compressor is the noisiest part of an air conditioner, and it is often put outdoors to keep the noise in the room down. A refrigerator has a similar structure to an air conditioner. The cold evaporator air circulates inside the box, and the hot condenser air is released into the room.

Rain is part of a process that acts like an air conditioner that cools the surface of the earth. When water evaporates at the surface, the air cools. Air movements take the water vapor up to a high altitude where it condenses to form rain. When it condenses the air warms. The rain returns to the ground to complete the cycle. The evaporative cooling power is significant. In the 2013 5th Assessment Report of the United Nations Intergovernmental Panel on Climate Change (IPCC), this cooling power is assessed as $84\,W/m^2$. In comparison, the carbon dioxide forcing from cumulative fossil-fuel burning is reckoned to be $2\,W/m^2$.

Air conditioning systems can be designed with a reversing valve at the compressor so that the refrigerant circulates in the other direction. The evaporating coils become condenser coils and vice versa. This heats the room, while outside air is cooled. It is as if we opened the inside of a refrigerator to outside air. This configuration is called a *heat pump*. The air conditioner/heat pump combination has major advantages over other heating and cooling systems. Nothing is burned, there is no smoke, and there is no danger of fire or carbon monoxide poisoning. Perhaps surprisingly, the heating power delivered to the room can be several times as large as the electrical power supplied to the compressor. We should think of the process as moving heat from the outside to the inside rather than generating heat. The same is true for the cooling power for an air conditioner. Engineers define the coefficient of performance COP as

$$COP = \frac{\text{Heating or cooling power delivered}}{\text{Electrical power supplied}}. \qquad 8.10$$

For heating systems based on combustion and electrical heating elements, this definition is equivalent to the efficiency, so it would be less than one. The wood stove in Figure 1.1 had an efficiency of 75%. Modern gas furnaces can be 90% efficient, and a portable electric resistance heater is 100% efficient, because all of the electrical power supplied to the heater is converted to heat. In contrast, the values for the COP for air conditioner/heat pump systems are from 4 to 6. In practice, the coefficient of performance will vary depending on the details of the installation and the outside temperature and humidity. If the ducts are long, then the temperature of the air will change on the way to the room. The COP for air conditioners is lower if it is hot outside and the COP for heat pumps is reduced if it is cold outside. Of course these are precisely the conditions when they are most useful. A representative value for the COP for air conditioner/heat pumps in practice is 3. This means that a heat pump can be competitive on a primary energy basis with a natural gas furnace when the electricity is generated by burning natural gas.

Air conditioners can also be used to dehumidify the air. The metal fins that surround the evaporator coil are cold and water condenses on them. When the unit is operating as an air conditioner, the water collects in a pan and drains outside. This process lowers the humidity in the room. The application of the first air conditioner was more to control the humidity than to control the temperature. In 1902, the Sackett-Wilhelms Lithography Company of Brooklyn, New York was having trouble with its color printing. When paper absorbs water, the fibers in the paper get fatter and the paper becomes longer in the direction across the grain of the paper. In color printing, the different colors are applied in different steps. When the paper swells the alignment of the colors is ruined. Sackett-Wilhelms approached the Buffalo Forge company of Buffalo, New York about their problem. They asked for a system that could maintain a relative humidity of 55% throughout the year. Buffalo Forge assigned the project to a young engineer, Willis Carrier, who had graduated from Cornell University the year before. Carrier delivered a system for Sackett-Wilhelms and then went on to found the Carrier Engineering Corporation in 1915. Carrier is still a major supplier of air conditioning and heating systems a hundred years later.

We can relate the heating and cooling power to the thermal resistance and the temperature rise and fall. We rewrite Equation 1.7 that defines the thermal resistance as

$$P = T / R_t \tag{8.11}$$

where P is the the heating or cooling power in watts, T is the temperatue rise or fall in K, and R_t is the thermal resistance in K/W. We can reduce energy consumption if we accept lower indoor temperatures in the winter and higher indoor temperatures in the summer. However, the trend has been in the other direction. To illustrate, the late David McKay noted in his book, *Sustainable Energy:Without the Hot Air*, that in 1970 the average temperature of British homes in the winter was 13 °C. British homes are not that cold today and neither are the homes in any other OECD country. We can also reduce energy consumption by increasing the thermal resistance. It is often practical to double the thermal resistance in an existing home by adding insulation in the attic and in the walls and by changing to multi-glazed windows. The larger increase comes from adding insulation, and this is also cheaper than new windows. That said, extremely sophisticated multi-glazed windows are now available that are coated to reflect infrared radiation and filled with argon gas. Argon has a thermal resistance that is 50% higher than air. Insulation and multi-glazed windows are typically required in new construction by building codes. Existing homes can be retrofitted, but this is expensive, and depending on the weather, the energy prices, and the housing markets, the homeowners may or may not get their money back in a reasonable time. An increased thermal resistance does increase the time constant of the home and this reduces the daily temperature variation. This makes the home more comfortable even when the heater and air conditioner are not turned on at all. Finally, the Jevons Paradox is lurking. Increasing the thermal resistance of the home makes it less expensive to maintain high indoor temperatures in the winter and low temperatures in the summer, and many people do.

8.4.5   Inline Water Heaters

In addition to heating and cooling, buildings need to supply hot water for bathing and shaving, and washing clothes and dishes. A traditional way to provide hot water is a tank of water with a heater. More recently companies have developed inline water heaters that heat the water on the fly. Figure 8.25 is a diagram of an inline heater made by the Takagi Corporation. When the hot water tap is turned on, a flow sensor in the unit detects the flow and starts the natural gas burner that heats the water through a heat exchanger. Thermistors monitor the water temperature so that water can be delivered at the correct temperature.

Inline water heaters have many advantages. They eliminate the heat losses from the tank, which improves the efficiency about 20%. Inline heaters do not run out of hot water. A large family returning from a camping trip can take showers one after the other. Inline heaters take less space than a tank heater and they can be mounted on an outside wall so that the combustion is outdoors. On the other hand, tankless water heaters are more expensive than tank water heaters. In addition, in areas with hard water, mineral deposits may clog the heat exchanger. Natural gas burners also have advantages over electrical resistive heaters. Electricity typically costs several times as much as natural gas on an energy basis. In addition, the power capacity is larger for gas. The gas burner in the water heater in Figure 8.25 is rated at 56 kW. This is larger than the entire electrical supply for most residences. Many chefs prefer a gas stove over an electric range for the same reason, higher heating power.

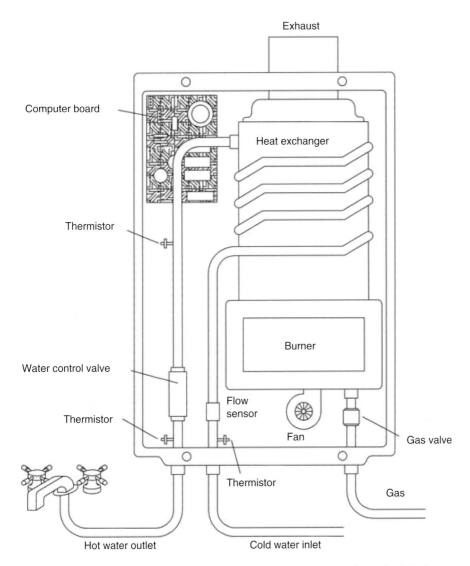

Figure 8.25 Takagi TK-3-OS inline residential water heater that uses natural gas. The default water temperature is 55 °C. The manufacturer rates the efficiency of the unit as 83%. From the owner's manual.

The discussion so far has focused on heating and cooling of individual buildings, However, sometime it is advantageous to use *district heating*, where a single heat source supplies many houses through a pipeline network that circulates hot water or steam to the homes. Through a heat exchanger, this primary hot water loop warms water for a secondary loop that circulates inside each home. For example, in places where geothermal energy is available for the primary hot water loop, as in Iceland, district heating makes sense (Section 7.6). Another good reason to use district heating is when the fuel is wood, coal, and peat. These are inconvenient to stock at a home, and it is also easier to add pollution control equipment at a central furnace. As an example, we will consider the district heating system of Finland. Information on district heating is available from the trade group Finnish Energy. The district heat supply in Finland in 2017 was 3.1 Mtoe. This is 11% of the total primary energy consumption of the country. Biomass, including wood fuel, industrial wood waste, and municipal waste, supplied 36% of the district heat. Coal was 23% and peat 14%. Industrial waste heat supplied 9%. District heat supplied 46% of the energy used for space heating in the country. This compares with 17% for electrical resistive heaters, 15% for heat pumps, 13% for wood stoves, and 8% for fuel oil. Resistance heaters and wood stoves also supply heat for the saunas that are universal in Finnish homes.

8.4.6 Split-Zone Systems

In general, however, heating and cooling systems work best at the lowest level. It takes less energy to heat or cool a room than a house, and an office than a building. In a single room the temperature can be tuned to a particular activity, cooler for sleeping and exercising, warmer for a restroom and an office. Figure 8.26 is

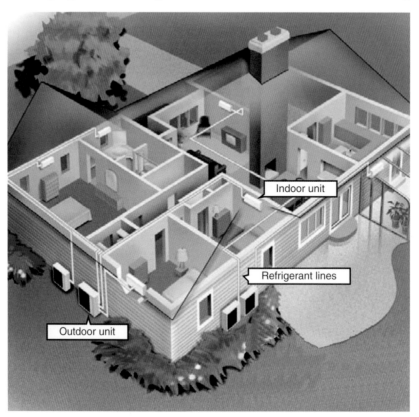

Figure 8.26 A split-zone system. From the Mitsubishi Electric HVAC Advanced Product Division. HVAC stands for Heating, Ventilating and Air Conditioning.

a diagram of a split-zone system. These were made popular by the Mitsubishi Corporation, but they are now available from other vendors. They can operate both as an air conditioner and a heat pump. There are indoor units in individual rooms. The split-zone systems are extremely flexible, with a separate remote control in each room. The indoor units have only a fan, so they are quiet. The noisy compressors are in the outside units. In addition to heating and cooling, they filter and dehumidify the air. The inside and outside units are connected by hoses that carry the refrigerant as a liquid. This eliminates the bulky air ducts in a traditional forced-air system.

8.5   Information Technology

The transistor was invented in 1947 by John Bardeen, Walter Brattain, and William Shockley of the Bell Telephone Laboratories in New Jersey. They received the Nobel Prize for this invention in 1956. The transistor made electronic computer logic and memory circuits feasible. Before the transistor was available, people had made logic and memory circuits with electromagnetic relays and vacuum tubes, but these were expensive and unreliable, and they consumed enormous amounts of power. A key advantage of the transistor was that people developed ways to make complete transistor circuits in an integrated fashion on a wafer of silicon through photographic lithographic processes. In the early years, these *integrated circuits* contained only a few transistors, resistors and capacitors. However, the technology has improved continuously, and now integrated circuits have billions of transistors.

The development of the Internet was initiated by the Advanced Research Projects Agency (ARPA) of the United States Defense Department. It was known as ARPANET at first. The first Internet message consisting of the letters "L" and "O" was sent by Professor Leonard Kleinrock's group from a classroom at the University of California at Los Angeles to the Stanford Research Institute in 1969. From that beginning Internet use has expanded dramatically. By the end of 2017, the United Nations International Telecommunications Union (ITU) estimated that 48% of the world's population had access to the Internet.

8.5.1   The Jevons Paradox in Information Technology

The dominant electronics technology has been CMOS (complementary metal oxide semiconductor). For CMOS, the conventional measure of efficiency is the switching energy, which is calculated for the change of a binary digit between 0 and 1. This is done in a nominal way as the capacitive energy that is stored in a gate of the transistor when a transistor is turned on and dissipated when the transistor is turned off. We write the switching energy E as

$$E = CV^2 / 2 \qquad\qquad 8.12$$

where C is the gate capacitance in farads and V is the supply voltage. Students who have had physics will recognize this formula as the electrostatic energy stored in a capacitor. Figure 8.27 shows the switching energies for different Intel fabrication processes from 1996 to 2014. The individual processes are called *nodes*. The nominal minimum feature size in nanometers is shown for the 1996 and 2014 nodes. In successive nodes, the gate capacitance and the supply voltage are lowered so that the switching energy becomes smaller. The y-axis is logarithmic and the slope of the regression line is for a decline of 22%/y.

Computers based on a transistor technology with a smaller switching energy will use less power. However, because of the new applications of information technology that are being developed, the energy consumed by information technology has not been decreasing. Figure 8.28 shows the International Energy Agency's estimate of the electricity consumed by network connected devices. Overall energy

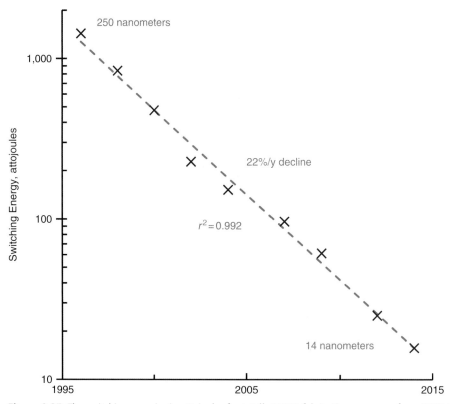

Figure 8.27 The switching energies in attojoules for Intel's CMOS fabrication processes from 1996 to 2014. An attojoule is 10^{-18} J. The scale is logarithmic. The data are from the Stanford Digital Repository, Chi-Shuen Lee, Jieying Luo, Thomas Theis, and Philip Wong, 2017, "CMOS Technology Scaling Trend." Available at: http://purl.stanford.edu/gc095kp2609

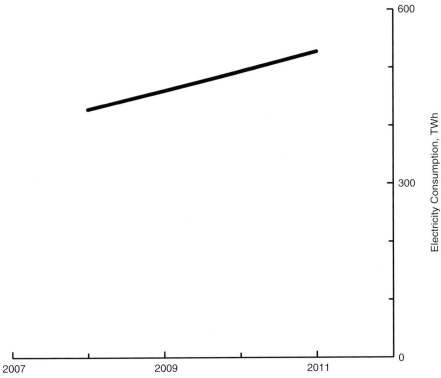

Figure 8.28 World electricity consumption for network connected devices. The 2011 consumption of 527 TWh was 1% of world primary energy demand that year. The data are from the IEA, 2012, *More Data, Less Energy.*

use has been increasing at 7%/y, in spite of the 22%/y decline in switching energy. This is a dramatic example of the Jevons Paradox.

8.5.2 Bitcoin Mining

An interesting component of information technology energy use is the data center. Data centers contain large numbers of specialized computers called servers that provide services to outside clients. These centers are connected to the outside world by fiber optic lines. A server can dissipate hundreds of watts and there can be tens of thousands of servers in a data center. The electricity to provide cooling can be comparable to that provided the servers, so the total center demand can be tens of megawatts. From this perspective, a northern country like Iceland is attractive for a data center so that cold outside air can be used for cooling.

A bitcoin mine is a particular type of data center (Figure 8.29). The miners are custom computers that do calculations called *hashes* that validate bitcoin transactions. The mining companies receive bitcoin as pay for their hashes. An example is the Antminer S9 manufactured by Bitmain and advertised on Amazon for $600. This computer uses specially designed integrated circuits with a 16-nanometer

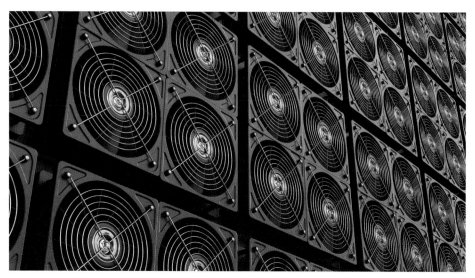

Figure 8.29 Rendering of a bitcoin mine. Each miner is cooled by a fan. Credit: Imstockwork/Shutterstock.

process. It is rated at 47 PH/h, where H is the abbreviation for hash. The P prefix is *peta*, for 10^{15}. The power consumed is substantial, 1.3 kW. As of mid-2018, 1.2 ZH were required for each bitcoin. The Z prefix is *zetta*, for 10^{21}. This is a stupendous number. We can calculate the electrical energy E consumed by an Antminer S9 machine to earn a bitcoin as

$$E = \frac{1.2 \text{ ZH/bitcoin}}{47 \text{ PH/h}} \; 1.3 \text{ kW} = 33 \text{ MWh/bitcoin}. \qquad 8.13$$

In mid-2018, 100 bitcoins were being mined per hour around the world. At the 33 MWh/bitcoin required by the Antminer S9, the electricity consumption would be 29 TWh/y. This is 5% of the electricity consumed for information technology in Figure 8.28. Bitcoin mining can be profitable. According to the IEA's 2017 *Electricity Information*, the average price for industrial electricity in the United States in 2017 was $69/MWh. At that price, the cost of electricity to earn a bitcoin is $2,300. The price of a bitcoin varies over time, but in mid-2018, it was $7,400. In Iceland, industrial electricity prices are less than half of those in the United States, and it has become a major bitcoin mining center.

It is interesting to compare bitcoin with gold. Unlike gold, bitcoin is used for purchases, 200,000 per day in mid-2018. On the other hand, gold is used in electrical products. Gold does not tarnish and it is a superb conductor. It is ideal for electrical connectors. However, much more gold is used in jewelry and formed into coins and bars to sell to investors than in industrial applications. The gold that is sold to investors acts as a store of value, like bitcoin. Many people buy gold as a hedge against severe inflation. Bitcoin has served a similar purpose in Venezuela. Gold production is a major source of mercury pollution in water. Small artisanal miners pan for gold in streams and they use mercury to purify the gold. Mercury forms an amalgam with gold. The mercury is removed from the amalgam in an acid solution, which often gets into the stream. Bitcoin mining is usually done with hydroelectric power, so it does not have this problem. Large gold miners actually use a lot of electricity. The amount varies from mine to mine, but 1 MWh per ounce is typical. Annual world gold production is a hundred million ounces per year, so the electricity that is used to produce gold for investors is comparable to the electricity used to mine bitcoin. Like gold, bitcoins have sometimes been lost or stolen. Many governments are opposed to bitcoin. One concern is that bitcoin can be used for criminal transactions. The bitcoin transactions are recorded anonymously in a distributed ledger called a blockchain. It is surely true that bitcoin is used for criminal transactions. So is cash. More fundamentally, governments wish to control the creation of money. A hundred years ago, they did not want to leave the creation of money up to gold miners. Today they do not want to leave the creation of money up to bitcoin miners. We will not speculate on the future of bitcoin. It may turn out to be a bubble, one of many going back to the tulip mania in 1637 during the Dutch Golden Age.

8.6   The Rosenfeld Effect

The shock of the 1973 Yom Kippur War and Arab Oil Embargo stimulated an interest in reducing energy demand. Arthur Rosenfeld, a professor of physics at the University of California at Berkeley, was a most influential advocate (Figure 8.30). He and his students invented energy-efficient appliances and efficient ballasts for fluorescent lighting. In addition, he developed architectural software for modeling energy losses in buildings. Rosenfeld also promoted product efficiency standards. He was influential in energy efficiency policy at the national level during the Clinton Administration. Today spending on energy efficiency is significant. The IEA estimated that 236 G$ was spent worldwide on energy efficiency in 2017.

8.6.1   Refrigerator Efficiency

One of Rosenfeld's interests was efficiency standards for refrigerators. Figure 8.31 shows his data for refrigerator volume and power consumption. Before the Arab Oil Embargo, the volume grew steadily, from $0.23 \, m^3$ in 1947 to $0.52 \, m^3$ in 1974. Power consumption increased by an even larger factor, from 40 W to 208 W. During this time the share of households that had refrigerators grew from 71% in 1947 to close to 100% in 1974.[6] From the perspective of social history, refrigerators played

---

[6] A good reference for historical data on the appliances in American households and the social impacts is Jeremy Greenwood, Ananth Seshadri, and Mehmet Yorukoglu, 2005, "Engines of liberation," *Review of Economic Studies*, vol. 72, pp. 109–133.

Figure 8.30 Arthur Rosenfeld (1926–2017), pioneer in product efficiency standards. Credit: Lawrence Berkeley National Laboratories.

a significant role in enabling women to work outside the home because they no longer had to buy food every day. After 1974, refrigerator power consumption was reduced by adding insulation and improving the seals. The volume in 1996 was 0.58 m³, 12% larger than in 1974. However, the power dropped all the way back down to 40 W. In considering energy consumption on a per-person basis there are two factors that offset these results somewhat. One is that the household size was shrinking. The US Census Bureau reckoned that the average number of people per household in 1996 was 2.65, down 12% from that in 1974. In addition, since 1996, an increasing number of people are buying two refrigerators. EIA surveys found that 30% of households owned a second refrigerator in 2015, compared with 15% in 1996. From a broader perspective, we could say that the mechanical engineers have given consumers refrigerators that were 2.5 times larger than in 1947 but consumed the same power.

The State of California credits Professor Rosenfeld's influence with flattening California electricity demand compared to the rise of the United States as a whole. Figure 8.32 shows the history of per-person electricity consumption compared with the United States as a whole. The rising slopes are similar before the shock of the Arab Oil Embargo. Both curves change slope, but the California curve actually levels out.

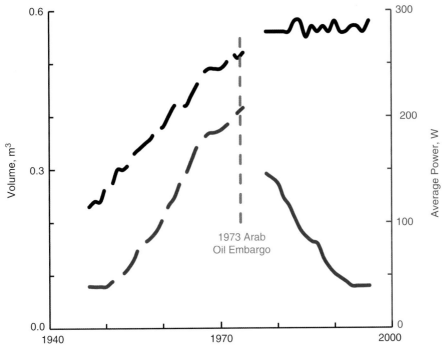

Figure 8.31 The evolution of American refrigerator size and average power from 1947 to 1996. The data are digitized from Arthur Rosenfeld, 1999, "The art of energy efficiency: Protecting the environment with better technology," *Annual Review of Energy and the Environment*, vol. 24, pp. 33–82.

There are factors besides efficiency policy that could have affected demand. California has evolved differently from the rest of the country from an economic and social perspective. The state has discouraged heavy industry in favor of software and entertainment. Much of Southern California's aerospace industry, which was world-leading in the 1950s and 1960s, has moved to other states. Software and entertainment have done well, but the income in these industries is highly concentrated among a few people, in contrast to the heavy industry that left. The US Census Bureau has determined by its supplemental poverty measure that California has the highest fraction of poor people of any state. Over the time range of the graphs in Figure 8.32, Americans have installed air conditioning in most of their residences. Air conditioning use is higher in hot, humid climates. In the Gulf Coast and Atlantic states climate is influenced by the warm Gulf of Mexico and the north-bound Gulf Stream. These states are hot and humid. California is dry, and it is cooled along its coast by the south-bound California Current.

8.6.2 A State Residential Efficiency Model

To get a better sense of how California residential electricity demand differs from other states and how much of this difference might be ascribed to efficiency policy, we will develop a model for state residential electricity demand in 2013 based on

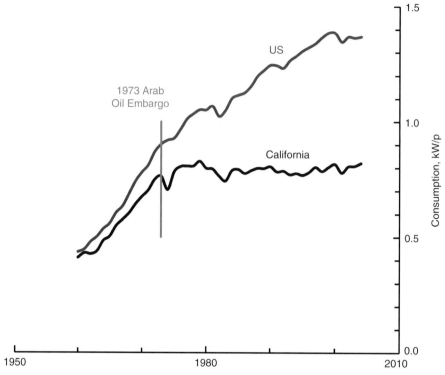

Figure 8.32 The Rosenfeld Effect. Per-person electricity consumption for California, compared with the United States as a whole. The data come from the California Energy Commission.

two variables, average residential prices in ¢/kWh, and annual cooling degree days. Cooling degree days are a common index for hot weather. Cooling degree days are calculated by accumulating the positive difference in degrees Fahrenheit between the daily average temperature and 65 °F. By convention, the units are °F·days. The National Oceanic and Atmospheric Administration (NOAA) calculates these for each state. Figure 8.33 plots residential electricity use per customer against the average residential price for each state. Note that the data are demand per customer, where the customers represent households rather than individuals. In addition to California, the data point for the second most populous state, Texas, is shown. There is a downward trend. Higher prices are associated with less electricity use. Texas has 30% lower electricity prices than California, 11¢/kWh versus 16¢/kWh and twice the residential demand, 1.6 kW/customer compared with 0.9 kW/customer. Figure 8.34 shows a plot of electricity use versus cooling degree days. There is an upward trend, with more electricity use in warmer states. Texas has three times the degree cooling days that California does, 2,900 °F·days versus 1,000 °F·days.

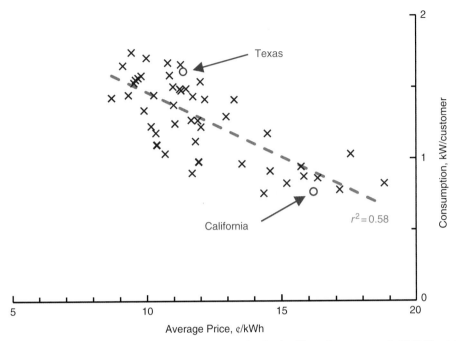

Figure 8.33 State residential electricity use versus price for the 48 contiguous states in 2013. The data come from the EIA, "Residential Electricity Use."

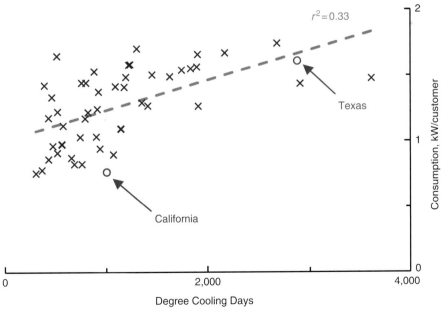

Figure 8.34 Per-person residential electricity consumption versus degree cooling days in 2013. The electricity data come from the EIA and the annual degree cooling days are from NOAA.

The model for electricity demand is of the form

$$d = d_0 + ap + bc \qquad\qquad 8.14$$

where d is in kW/customer, p is the price in ¢/kWh, and c is the number of cooling degree days. The constants d_0, a, and b are fitted by minimizing the sum of the squares of the residuals, which are the differences between the data and the model values. This gives $d_0 = 2.0$ kW/customer, $a = -77$ W/customer/(¢/kWh), and $b = 150$ mW/customer/(°F · day). The comparison between data and model is shown in Figure 8.35. The dashed line is the locus of points where the model matches the data. Texas is close to the line. In a statistical sense, its demand is explained by its low electricity prices and its hot weather. California is 170 W under the model.

Much of the difference between California's residential consumption and other states' use is explained in a statistical sense by its expensive electricity and mild summers. It is possible that California would come even closer to model if other

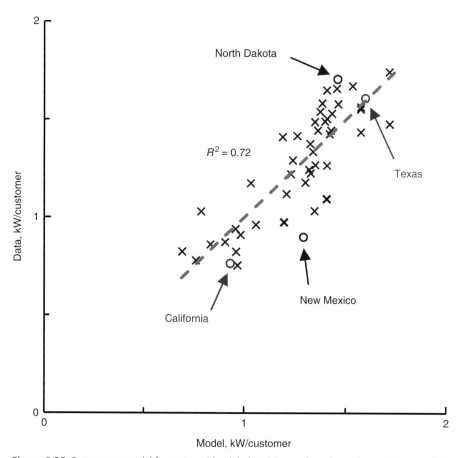

Figure 8.35 Data versus model for state residential electricity use based on price and the population-weighted degree cooling days.

variables were considered, like poverty, humidity, and heating degree days. The state where the residential demand is farthest below the model is New Mexico at 390 W/customer under. Like California, New Mexico is dry, and a high fraction of the population meets the census bureau poverty criteria. The state that is highest above the model is North Dakota, at 250 W/customer over. North Dakota is noted for its severe winters and we would expect electricity demand for heaters to be high, unlike California, which has mild winters.

8.7   Trends in Residential Energy Demand

We conclude the chapter with a look at trends in American residential demand. In the wake of the Arab Oil Boycotts, the Energy Information Administration (EIA) within the US Department of Energy started a series of surveys of American home energy use. These carry the acronym RECS, for Residential Energy Consumption Survey. Here we need to issue a caution. Surveys are hard. This is not as easy for the government as ordering coal companies to send in production numbers and adding them up. In surveys, it is a fundamental problem that we are making models from incomplete and inconsistent data. Over time, people respond in different ways and new components of demand arise. We will use the 11 surveys conducted from 1979 to 2009 to determine the trends. There was also a survey in 2015, but it had significant changes in the modeling approach for the estimation for electricity use so that is not compatible with the earlier surveys. Moreover, the six-year gap makes it difficult to distinguish the effects of the Great Recession, which had a large impact on housing construction. There was one more survey in 1978 that is not included. It did not include a question on wood use, and the time frame is covered by the 1979 survey.

Figure 8.36 shows American total residential energy consumption from 1979 to 2009. The regression line shows a slow decline at 2.7 kgoe/p/y. There is a significant rise in the electricity share, from 47% in 1979 to 67% in 2009. Figure 8.37 shows wood consumption. This is mostly for heating houses, with smaller amounts for heating water and cooking. Wood consumption rises when energy is relatively expensive, as in the early 1980s and after 2005. In 2009, the consumption was 41 kgoe/p, 3% of total demand. The decline in wood consumption is 2.6 kgoe/p/y. This is a good match to the overall decline of 2.7 kgoe/p/y.

The largest component of demand in the RECS is for heating. This is shown in Figure 8.38 from 1987 to 2009. Heating energy was 0.65 toe/p in 2009, 44% of the total. In contrast to the overall demand, there is a significant decline in heating demand at 10.3 kgoe/p/y. We would associate this with improved insulation in houses and more efficient heaters. During this time, the electricity share has risen from 19% to 28%. Electricity consumption is a modest part of the overall heating

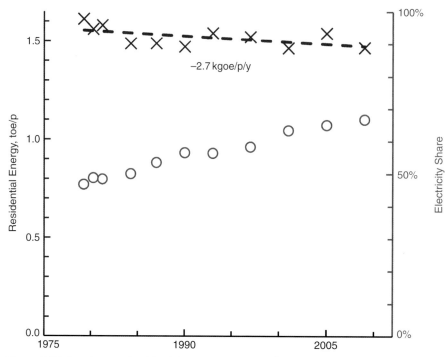

Figure 8.36 American total residential energy consumption from the EIA Residential Energy Consumption Survey (RECS). The electricity contribution is calculated on a primary energy basis. The population data come from the United Nations Population Division.

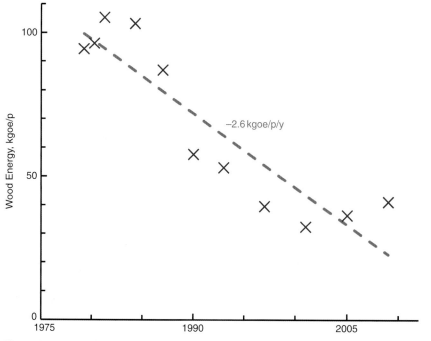

Figure 8.37 American residential wood energy consumption from the EIA Residential Energy Consumption Survey (RECS). The population data are from the United Nations Population Division.

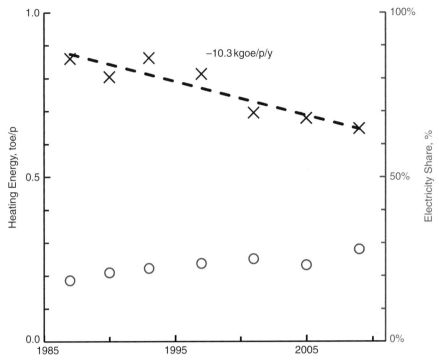

Figure 8.38 American residential heating energy consumption from 1987 to 2009. This includes space heating and water heating in the EIA Residential Energy Consumption Survey (RECS). The electricity contribution is calculated on a primary energy basis. The population data are from the United Nations Population Division.

demand, but it is nevertheless a significant electrical load, comparable to air conditioning, lighting, and refrigeration. Heating electricity by itself is actually rising, from 80 W/p in 1987 to 91 W/p in 2009.

The next largest component is appliances; see Figure 8.39. Appliances include cooking stoves, dishwashers, clothes washers and driers, fans, evaporative coolers, lighting, TVs, laptops, and audio systems. It does not include air conditioners and refrigerators, which the EIA treats as separate categories. In 2009, appliance energy amounted to 0.58 toe/p, 39% of the total. In contrast to heating energy, which is declining at 10.3 kgoe/p/y, appliance energy is rising at 9.5 kgoe/p/y. This rise offsets most of the heating energy decline. Appliance energy use is mostly electrical and the share is stable, with an 89% share in 1987 and 91% in 2009. Some natural gas and propane is used in cooking stoves and clothes driers.

In the RECS, the EIA includes lighting electricity in appliances. However, a recent Department of Energy study gives estimates for residential lighting electricity demand, and these are shown in Figure 8.40. This shows a decline of 2.2 W/p/y, or 4.4 kgoe/p/y on a primary basis. This reflects the ongoing shift from

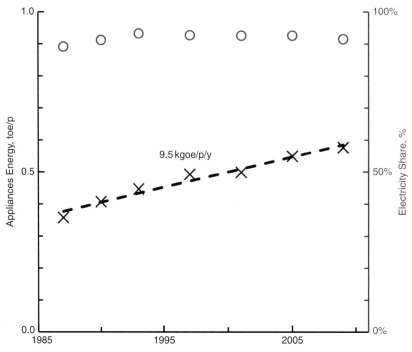

Figure 8.39 American residential appliance energy consumption from 1987 to 2009. The electricity contribution is calculated on a primary energy basis. The population data are from the United Nations Population Division.

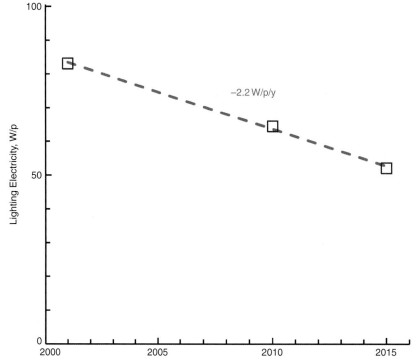

Figure 8.40 American residential lighting electricity in 2001, 2010, and 2015. The data come from the Department of Energy, 2017, "2015 U.S. Lighting Market."

incandescent bulbs to compact fluorescent lamps and LEDs. However, this decline is overwhelmed by the rise in electricity consumption by other appliances.

The smallest sector is cooling, including air conditioners and refrigerators, shown in Figure 8.41. Together they accounted for 17% of residential primary energy consumption in 2009. Air conditioners and refrigerators almost all run on electricity. There are a few refrigerators that use propane in off-grid homes. Only the electrical consumption is included in the figure. In 2009, the draw for air conditioners was 69 W/p and the draw for refrigerators was 53 W/p. The refrigerator demand is dropping at 0.7 W/p/y. This reflects the efficiency improvements that started after the 1973 Arab oil boycotts. However, air conditioner demand is rising at 1.1 W/p/y. The efficiency of air conditioners is also improving, but not enough to offset the increase in demand from new installations. The net change for cooling electricity is a rise of 0.4 W/p/y, or 0.8 kgoe/p/y on a primary energy basis.

Excluding wood, American residential demand has been flat for 30 years. The consumption in 2009 was 1.5 tgoe/p. We need to keep in mind that this is per-person demand. The US population grew at 1%/y during this time, so total demand

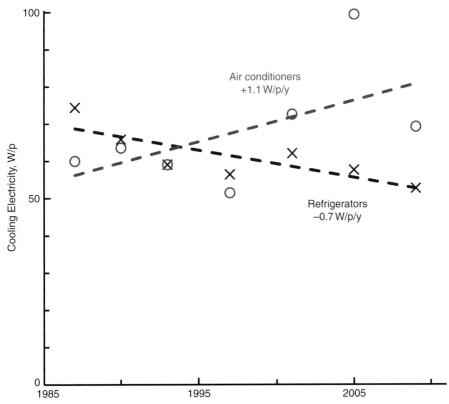

Figure 8.41 American residential cooling electricity consumption from 1987 to 2009. This includes space heating and water hearing in the EIA Residential Energy Consumption Survey (RECS). The population data are from the United Nations Population Division.

grew significantly. There have been large increases in the efficiency of individual appliances, particularly refrigerators, air conditioners, lights, and electronics, as well as improvements in home insulation. However, for appliances and air conditioning, demand has gone up, and this has offset the energy reduction in heating and refrigerators. The efficiency improvements do improve our quality of life, but the Jevons Paradox casts a long shadow over residential energy consumption. Possibly the most significant result from the surveys is the large increase in the electricity share of residential demand from 47% to 67% from 1979 to 2009, which bodes well for a future transition from fossil fuels to alternative energy sources that produce electricity.

Concepts to Review

- Trends in industrial demand and building demand
- Traditional building climate control
- Increased mortality from hot and cold
- Lighting history in the United Kingdom
- The development of light-emitting diodes (LEDs)
- Fundamental efficacy limits
- Hybrid lighting products
- A pre-electrical machine shop
- The universal motor
- Nikola Tesla and the invention of the induction motor
- The rotating magnetic field
- The comfort zone
- Heating and relative humidity
- Evaporative coolers
- Air conditioners and heat pumps
- District heating
- Split-zone systems
- Switching energies in electronics
- Information technology electricity consumption
- Bitcoin mining
- Arthur Rosenfeld and efficiency improvements in appliances
- Modeling residential electricity demand
- Trends in US residential energy consumption

Problems

Problem 8.1   Kettles

Figure 8.42 shows the electricity consumption in the United Kingdom during the European Cup quarter-final football (soccer in American English) match between England and Italy in Kiev on June 24, 2012. The game began at 1945 British summer time.

At halftime there was a spike in electricity consumption. Let us assume that the spike came from electric tea kettles. Estimate the number of tea kettles that people turned on. You may assume one liter of water per kettle and a temperature rise from $10\,°C$ to $90\,°C$ for each tea kettle.

Problem 8.2   The Jevons Paradox

Locate Stanley Jevons' 1865 book, *The Coal Question*, online and read Chapter 7, "On the Economy of Fuel."

a.   In your own words, summarize his explanation of the Jevons Paradox.
b.   Distinguish between rebound and backfire. Discuss an example of either from your personal experience.

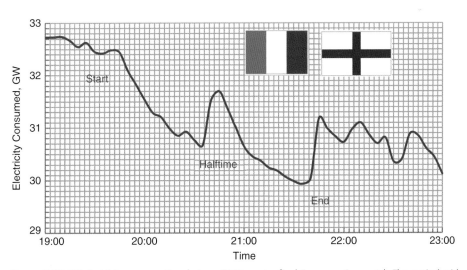

Figure 8.42 UK electricity consumption during a 2012 quarter-final European Cup match. The vertical grid lines are 3 minutes apart and the horizontal grid lines are 100 MW apart. The data came from Gridwatch.co.uk.

Further Reading

- Bernard Carlson, 2013, *Tesla, Inventor of the Electrical Age*, Princeton University Press. The technical descriptions are good, and there is a discussion of Galileo Ferraris' induction motor.
- Nick Holonyak, 2012, the inventor of the LED. An interesting YouTube presentation at www.youtube.com/watch?time_continue=69&tv=KKkzBVNozjI
- David MacKay, 2009, *Sustainable Energy: Without the Hot Air*, UIT Cambridge. The late David MacKay was a professor of engineering at the University of Cambridge and he was the scientific advisor to the UK Department of Energy and Climate Change. This book is fun to read and it is packed with interesting ideas for reducing stationary demand. The pdf version is available for free at www.withouthotair.com/download.html

9 Transportation Demand

I spent four lonely days in a brown LA haze
and I just want you back by my side.

Jimmy Buffet
"Come Monday"

9.1 Introduction

For most of human experience, for most people, land transportation meant walking. For raiders like the Mongols and Comanches, it meant horses. In some societies, the better off rode in carriages. With the coming of the steam locomotive in the 1800s, people could ride a train. However, the real agent of deliverance has been the motor car. Steam engines needed a firebox, boiler, and large cylinders and this made them too heavy for a car. The problem was solved by the invention of engines where the fuel burned inside the cylinders. These engines are called *internal-combustion* engines to distinguish them from the *external-combustion* steam engines. The most important car engine has been the Otto-cycle, 4-stroke engine invented by Nikolaus Otto in 1876 in Cologne, Germany. In Otto's engine, fuel mixed with air is first drawn into the cylinder by the motion of the piston. Next the piston compresses the fuel mixture, which is then lit by a flame. The heated gas expands and propels the piston. Finally, the piston reverses and pushes the mixture out of the engine and the cycle can begin again. Otto's engine burned coal gas, which does not have a high enough energy density to be used in cars. However, one of his engineers, Gottlieb Daimler, started a company to develop engines for cars that ran on gasoline. Since that time, the Otto engine has been subject to continuous development. The modern engines use spark plugs to light the fuel mixture, so the engines are often called *spark-ignition* engines.

The main competitor to the Otto engine for car motors has been the diesel engine invented by Rudolf Diesel in 1897 in Augsberg, Germany. In the diesel engine, fuel is injected into the cylinder after the piston has compressed the air inside to the point that it is so hot that the fuel ignites without a spark plug. Diesel's engines are called *compression-ignition* engines. Diesel engines produce higher temperatures inside the cylinders than spark-ignition engines and this gives them fundamental efficiency advantages. Diesel engines can be made with much higher power ratings than spark-ignition engines and they dominate freight trucks, trains, and ships. The fuels for the two engines are complementary in that gasoline comes from the lower-carbon-number fraction of crude oil, and diesel fuel comes from the higher-carbon-number fraction. Diesel fuel is less volatile than gasoline, and this reduces the fire risk. On boats where there may be no safe escape, diesel engines are often chosen for this reason alone.

The work of Otto, Daimler, and Diesel made Germany the home of the motor car. However, the German companies produced luxury cars. The next step forward was to develop a car that could be mass produced. Surprisingly, this happened in the United States, not Germany. Considerable scientific knowledge was involved in engine development and at that time Europe completely dominated scientific research. Only one American won a Nobel Prize in any scientific field before World War I.[1] Moreover, American roads were appalling – 90% of the roads were dirt, without a gravel or asphalt surface.

9.1.1   Henry Ford and His Model T

As with the electrical grid, the vision for an inexpensive car came from one man. Henry Ford (Figure 9.1) was born in 1863 on a farm near Detroit, Michigan. He was as an apprentice machinist as a teenager, but he did not have a formal technical education. In 1891, he began work at the Edison Illuminating Company in Detroit. He met Thomas Edison at the company in 1896 and he told Edison about his personal experiments building gasoline-powered vehicles. Edison encouraged him and in later life, Ford and Edison became close friends.

In 1899, Ford went out on his own to start a company to produce cars. There were several failures and Ford built and drove race cars to try to stay solvent. Finally, in 1903, he founded the Ford Motor Company, which made money. However, he had disagreements with his investors. Ford wanted to build an inexpensive car while they preferred luxury cars. He finally got his way with the Model T, which was introduced in the fall of 1908. He expressed his ideas in a speech to his employees the following year:

> I will build a motor car for the great multitude. It will be large enough for the family but small enough for the individual to run and care for. It will be constructed of the best materials, by the best men to be hired, after the simplest designs that modern engineering can devise. But it will be so low in price that no man making a good salary will be unable to own one—and enjoy with his family the blessing of hours of pleasure in God's great open spaces.[2]

The design was revolutionary. Ford used several types of vanadium steel to keep the weight down to 550 kg, which was much lighter than other cars, then and now.

---

[1] This was Albert Michaelson in 1907 for his measurements on light.

[2] Quotation from Henry Ford and Samuel Crowther, 1922, *My Life and Work*, Garden City Publishing.

Figure 9.1 Henry Ford (1863–1947), the developer of mass production of cars. This photograph was taken in 1919. Credit: United States Library of Congress/public domain.

The engine was an inline four with a capacity of 20 hp. The block for the four cylinders was a single casting. Most cars at that time used separate castings for the cylinders. The engine rating of 20 hp is considered small today, but because of its light weight, the Model T was a lively car for its time. The suspension system allowed the axles to move independently at large angles to allow driving on rough terrain. The clearance, 25 cm, was relatively high to help the car get through mud. It was a true off-road vehicle that could be driven across a farmer's field. Later, attachments were developed for the Model T that allowed it to saw wood, mill wheat, and pump water.

The price for the base model of the Model T was $825, or $16,000 in 2017 dollars. The car was wildly successful. In the 1909 fiscal year, 10,666 Model Ts were sold. In comparison, total car production in Germany in 1909 was 7,300. And this was just the beginning. Ford developed a moving assembly line where the cars are pulled along the line and the parts are added along the way. He got the idea from the way carcasses were handled in meat packing plants. Figure 9.2 shows a Model T assembly line. The efficiency measures that were adopted eventually improved manufacturing productivity by a factor of four. This allowed Ford to lower the prices of his cars and at the same time to raise the pay of his workers to $5 a day, double the prevailing wage. This cut employee turnover and reduced training costs. Production reached a peak of 2,011,125 in 1923. The price was $364, or $7,000 in 2017 dollars. Half the cars made in the world that year were Model Ts. No car since then has remotely approached that share.

Figure 9.3 shows the early production figures for Germany and the United States. By 1910, the American share of world car production was 80%. The share peaked at 90% in 1924. In 1938, on the eve of World War II, American car production was still seven times the German production. The difference in production capacity between the two countries was a critical factor in how the war developed, because automobile assembly lines could be retooled to produce tanks and airplanes. In the aftermath of the Japanese attack on Pearl Harbor, Hawaii, on December 7, 1941, the German chancellor, Adolf Hitler, declared war on the United States. Hitler's decision is not easy to explain, because he was aware of the difference in industrial capability between the two countries. During the war, the United States, in

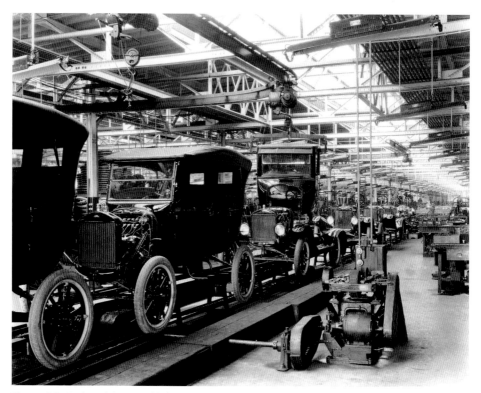

Figure 9.2 Ford Model T assembly line (www.american-automobiles.com/Ford/1913-Ford.html). Photographer unknown/public domain.

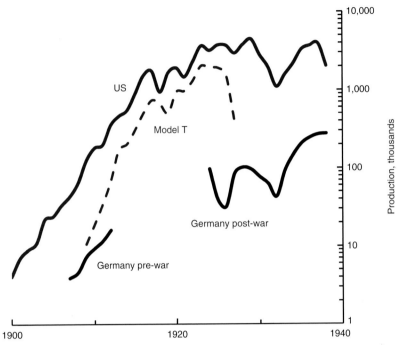

Figure 9.3 Early car production for Germany and the United States. The scale is logarithmic. National car production data are from Brian Mitchell, 2007, *International Historical Statistics*. The Model T production data come from R.E. Houston, Ford Production Department, available at www.mtfca.com/encyclo/fdprod.htm

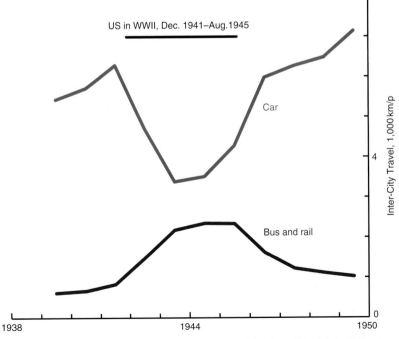

Figure 9.4 Inter-city travel in the United States around the time of World War II. The data come from Richard Gilbert and Anthony Perl, 2010, *Transport Revolutions*, New Society Publishers. The data points are plotted half-way through each year so that the alignment with the war timeline is accurate.

addition to equipping its own army with tanks, vehicles, and airplanes, delivered 500,000 vehicles to the Russian front where most of the land fighting in World War II took place. At the same time, the German army was handicapped by lack of vehicles. Its tanks dominated newsreels, but its infantry walked from the railhead and its supplies moved by horse and wagon.

World War II also performed an interesting natural experiment on American travel preferences. During the war, gasoline for cars was rationed, primarily because there was not enough rubber for tires. Some people were given larger allowances, like physicians, who made house calls in those days. Figure 9.4 compares the travel in cars with buses and trains. Travel by car dropped during the war, and there was a corresponding increase in bus and rail travel. However, by 1946, the first year after the war, car travel is up to the pre-war level. This indicates a clear preference for cars.

9.1.2   Transportation Energy Demand

Turning now to recent years, Figure 9.5 shows the history of world transportation energy demand on a per-person basis. Demand has been rising at a 10-y annualized rate of 0.8%/y, reaching 379 kgoe/p in 2017. The transportation share of

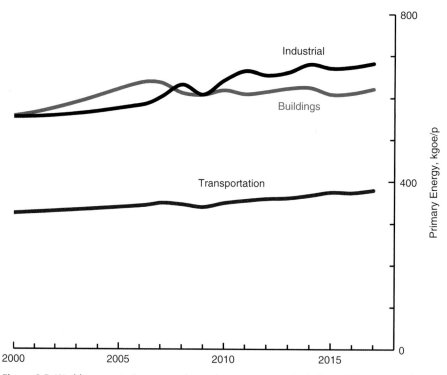

Figure 9.5 World transportation energy demand on a per-person basis from 2000 to 2017. The energy data come from various editions of the IEA *World Energy Outlook*. The population data are from the UN Population Division. The calculations are done on a primary energy basis. The corresponding plots for buildings and industrial demand from Figure 8.1 are added for comparison.

overall primary energy demand has been stable. It was 23% in 2000 and it was 23% in 2017.

Figure 9.6 compares the OECD and the non-OECD transportation demand. The OECD transportation demand has been almost flat, declining at a 10-y annualized rate of 0.1%/y from 2007 to 2017. It was 979 kgoe/p in 2017. On the other hand, the non-OECD demand is increasing steadily at 2.5%/y, reaching 254 kgoe/p in 2017. However, the OECD transportation demand is still 3.9 times larger than the non-OECD demand.

Electricity has had little impact on transportation so far. Figure 9.7 shows the electricity share of transportation demand. The share was 2.5% in 2007 and 3.0% in 2017. Electric passenger trains have been successful in some countries. In a train, the electricity can be supplied by the grid through overhead wires and power can be returned to the grid by regenerative braking. Only a tiny fraction of freight moves by electricity. While battery electric vans and trucks are technically possible, the cost of batteries has kept them off the market so far. Similarly, the market for electric boats has been limited to small craft that take passengers around a harbor and ferries with short routes. The low energy density of batteries compared to

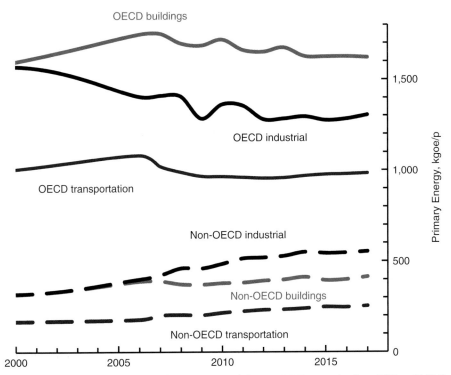

Figure 9.6 Transportation demand for the OECD and the non-OECD countries from 2000 to 2017. The data come from various editions of the IEA *World Energy Outlook*. The corresponding plots for buildings and industrial demand from Figure 8.5 are included for comparison.

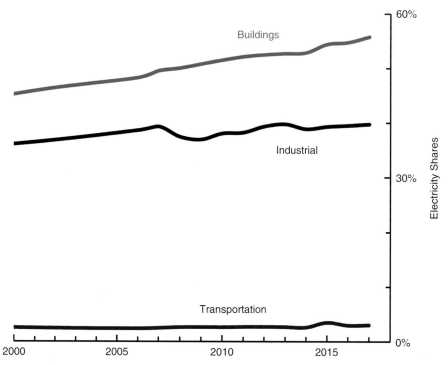

Figure 9.7 Electricity share of transportation demand from 2000 to 2017, calculated on a primary energy basis. The data come from various editions of the IEA *World Energy Outlook*. The corresponding curves for buildings and industrial demand from Figure 8.6 are repeated here for comparison.

aviation fuel limits electric airplanes to short ranges, but unmanned electric drones have been very successful, particularly for aerial photography.

9.1.3 Electric Vehicles

Many governments are encouraging electric vehicles (EVs). EVs are fun to drive. They are quiet and drivers love the instantaneous torque that is available from electric motors. EVs are designed with the batteries at the bottom of the frame. This gives them a low center of gravity and excellent handling. There is no transmission and no clutch in an electric car. Some Chinese cities with severe air pollution make it difficult for owners of gasoline cars to get license plates, but allow electric cars because they have no local emissions. For good performance in snow and mud, designers put an electric motor at every wheel. A major limitation is that the energy density of batteries is much less than that of oil. For example, the type-18650 lithium-ion battery often used in electric vehicles has an energy density of 800 MJ/t. This compares with the tonne-of-oil energy equivalent of 42 GJ/t. In addition, it is much faster to fill a tank with gasoline than it is to charge an electric vehicle battery. A gasoline car can be refueled in a minute. Expressed as a power, this is 15 MW. In comparison, the Tesla home charging station delivers 20 kW.

Norway has the highest EV market share, 39% of light vehicles in 2017. Several factors make Norway a special case. The country has low residential electricity prices, high gasoline prices, high import tariffs that are waived for electric vehicles, and narrow winding cliff roads that exhaust drivers and discourage them from long-distance driving. In the United States, Elon Musk's Tesla Model 3 sedan (Figure 9.8) achieved a breakthrough in 2018, with sales of 140,000. This was 59% of the US EV market that year. Figure 9.9 shows the sales history for the different models. Tesla's other models, the S sedan and the X SUV have never exceeded 30,000 in a year. The S and the X are more expensive cars, priced at around $100,000, whereas the Model 3 is closer to $50,000. It should also be noted that the federal subsidies for people who buy Teslas are expiring and this may reduce future sales. What distinguishes the Tesla EVs is that they are full-range vehicles with 500 km between charges. Their competitors have ranges of less than 400 km. Range for electric vehicles is a particular issue in cold weather. Cars with internal combustion engines use waste heat from the engine to warm the passenger compartment. In an EV, electricity is consumed in heating both the passenger compartment and the batteries themselves. Cumulative American EV sales through 2018 were 630,000. This could be compared with the total number of registered light vehicles in the United States, which was 243,000,000 in 2015.

Figure 9.8 The 2017 Tesla Model 3 sedan. One thing that makes this EV look different from a gasoline car is that the front cooling radiator is smaller. The reason is that electric motors are more efficient than gasoline engines and generate less heat. Photograph by Dale Yee.

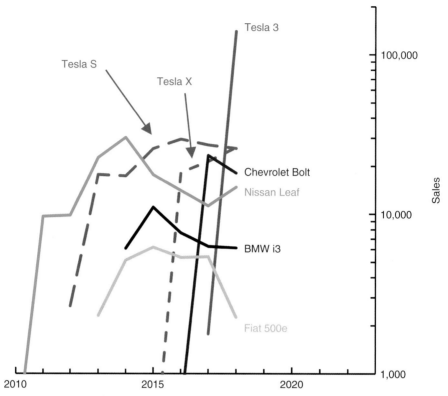

Figure 9.9 Electric vehicle sales in the United States. The scale is logarithmic. These are the battery electric vehicle models (not hybrids) with sales above 2,000 in 2018. The BMW i3 numbers include some cars with a range extender, a small gasoline generator that charges the battery to boost the range. The data come from the Inside EV monthly plug-in and EV sales scorecard at https://insideevs.com/monthly-plug-in-sales-scorecard/

9.1.4   Transportation Oil Demand

While transportation is a modest part of overall energy demand, it does take a lot of our oil. This is shown in Figure 9.10. The share has increased from 37% in 1971 to 58% in 2017. This reflects both the ability to substitute natural gas for oil in stationary demand and the difficulty of substituting anything for oil in transportation demand.

For transportation oil, shares for land, sea, and air travel have been quite stable over the years. Table 9.1 gives the percentages for 2000 and 2015. In 2015, 47% of the transportation oil demand was for passengers by land, while 32% was for land freight. Air transportation, which is dominated by passenger travel, used 11% of transportation oil. For sea transportation, which is mostly freight, the share was 10%.

9.1.5   Commuting

Almost half of the transportation oil demand is for passengers by land, and for many people, the most important land travel is going to work. Table 9.2 shows how Americans have commuted to work since 1960. The biggest change has been

Table 9.1 World transportation oil demand shares. From the IEA 2016 *World Energy Outlook.*

	2000	2015
Passengers by land	47%	47%
Freight by land	31%	32%
By air	12%	11%
By sea	10%	10%

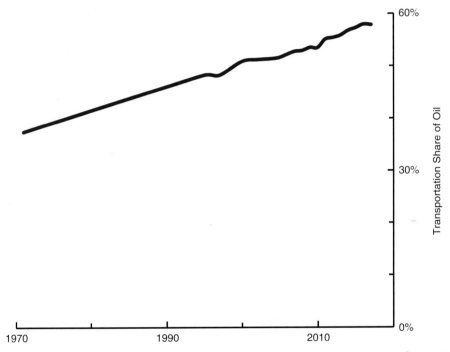

Figure 9.10 Transportation share of world oil demand from 1971 to 2017. The data come from various editions of the International Energy Agency (IEA) *World Energy Outlook.*

the increase in the share of people who travel by car, up from 66.9% in 1960 to 87.9% in 2000, falling back slightly to 85.3% in 2016. Even this modest decline has been completely offset by the fact that the fraction driving alone has been rising. The first survey to ask whether people were alone in their cars was in 1980, when 77% of the car commuters drove alone. This increased to 86% in 2000, and to 89% by 2016. The problem with carpools for commuters is that they are less flexible than driving alone.

Two categories show large drops: transit and walking. Transit fell from 12.6% in 1960 to 5.1% in 2016. The walking share dropped from 10.3% in 1960 to 2.7% in 2016. The cycling share has been steady, 0.5% in 1980 and 0.6% in 2016. During this time, many cities have added bicycle lanes that make the commute more

Table 9.2 Commuting to work in the United States over the years.

	Car	Transit	Walk	Bicycle	Other	Home
1960	66.9%	12.6%	10.3%	na	2.6%	8.0%
1970	77.7%	8.5%	7.0%	na	2.9%	4.0%
1980	84.1%	6.2%	5.6%	0.5%	1.3%	2.3%
1990	86.6%	5.1%	3.9%	0.4%	1.1%	3.0%
2000	87.9%	4.6%	2.9%	0.4%	1.0%	3.3%
2010	86.3%	4.9%	2.8%	0.5%	1.2%	4.3%
2016	85.3%	5.1%	2.7%	0.6%	1.2%	5.0%

Data for 1960 to 1990 from the US Census. Data from 2000 to 2016 from the American Community Survey. Transit includes buses, rail, and ferries. The categories have changed over the years. From 1960 to 1990, taxi services were included in transit. In the more recent American Community surveys, they were counted as "other."

pleasant. The "other" category lumps together motorcycles, planes, and personal boats, and since 2000, taxis. This category has been stable, 1.3% in 1980 and 1.2% in 2016. Working at home shows the most interesting pattern. In earlier times many wives did piecework sewing at home, but beginning in the 1960s, women joined the work force outside the home. However, in the early 1980s people began telecommuting with personal computers, and since then working at home has been the fastest growing category, rising from 2.3% in 1980 to 5.0% in 2016.

Enormous sums have been spent on transit projects in the United States to try to get people out of their cars. Transit does have the advantage that a parking space is not needed at the work place. However, the fundamental problem with transit in the United States is that it takes too long, half an hour longer than driving each way. The only place in the United States where the transit dominates is the island of Manhattan, where three-fourths of the commutes are by transit. Manhattan is a special case. Parking is more expensive than anywhere else in the United States, and congestion in Manhattan is so severe that *walking* can be as fast as driving. It is also important to realize that many of the jobs in the New York metropolitan area are in the surrounding boroughs rather than in Manhattan. In these areas, the majority of the people drive to work.

Figure 9.11 shows fuel consumption over time for different types of passenger travel, expressed as passenger-kilometers per kgoe. We can think of this as the distance in kilometers that one passenger moves on 1 kgoe of fuel. Trains are the best, at 31 pkm/kgoe inside the city and 30 pkm/kgoe between cities. The fundamental reason for this is that rolling steel wheels on steel rails have low friction losses. Next is air travel at 26 pkm/kgoe. The improvement in fuel consumption for air travel has been remarkable. The annualized improvement is 3.2%/y. In 1970, it

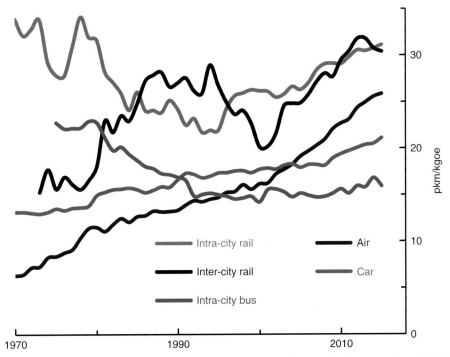

Figure 9.11 Fuel consumption for passenger travel in the United States. The data come from Oak Ridge National Laboratory, 2018, *Transportation Energy Data Book*. The average number of passengers is 1.6 for cars. Calculations are on a primary energy basis because some trains are electric.

was the worst in fuel economy, but by 2015 it was closing in on trains. Aside from speed, air has other advantages. The routing is flexible and the infrastructure is localized. There is no need to secure rights-of-way over thousands of kilometers. Cars follow at 21 pkm/kgoe. Cars have shown modest improvement from 1970 to 2015, with an annualized increase of 0.4%/y. At the bottom is intra-city buses at 16 pkm/kgoe. The problem with the buses is that most of the seats are empty; there are only nine passengers on average. The emphasis is on providing transportation for people who do not have it, rather than efficient fuel consumption.

One thing that makes transportation demand different from stationary demand is that many people are killed traveling. It is a rare adult who did not know someone who died in a car wreck. Passengers choose the way they want to travel on the basis of price, speed, fuel consumption, flexibility, and exercise value. There are also differences in safety. Table 9.3 shows the death rates for the different types of passenger travel in the European Union, sorted from the safest to the most dangerous. Trains and commercial airlines are the safest. Driving is twenty times more dangerous than trains and commercial airlines, and walking and cycling are an order of magnitude more dangerous than driving. Motorcycles are the worst, which would not come as a surprise to any emergency room physician. Buses are

Table 9.3 Hazards of passenger travel.	
	Fatalities/100 million pkm
Trains	0.035
Airlines	0.035
Bus	0.07
Ferry	0.25
Cars	0.7
Bicycles	5.4
Walking	6.4
Motorcycles	13.8

The data come from Table 1 in the European Transport Safety Council, 2003, "Transport Safety Performance in the EU, a Statistical Overview."

ten times safer than cars. Part of the reason for this is that buses are larger than cars. Similarly, large cars are safer than small cars. However, fuel consumption is lower for small cars and small cars are cheaper. The customer makes a trade-off between price and fuel consumption on the one hand and safety on the other.

9.1.6   Freight

Now we turn to freight. Table 9.4 below shows freight modal shares in different countries. The second column is traffic volume normalized to the population and the square root of the area. China's population is 40 times that of Canada, and Canada's area is 26 times that of Japan. Canada has the largest normalized freight traffic. This reflects the fact that Canada is a high-income country with a relatively large grain and mineral rail traffic. China has the lowest normalized freight traffic. China is a poorer country than the others. The American water freight system is weak. The problem is that there are three major mountain ranges that run north–south in the United States. These are the Appalachians in the East, the Rocky Mountains in the West, and the Sierras along the eastern border of California. Before the development of railroads and the building of the interstate highway system, these mountain ranges were a serious hindrance to American economic development.

The most interesting aspect of the table is that while the EU and Japan have the best passenger rail networks, they also have the weakest freight rail systems. On the other hand, Canada and the US have weak passenger rail networks, but excellent freight railroads. Freight and passenger rail are fundamentally incompatible. Passenger rail service follows a fixed schedule. In contrast, freight trains do not have a rigid time schedule. Rather, the dispatchers distribute the trains in space along the rails. Freight trains spend much of their time waiting at signal lights, which is unacceptable in a passenger network.

Another factor that helps determine the freight shares in Table 9.4 is the fuel consumption. Table 9.5 expresses this in units of tkm/kgoe, sorted from worst to best. We can think of this as the distance a tonne of freight can be moved with 1 kgoe of energy. Air freight has by far the highest fuel consumption, and that limits its cargo to items that have a relatively high value per kilogram. Trucks use considerably more energy than rail or pipelines. On the other hand, trucks can pick up from a seller and deliver to a customer. For this reason, trucks dominate at distances shorter than 500 km. This factor also helps explain the importance for trucks in the European Union and in Japan, where the typical distances are shorter than in China, the US, and Canada. Rail and pipelines are comparable in efficiency, but pipelines are safer. The best freight efficiencies are achieved by large ships. Ships have fundamental efficiency advantages for carrying freight as they

Table 9.4 Freight modal shares for selected countries and the EU.

	Volume, $\dfrac{\text{tkm}}{p\sqrt{A}}$	Road	Rail	Water	Air
EU 25 (2005)	4.1	46%	10%	44%	0.1%
Canada (2004)	8.2	29%	39%	32%	0.4%
US (2003)	5.6	36%	46%	17%	0.4%
China (2004)	1.7	11%	28%	60%	0.1%
Japan (2005)	7.3	59%	4%	37%	0.2%

The total freight traffic in tkm is normalized by dividing by the population and the square root of the area in km^2. The country areas are taken from the United Nations FAOSTAT online database. The data in tkm/p come from Richard Gilbert and Anthony Perl, 2010, *Transport Revolutions*, New Society Publishers, Gabriola Island, British Columbia.

Table 9.5 Freight fuel consumption.

	tkm/kgoe
Air	7
Heavy trucks	43
Rail	196
Pipeline	200
Container ship	233
Oil tanker	689

The data for US rail come from Oak Ridge National Laboratories, 2014, *Transportation Energy Data Book*. The estimates for ships and airplanes come from David MacKay, 2009, *Sustainable Energy: Without the Hot Air*, UIT Cambridge. The data for US heavy trucks come from the IEA, 2017, *The Future of Trucks*. The estimate for US pipelines comes from J. N. Hooker, 1981, "Oil pipeline energy consumption," Oak Ridge National Laboratory.

are made larger. Surface friction is proportional to the hull area, which increases as the square of the length, while freight volume increases as the cube of the length.

In this chapter, we will review the major transportation modes: roads, rails, sea, and air. The most important change in recent years has been the invention of the standardized shipping container by Malcom McLean. This has revolutionized the loading and design of trains and ships, and enabled the development of intermodal transportation, where containers move between ships and trains and trucks. We will conclude with a speculative look at the future of transportation, focusing on self-driving vehicles.

9.2   Roads

A nation's road network is critical infrastructure. A road network is expensive to build and to maintain, but it contributes enormously to the economy and to the quality of life of its citizens. Roads give drivers an exhilarating sense of freedom. For many people, their time in the car is a time of privacy that they do not feel at the office, on the bus, or even at home. Roads allow cars and trucks to go door to door. The backbone of a road system is its expressway network. In the United States, these are the interstates, while in the UK they are the motorways. Expressways are safer than other roads because the two directions of traffic are separated. Expressways do not have traffic lights, so the average speeds are much higher than on ordinary streets. However, they are expensive because of the crossovers. Figure 9.12 shows the intersection of two interstate highways in Los Angeles. One thing that makes this intersection complicated is that there are separate connections for the carpool lanes. The Interstate Highway system has 80,000 km of expressways, while the British motorways total 4,000 km. The differences in length reflect several factors. The most important is the size of the countries. However, another factor is that the British system primarily connects cities. The United States system also connects cities, but in addition it provides extensive access into cities. In many cities in other countries, access into the city center is slow and tedious. Finally, Americans drive more than the British do.

The Romans built an excellent network of roads paved with stones. Their primary interest was military as there was almost continuous fighting on the imperial marches. The roads let the legions move from one region to another to counter attacks. In addition, the roads allowed communications and the movement of supplies. After the collapse of the empire, the Roman roads deteriorated and there was little improvement in road technology until the early 1800s, when a Scot, John McAdam, invented a superior road. McAdam's design had two layers of carefully sized stones. The coarse layer at the bottom with 8-cm stones was 20 cm thick. The fine layer on top was 5 cm thick with 2-cm stones. The road had a 10-cm crown

Figure 9.12 The intersection of Interstate 110 and Interstate 105 in Los Angeles. The 110 runs up and down, while the 105 runs left and right. Credit: trekandshoot/Shutterstock.

and the water drained through the coarse bottom layer into side ditches. The attention to drainage helped prevent damage from frost. With use, the stones locked together and formed a stable, durable surface.

The McAdam roads were fine for horses and carriages, but cars kicked up dust, so a layer of asphalt was added on top of the stones. Asphalt is a mixture of bitumen and sand or gravel. Bitumen occurs naturally and it is also produced in large quantities as the viscous residue of petroleum refining. Informally, asphalt is called *tar*, so the asphalt covered McAdam road became the "tarmac." When properly maintained, these roads are quite smooth and they allow high speeds. On some German highways there are no speed limits and people routinely drive above 200 km/h on them. More recently, Portland cement mixes have been developed. The cement surfaces are not as smooth as asphalt, but they can withstand heavy loads. There are now millions of kilometers of asphalt and cement roads around the world.

9.2.1 The American Interstate Highway System

At the beginning of the twentieth century, American roads were terrible. One of the attractions of the Model T was that it had a high clearance and a flexible suspension to allow it to go on rough roads. Figure 9.13 shows a Model T on a muddy road in North Carolina. By the 1920s, US car production exceeded that in the rest of the world combined, and as a result, Americans began to support

Figure 9.13 A Model T on a muddy road in Johnston County, North Carolina in 1909. Credit: Albert Barden, State Archives of North Carolina, N_53_15_7070.

road construction. There had been limited programs to build paved roads dating back to the late 1800s, primarily led by cyclists, who need smooth surfaces. Some highways were built in the 1920s and 1930s, but construction stopped during World War II. The most important legislation was the Federal-Aid Highway Act of 1956 during the Eisenhower Administration that initiated the construction of the Interstate Highway System. President Eisenhower had been the commander-in-chief of allied forces in Europe during World War II, and he was impressed with the German autobahns. He envisioned roads like the autobahns to connect cities. Eisenhower was painfully aware that American roads were not up to the German standard. In 1919, he had participated in an army truck convoy that travelled across the United States. That trip took two months. "Federal-Aid" in the title of the legislation refers to the financing. The federal government provided 90% of the money and the states paid 10%. The federal money came from taxes on gasoline and diesel fuel.

By the time the Federal-Aid Highway Act was passed, engineers in the Federal Bureau of Roads had been thinking about a national highway system for many years. From their surveys they were aware that there was much more intra-city traffic than inter-city traffic. In their hands, the Interstate Highway system became quite different from Eisenhower's idea of an American autobahn. Construction continued 36 years from 1956 until 1992, when engineers blasted through the Rocky Mountains west of Denver to finish Interstate 70. Today this section of I-70 is one of the most spectacular drives in the entire system. The sacrifices Americans citizens made for the construction were remarkable. An inconceivable total of three-quarters of a million buildings were demolished for the highways. There are some gaps. I-70 stops on the outskirts of Baltimore. This allowed some residents to protect their neighborhoods from intrusion. On the other hand, Baltimore is a port city and ports need good highway connections to be competitive.

Today, the intra-city interstates are dominated by commuters, while the inter-city interstates primarily carry truck freight and tourist traffic. Figure 9.14 is a map of the inter-city truck traffic in the United States. The interstate traffic is shown in red. The non-interstate freight is shown in gray. There is significant traffic on non-interstate roads in only two places. One is the central valley of California, where California State Highway 99 supplements Interstate 5. The other is in south Texas, where Interstate 35 meets the Mexican border at Laredo, while US Highway 281 goes into McAllen. The Texas crossings are the two most important ones for truck freight along the entire Mexican border. The Interstate system is remarkable. To go from San Francisco to New York, the instructions are to get on I-80 and drive 4,700 km. There are no traffic lights and few tolls.

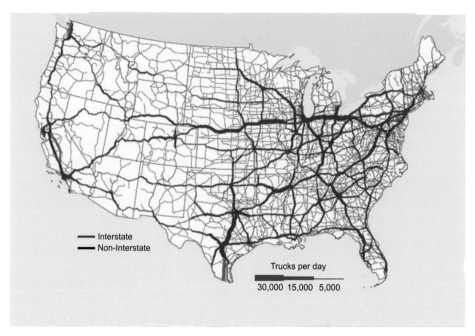

Figure 9.14 Truck traffic in the United States. Credit: US Department of Transportation.

9.2.2 Vehicle Fuel Efficiency

One way to get a sense of where vehicles are headed is to consider how fuel efficiency and horsepower have evolved. Figure 9.15 plots horsepower against fuel efficiency by model year from 1975 to 2017. During this time the average weight of vehicles stayed almost constant, 1,842 kg in 1975 and 1,834 kg in 2017. The most dramatic change in efficiency is the shock of the 1979 Iranian Revolution, when it jumps from 6.8 km/l (15.9 miles per gallon) in 1979 to 8.2 km/l (20.5 miles per gallon) in 1980. However, there is a turning point in 1987. This followed a period when the North Sea oil was coming in and the price of oil was falling. From 1987 until the next turning point in 2004, horsepower rose and efficiency fell. By 2004, prices were rising and there were concerns about Peak Oil again. Since 2004, the fuel efficiency recovered and by 2017, it was 15% above the 1987 level. This is not impressive, less than the 29% jump in one year after the Iranian Revolution. The problem is trucks. The light truck share of sales increased from 19% in 1975 to 42% in 2017. Many trucks are bought with powerful engines for towing trailers. For example, to tow a typical 3-t recreational vehicle (RV) with the popular Ford F150 pickup, one might choose a 325 hp engine.

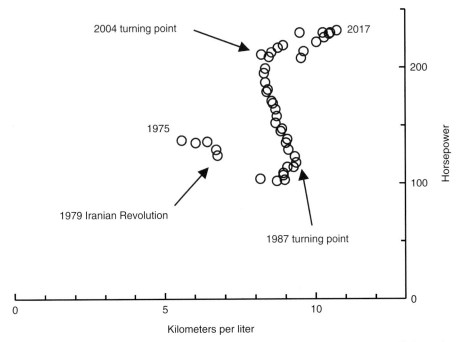

Figure 9.15 Horsepower vs. fuel efficiency in kilometers per liter for American cars and light trucks for the model years from 1975 to 2017. The data come from the EPA report, "Light-Duty Automotive Technology, Carbon Dioxide Emissions, and Fuel Economy Trends."

From an energy perspective there are two losses associated with driving on the highway, the drag force, given by Equation 7.34, and the rolling resistance force. The rolling resistance is expressed in the form

$$F_r = C_r F_g \qquad\qquad 9.1$$

where F_r is a force retarding the vehicle, C_r is a rolling resistance coefficient, and F_g is the gravitational force. Like the drag coefficient, the rolling resistance varies somewhat with speed. There are several factors that contribute to rolling resistance. As the wheels turn, the tires flatten and recover. In the cycle of flattening and recovery, heat is generated in the tire. This effect is called *hysteresis*. It is important for rubber tires, but less significant for the steel wheels on trains. Another factor is that the surface depresses as the vehicle rolls over it. This means even when a vehicle is on level ground, it is effectively rolling slightly uphill. This is an important factor for trains. The rail become stiffer as the wheel digs into it, so that C_r goes down as the mass of the freight car increases.

For cars, a representative value of C_r is 0.01. As an example, we will estimate the losses for the Ford 150 pickup truck. The truck parameters in the calculations are given in Table 9.6. Ford is a bit coy about the aerodynamic properties of its trucks, so they are estimated for the calculation. Figure 9.16 shows the results

Table 9.6 Estimated loss parameters for a Ford F150 pickup rolling on level ground. This pickup has been the best-selling vehicle model in the United States for many years.

Rolling resistance coefficient C_r	0.01
Mass, t	2
Drag coefficient C_d	0.4
Area, m^2	3

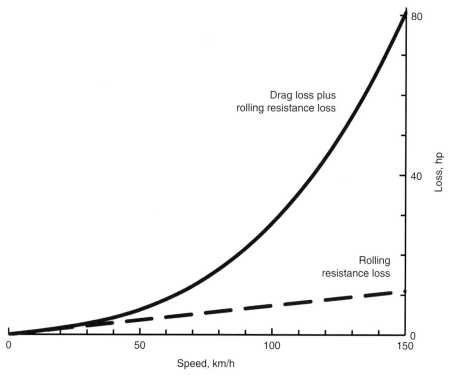

Figure 9.16 Calculated drag and rolling resistance losses for a Ford F150 pickup truck on level ground. The parameters for the calculation are given in Table 9.6.

as a function of speed. The power is expressed in horsepower so that it can be compared with engine horsepower. There are two components of the loss. The rolling resistance force is given by Equation 9.1 and in this approximation, the loss increases linearly with the speed. The drag force is quadratic in the velocity, so the loss varies as the cube of the velocity. The drag force dominates at high speeds. The crossover speed, where the rolling resistance force is the same as the drag force, is 59 km/h (37 miles per hour). The horsepower losses in the figure are lower than the power capacity of typical engines. It is easy for a car to maintain a speed on level ground. The higher power capacity of engines gives a margin for maintaining speed going uphill, accelerating onto an expressway, driving at high altitudes, and pulling a trailer.

9.3   Rails

Trains have large capacities compared to road vehicles. The Japanese *shinkansen* carry 1,300 people and freight trains in the United States routinely pull 300 forty-foot containers. The rolling resistance is an order of magnitude better than cars – a C_r of 0.001 is typical. In addition, drag is a relatively small effect in trains because the masses are so much larger. These factors make trains fundamentally more efficient than road vehicles. A more surprising consequence of the low rolling resistance for trains is that they will run away on even a slight grade if the parking brakes are not set properly.

9.3.1   The Japanese Shinkansen

The genesis of the high-speed passenger train was in Japan. After the Battle of Sekigahara in 1600, the Tokugawa Clan that ruled Japan severely limited contact with Westerners, allowing only a few Dutch ships each year at the port of Nagasaki. The Tokugawas were defeated in a civil war in 1868. The young Meiji Emperor acquired increased power and he moved his residence from Kyoto to Tokyo, formerly called Edo. The new government built a navy, established a telegraph service, and started a postal system. In addition, it encouraged railroad development both by the Imperial Japanese Government Railway and by private companies. From 1904 to 1905, Japan fought a major war with Russia. Japan won the war, but the military complained that the support from the railroads was not good enough, and the railroads were nationalized the following year. By the end of the Meiji Era in 1912, the country had a comprehensive national network of 5,000 route kilometers. In addition, it was making its own rails, rolling stock, and locomotives. Japan is a mountainous country and the country started with a narrower gauge than in other countries to keep costs down. However, it was appreciated in the Railway Ministry that the country would eventually need the larger standard-gauge trains and that it would be possible to connect Japan's four most important cities, Kyoto, Nagoya, Osaka, and Tokyo by a single 550-km long line. In the 1930s, plans were made for a high-speed line connecting these cities. Several tunnels were built for the line but construction was interrupted by World War II. After the war, construction resumed in the 1950s and the new standard-gauge, high-speed line opened in October 1964 (Figure 9.17). The train was called the *Tokaido Shinkansen*. Tokaido, literally "eastern sea route," was the name of the coast road that connected Kyoto and Tokyo. Shinkansen means "new main line." The name is also used for the trains themselves.

The shinkansen was successful from the start. By 1970, the Tokaido Shinkansen had 233,000 riders per day (Figure 9.18). The system was re-privatized in 1987. By 2004, there were 360,000 passengers per day. The passenger traffic on a shinkansen

Figure 9.17 The classic postcard photograph of the 0 Series Tokaido Shinkansen crossing in front of Mount Fuji. Credit: RogerW/CC BY-SA 2.0, https://creativecommons.org/licenses/by-sa/2.0/

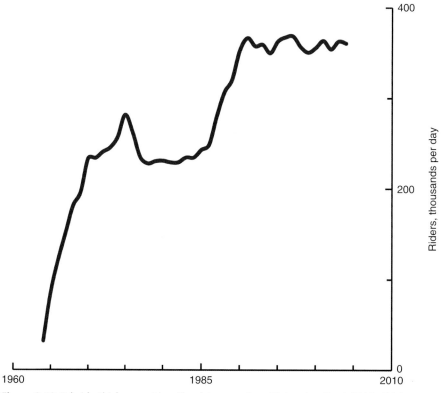

Figure 9.18 Tokaido Shinkansen riders. The data come from Christopher Hood, 2006, *Shinkansen*, Routledge.

line is comparable to that of a major expressway. The route length has continued to expand. In 2014 there were 2,848 route kilometers in the shinkansen network. The maximum speed in 2018 was 320 km/h on the Tohoku Shinkansen that runs from Tokyo through Sendai to Aomori. New train series have been introduced as power electronics has improved. The trains are sets of cars that are permanently attached to each other. There is an engineer cab at each end of the train so that it can be operated in either direction. The Tokaido Shinkansen has 16-car train sets. Rather than a single locomotive pulling passenger cars, in a shinkansen most of the cars have AC induction motors under the floor. This gives excellent track adhesion and acceleration and the leading and trailing cars can carry passengers. Electricity is supplied through pantographs above the cars that make contact with catenary wires suspended from towers over the tracks. The trains use regenerative braking where the motors act as generators and return electrical power to the grid. Mechanical brakes dissipate only a few percent of the kinetic energy of the train. The shinkansen are punctual, with average schedule delays of seconds. The safety record is inconceivably good, with no deaths due to derailments. Japanese cities have superb connecting networks of airports, ordinary trains, subways, buses and taxis, so the shinkansen can be used for both inter-city travel and commuting.

The shinkansen do have disadvantages. There is no checked luggage, so passengers must engage a shipper for large bags. The curves must be gentle to allow high speeds and this means that are long runs inside tunnels, where the sounds and pressure changes are unpleasant. The shinkansen are also expensive. A reserved ticket on the Hikari Superexpress from Tokyo to Osaka costs 14,140 yen, or $125. The trip takes three hours. This fare could be compared with the Southwest Airlines $49 Getaway fare going the same distance from a Southern California airport to the San Francisco Bay Area. The smaller airports have close-in parking and short security lines. The scheduled flight time is half the shinkansen travel time. Moreover, it is inexpensive to drive a car from Los Angeles to San Francisco. The cost for gasoline is less than the Southwest fare and there are no tolls. In contrast, in Japan, it would cost more to drive from Tokyo to Osaka than to take the shinkansen because of the tolls.

9.3.2 American Rail Freight

One of the factors in the success of the shinkansen is that the trains had their own dedicated tracks so they did not have to work around freight trains. Passenger trains also hinder freight trains because they need track priority to maintain their schedules. In the 1960s and 1970s, American railroads were hobbled by being forced by regulators to maintain passenger services that were losing money because of competition from airplanes and automobiles. Several major railroads went bankrupt, and others did not have the ability to make investments for tracks and to buy locomotives. However, in 1980, near the end of the Carter

Administration, the Staggers Act deregulated the railroads. This gave the railroads flexibility to set rates and to cancel services. The consequences, shown in Figure 9.19, were remarkable. Freight volume almost doubled and rates were cut in half. Productivity more than doubled. Many railroads merged and an integrated North American railway system including Canada and Mexico has emerged. There are now ten major North American railroads that the US Surface Transportation Board rates as Class 1 based on their revenue. These include two national passenger railroads, the American Amtrak and the Canadian Via. The other eight are primarily freight lines. The two largest railroads, close in size, are Union Pacific and BNSF. Norfolk Southern and CSX are freight lines that operate only in the United States. Canadian Pacific, Canadian National, Norfolk Southern, and CSF operate both in Canada and the United States. Kansas City Southern operates in both the United States and Mexico, while Ferromex only has trackage in Mexico. Unlike the other parts of a national transportation network, freight railroad companies can make money without significant subsidies. This gives them the revenue to improve tracks, to buy new locomotives to satisfy tighter pollution limits, and to buy new safer tank cars.

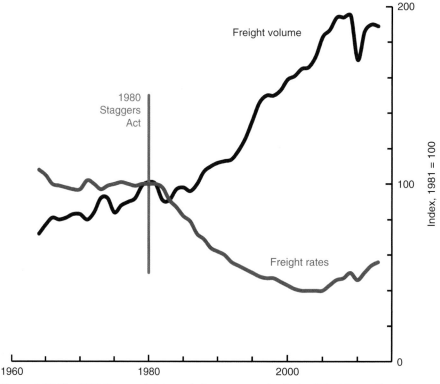

Figure 9.19 The 1980 Staggers Act deregulating transportation in the United States. The data come from Robert Gallamore and John Meyer, 2014, *American Railroads*, Harvard University Press.

There have also been technical improvements. The most important is the development of the stacking cargo container. This idea was due to American trucker Malcom McLean, who wanted to make loading and unloading of ships more efficient. The containers were also made compatible with trains, where it meant that individual boxcars no longer needed to be loaded and unloaded. The freight services are called *intermodal* to reflect the fact that the same container can go by ship, train, and truck. Figure 9.20 shows an intermodal train with double-stacked containers and truck trailers mounted on flatcars. Another advance is the ability to have multiple locomotives on the same train all controlled by engineers in the lead locomotive. The locomotives can be distributed among other parts of the train and the power and brakes for the distributed locomotives can be adjusted independently to minimize the stress on the couplers. When a long train is passing over hills and valleys, some locomotives may be pulling at the same time that others are braking.

The locomotives in the figure are diesel electrics. There is a primary diesel engine that drives an electric generator. The electricity that is generated is then converted electronically to provide the voltages for the motors at each axle of the locomotive. The locomotives do not have a grid connection so there is no way to do regenerative braking. However, there is an interesting electrical braking system that acts like regenerative braking in that it does not require friction brakes. It is

Figure 9.20 BNSF intermodal freight on the Tehachapi Loop in Southern California. Any train longer than 1200 m will cross over itself on the loop. The purpose of the loop is to gain height. The grade is 2%. Credit: David Brossard/CC BY-SA 2.0, https://creativecommons.org/licenses/by-sa/2.0/

called *dynamic braking.* In dynamic braking, the connections to the motors are switched so that they act as generators rather motors. The electricity is dissipated in a resistor with a large fan. These resistor-fan modules can each dissipate 1 MW of power to slow the train.

There are two types of yard facilities for sorting freight. The first is a classification yard, shown in Figure 9.21. A classification yard breaks up a set of cars and reassembles them into new attached sets. The attached sets are called *consists.* The cars are pushed by a switching locomotive to the top of a gentle slope. The cars are decoupled and roll down the slope one by one. In the photograph, there are two of these rolling cars in the lower right corner. An operator in a tower sets switches to steer the cars to the correct consist. As each car rolls into its consist, the couplers attach automatically. The second kind of facility is an intermodal yard, shown in Figure 9.22. Here there are cranes that load and unload the containers from railroad cars and trucks and stack and unstack them.

In Section 4.7.1, we discussed the importance of the railroads in transporting coal. In recent years, trains have also played a role in the fracking revolution. For a while, oil production grew more quickly than pipeline capacity and oil trains picked up the slack. However, shipping oil by pipeline is cheaper, and the railroads lose the business when the pipelines catch up. Figure 9.23 shows the history of crude-oil movements by pipeline and rail in the US during the recent growth in

Figure 9.21 The Union Pacific classification yard in Fort Worth, Texas in 2014. Photograph by Dale Yee.

Figure 9.22 Intermodal freight terminal in Stourton, UK. Credit: Ian Kirk/CC BY-SA 4.0, https://creativecommons.org/licenses/by-sa/4.0/

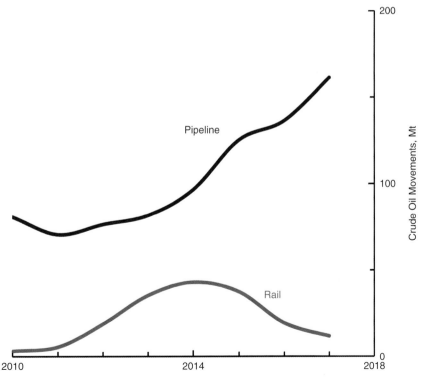

Figure 9.23 Crude oil movements by pipeline and rail in the United States. This is the crude oil that crosses the boundary of one of the five Petroleum Administration for Defense Districts (PADD). The data come from the EIA.

oil production. Rail shipments peaked in 2014, when they had a 31% share of the combined movements. However, by 2017, the rail share had fallen to 7%.

9.4 Impacts

9.4.1 The Wreck of the MMA-002 Oil Train

On July 5, 2013, the eastbound Montreal, Maine, and Atlantic freight train, MMA-002, pulled into Nantes, Quebec at 10:49pm to stop for the evening.[3] The engineer, Tom Harding, had left Farnham, Quebec at 1:30pm, a half hour late because of personal errands. He was having trouble with his lead locomotive. It was smoking and leaking oil, and he complained about it to his rail traffic controller in Farnham, Richard Labrie. Harding had been working for railroads for 33 years, starting with Canadian Pacific in 1980. He had driven trains down this line hundreds of times. Figure 9.24 shows the company route map. The train was carrying 6.3 kt of crude oil from the Bakken Oil Field in North Dakota to the Irving Refinery in St. John, Nova Scotia. This refinery is the largest in Canada, with a capacity of 40 kt/d.

It happened that Richard Deuso, a fireman from Newport, Vermont snapped a photograph of MMA-002 that afternoon near Brookport, Quebec. Deuso's photograph is shown in Figure 9.25. In the photograph, five locomotives are pulling a consist of tank cars. There is a caboose after the lead locomotive. This caboose held electronics to control the four following locomotives. Between the last locomotive and the first tank car there is a white car filled with gravel. This is a buffer car, meant to provide protection for the locomotives from burning crude oil.

Shutting down a freight train is a complex process. There are three braking systems. First, there are air brakes on each car of the train. The air pressure for the brakes is supplied by a compressor on each locomotive, and there are hoses that provide connections between each car. This system is called the *automatic* brakes because there are reservoirs on each car that engage the brakes if the pressure in the hoses is lost suddenly. That would happen if a coupler broke while the train was moving. The automatic brakes are also used for stopping the train. In addition, there are individual air brakes on each locomotive that are used as parking brakes. Finally, each locomotive and car has a manual parking brake. Harding followed the procedure that was standard for the railroad. He left the individual air brakes on the locomotives engaged and he set the hand brakes for the five locomotives, the caboose, and the buffer car. He left the lead locomotive running so that its compressor could keep the individual air brakes engaged and he took a taxi to a hotel.

---

[3] The chronology is taken from the Transportation Safety Board of Canada "Railway Investigation R13D54." One will see small differences in other accounts. In many cases precise train information is available from the Locomotive Event Recorder, which is like the black box in an airplane.

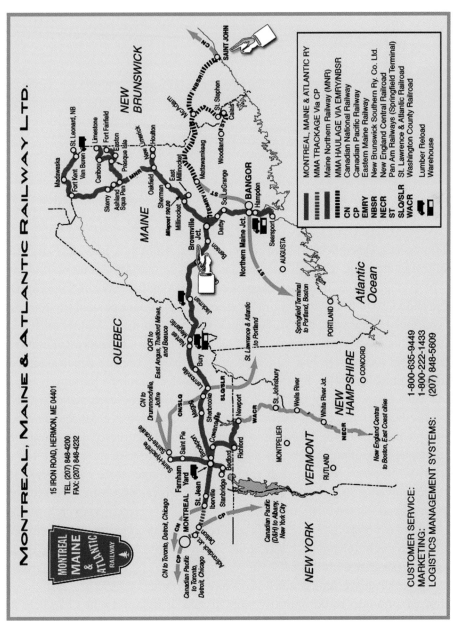

Figure 9.24 Montreal, Maine, and Atlantic Railway route map. From www.oil-electric.com/2013/07/a-night-to-remember.html

Figure 9.25 Montreal, Maine, and Atlantic oil train MMA-002 near Brookport, Quebec, on July 5, 2013. Courtesy of Richard Deuso.

Harding was scheduled to drive the westbound MMA-001 back to Farnham the following morning, while an American crew would take MMA-002 on into Maine.

At 11:40pm, the Sûreté du Québec, the provincial police force, received a 911 call reporting a train fire. With admirable efficiency, the police phoned the Nantes Fire Department, Richard Labrie in Farnham, and the Montreal, Maine, and Atlantic yard in the town of Lac Mégantic, 10 km down the line to the east. The fire was in the lead locomotive that had been giving trouble. The fire department was as competent as the police. By 11:58pm, the fire was out and the burning locomotive was shut down. At 12:35am the track foreman from the yard arrived, met with the fire fighters and called Labrie. Labrie was a locomotive engineer himself, and he gave permission for everyone to leave. At this point, there was no compressor operating to maintain the brake air pressure. The pressure in the hose began to drop slowly as air leaked out of the joints. The contact force of the brake shoes in the individual brakes fell, but the pressure drop was not fast enough to engage the automatic brakes. The track where the train was parked was on a grade. By 12:58am, the brake pressure had fallen from 70 psi to 27 psi, and the train started to roll downhill towards Lac Mégantic. When it arrived at the town at 1:15am, it was travelling at 65 miles per hour. Where the tracks crossed the Rue Frontenac in town, the speed limit was set at 10 miles per hour because of a switch, which allowed a train to go left into the railroad yard or right over the Chaudière River. At this point the coupler connecting the buffer car and the locomotives failed. The locomotives continued along the track while the buffer car and tank cars derailed. The tanks ruptured and the crude oil caught fire (Figure 9.26). The fires burned out downtown Lac Mégantic and killed 47 people. It was the worst freight rail accident in Canadian history.

Table 9.7 Masses in tonnes for the Lac Mégantic runaway train. The data come from the Transportation Safety Board of Canada, 2014, "Railway Investigation R13D54."

Locomotive (5)	177
Caboose	27
Buffer car filled with gravel	95
Tank car (73)	116
Train total	9,475

Table 9.8 Forces in kN for the Lac Mégantic runaway train. The data come from the Transportation Safety Board of Canada, 2014, "Railway Investigation R13D54."

Force of gravity on the train	92,918
Rolling resistance force ($C_r = 0.001$)	93
Required braking force to hold the train (0.92% grade)	762
Holding force for the hand brakes that were set	216
Additional hand brake holding force needed without air brakes	546
Single tank car hand brake holding force	38

Figure 9.26 The wreck of the Montreal, Maine, and Atlantic MMA-002 oil train, Lac Mégantic, Quebec, July 6, 2013, taken from a Sûreté du Québec helicopter. Credit: Sûreté du Québec/CC BY-SA 1.0, https://creativecommons.org/licenses/by-sa/1.0/

The report from the Transportation Board of Canada gives enough information to analyze the accident. Table 9.7 shows the masses of the locomotives and cars and Table 9.8 shows the forces. The force of gravity on the train, 93 MN, is found by multiplying the mass of the train by the acceleration due to gravity, $9.8 \, \text{m/s}^2$. In the report, the rolling resistance coefficient C_r was taken to be 0.001, giving a rolling resistance force of 93 kN. The grade where the train was parked was 0.92%, so the component of the gravitational force down the track, net of the rolling resistance force, is 762 kN. This is the force that must be provided by brakes to hold the train.

The investigators recovered the locomotives and the caboose and several undamaged tank cars and measured the holding force of the brakes. They estimated that the total holding force of the hand brakes that Harding set was 216 kN and that the hand brake in each tank car could have provided an additional 38 kN. This means that 546 kN of additional braking force was needed to keep the train from moving. Initially this force was provided by the individual locomotive air brakes, but as the air leaked out of the lines, the brake force dropped below this level. To provide the 546 kN entirely from hand brakes on the tank cars would have required him to set the brakes on 14 tank cars. For a comparison, the two largest North American railroads, Union Pacific and BNSF, require 22 brakes to be set for a 10,000 t train on a 1% grade.

There were other contributing factors. Diesel engines are extremely reliable, but they must have competent service. There was a siding available at Nantes to park a train that had a safety derailing device that would have prevented the runaway. However, the railroad had rented out the siding track to store flat cars belonging to a factory in Lac Mégantic. The train could have been safely parked on level ground at Vachon, the next town beyond Lac Mégantic, except that would have made the trip take longer than the safety agency, Transport Canada, allowed Tom Harding to drive. Likewise, if he had returned to the train that night to start another locomotive after the fire was out, that would have extended his work day, and the agency rules would not have allowed him to drive the next day. That would have stranded the west-bound MMA-001 train.

As a result of the accident, the Montreal, Maine, and Atlantic line went bankrupt. Harding and Labrie were put on trial and they were acquitted by the jury. A new rail line is being built to bypass the town. An additional result of the accident is that North American railroads are shifting to sturdier tank cars. Fundamentally, however, as we saw with the passenger modes, there are differences in safety in the freight modes. Shippers of hazardous materials in the United States are required to report incidents to the Department of Transportation. Normalized by the number of tonne-kilometers, there 32 times as many road incidents as rail incidents, and 34 times as many rail incidents as pipeline incidents.[4] If safety is the most

---

[4] These results were compiled in a report written by Diana Furchtgott-Roth and published the month before the Lac Mégantic disaster. The hazmat statistics for pipelines are for liquids. Her report, "Pipelines are safest for transportation of oil and gas," is available online at www.manhattan-institute.org/html/pipelines-are-safest-transportation-oil-and-gas-5716.html

important criterion, choose trains over trucks and pipelines over trains. There is a final factor that makes railroads more dangerous than pipelines. Because the railroads carried passengers in their early days, the tracks go through towns. Long-distance pipelines generally avoid them.

9.4.2 "A Brown LA Haze"

Air pollution from burning coal has been a problem for hundreds of years (Section 4.4.2). California has little coal, so it was a surprise when smog developed in Los Angeles. Caltech Chemistry Professor Arie Haagen-Smit described the problem in 1952.

> Air pollution in the Los Angeles area is characterized by a decrease in visibility, crop damage, eye irritation, objectionable odor, and rubber deterioration.[5]

Figure 9.27 shows a view of downtown Los Angeles through the smog. Long-time Los Angeles residents will appreciate that this was actually not a bad day. On a bad day the buildings would not be visible at all and workers in a high-rise office would not see anything but haze. The "brown LA haze" in the Jimmy Buffet epigraph at the beginning of the chapter was particularly visible flying into the

Figure 9.27 Smog in Los Angeles, September 1973. Credit: Gene Daniels/US National Archives and Records Administration/public domain.

---

[5] A. J. Haagen-Smit, 1952, "Chemistry and physiology of Los Angeles smog," *Industrial and Engineering Chemistry*, vol. 44, pp. 1342–1346.

city. There were many people who flew in for job interviews, saw the pollution from the plane, and turned down the offers.

Professor Haagen-Smit diagnosed the primary problem as car emissions.

> These effects are attributed to the release of large quantities of hydrocarbons and nitrogen oxides to the atmosphere. The photochemical action of nitrogen oxides oxidizes the hydrocarbons and thereby forms ozone, ...

Ozone (O_3) is a strong oxidizer that damages animal lungs and plant cells. It is not emitted directly. Rather it is a secondary pollutant formed from chemical reactions. The concentrations of ozone were greatest in the middle of the day because ultra-violet radiation from the sun drove the reaction. Exercising at that time was a bad idea because it hurt to breathe deeply. There were also local factors that made the problem worse. There are steep mountains to the north and to the east that reduce the flow of air out of the basin. In addition, there are often inversions where a layer of warmer air traps cooler air below it, along with the reactants for a witches' brew of pollutants. Figure 9.28 shows the results of a pollution model developed by a group of researchers led by Caltech Professor John Seinfeld. Air pollution is a complex problem, and this is a complex figure. The modest peaks at night of nitrogen oxides and reactive hydrocarbons are from refineries on the coast. The large concentrations inland of carbon monoxide and reactive hydrocarbons from 3pm to 6pm are from cars and trucks.

The city was saved by a remarkable invention, the catalytic converter. The inventors were chemical engineers Carl Keith and John Mooney at the Engelhard Corporation. The catalytic converter removes nitrogen oxides, carbon monoxide, and reactive hydrocarbons from vehicle exhaust. What makes this so impressive is that the reactions are quite different for the different pollutants. The NO_x is reduced to N_2, while the CO and hydrocarbons are oxidized to form carbon dioxide and water. This is shown in Figure 9.29.

Figure 9.30 shows the progress in reducing ozone levels in Los Angeles. The Southern California Air Quality Management District established a system of smog alerts. A Stage-1 alert was for an ozone level above 0.2 ppm. At this level people who were sensitive to pollution were urged to stay indoors. A Stage-2 alert was for a level above 0.35 ppm. This air is irritating to everyone. In a Stage-2 alert, the Air Quality Management District had the power to stop restrict commuting to work. In the early 1980s, there were over a hundred Stage-1 alerts each year, and Stage-2 alerts were not uncommon. The plot shows the maximum 1-hour ozone levels recorded each year. The trend line is for a reduction of 3.5%/y. The maximum level in 2012 was 70% below the level in 1980. The last second-stage alert occurred in 1988, and the last first-stage alert happened in 1998. The worst pollution days in Los Angeles now are not from cars, but from fires, which are a fundamental component of the forest ecology of Southern California.

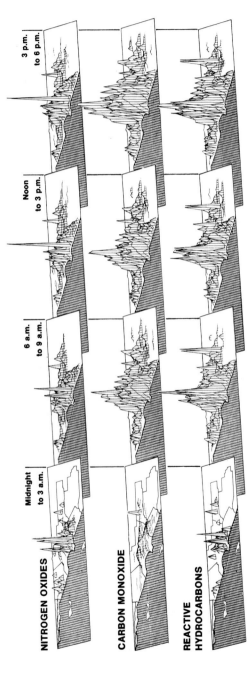

Figure 9.28 Modeling air pollution in Los Angeles in the 1970s. The model and the validation are described in two papers, Gregory McRae, William Goodin and John Seinfeld, 1982, "Development of a second-generation mathematical model for urban air pollution, part 1 Model formulation," *Atmospheric Environment*, vol. 16, pp. 679–696, and Gregory McRae and John Seinfeld, 1983, "Development of a second-generation mathematical model for urban air pollution, part 2 Evaluation of model performance," *Atmospheric Environment*, vol. 17, pp. 501–522. Personal communications from John Seinfeld.

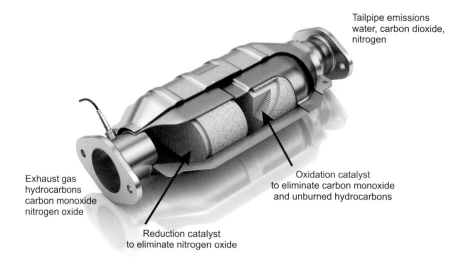

Figure 9.29 A rendering of the catalytic converter with labels added. Credit: Slavoljub Pantelic/ Shutterstock.

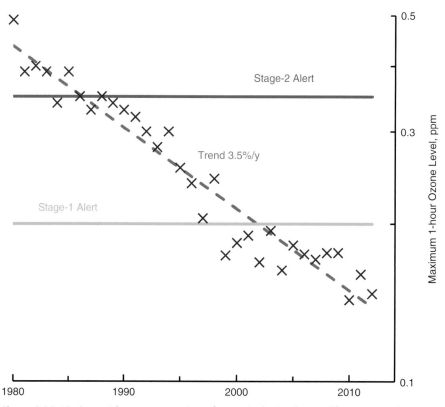

Figure 9.30 Maximum 1-hour concentrations of ozone in the Southern California Air Quality Management District. The scale is logarithmic. The data come from various editions of the California Air Resources Board (CARB) *Annual.*

9.4.3   The Volkswagen Diesel Scandal

The weakness of diesel engines is that the same high temperatures that give it high efficiency create nitrogen oxides in abundance. This dilemma led to the greatest scandal in automotive history. Beginning in the 2000s, several car manufacturers wrote firmware into the onboard computers to recognize when cars were being tested on a dynamometer. A dynamometer is a device that allows the wheels to be driven against resistive rollers. The car does not go anywhere, and an onboard computer can tell it is a test because among other things, the steering wheel does not change position. The engine parameters were then adjusted to give low nitrogen oxide emissions. This was only for the test, because it hurt fuel efficiency. The fraud was discovered in 2013 when Professor Gregory Thompson and his colleagues at West Virginia University measured emissions from two diesel cars on road tests. Table 9.9 shows their results. Thompson did not identify the cars in his report, but it was revealed later that the cars were a Volkswagen Jetta and a Volkswagen Passat. Both cars passed the dynamometer test easily. However, they failed road tests utterly when the engine firmware could no long recognize that the cars were being tested. The Jetta was 35 times the limit in the mountains and the Passat was 19 times the limit in urban Los Angeles. After these results were published, Volkswagen admitted that 500,000 of their cars in the United States had the deceptive firmware. Volkswagen has been fined billions of dollars and it undertook an enormous buyback program for these cars. Professor Thompson's group also tested a BMW and it was clean. BMWs are more expensive than Volkswagens and this made it easier for them to absorb the cost of the pollution control that was needed to pass the tests honestly. The same is true for diesel trucks, locomotives, and ships.

This episode raises questions. Why was it that an American university research group brought the cheating to light? What about Volkswagen's competitors? Didn't they realize that Volkswagen was cheating? Why didn't they report it?

Table 9.9 NO_x emission measurements for two Volkswagen diesel cars by Gregory Thompson's group at West Virginia University. The units are grams per kilometer.

	EPA limit	CARB dynamometer test	Southern California highway	Urban Los Angeles	Southern California mountains	Urban San Diego
Jetta	0.043	0.022	0.61	0.99	1.51	1.48
Passat	0.043	0.016	0.34	0.81	0.67	0.68

The limit is set by the US Environmental Protection Agency (EPA). The dynamometer tests were done at the California Air Resources Board (CARB) facility in El Monte, California. From Gregory Thompson, 2014, "In-Use Emissions Testing of Light-Duty Diesel Vehicles in the United States," prepared for the International Council on Clean Transportation.

The answer is that Volkswagen was not the only cheater. Moreover, Volkswagen was not just cheating in the United States. The problem was worse in Europe, where diesel cars are quite popular. As a result of the scandal, many European cities are banning older diesel cars and some car companies are cancelling plans to develop new diesel cars in favor of electric vehicles.

9.5 By Sea

9.5.1 From Sails to Engines

Historically cargo ships had one energy transition, from sail power to engine power. Figure 9.31 shows the number of ships in the British registry broken out between sailing vessels and ships with engines. The number of engine ships in the registry passed the number of sailing ships at a surprisingly late date, 1905, when there were 10,000 ships of each type. There were several factors that slowed the transition. The fuel for the powered ships was coal. Large numbers of firemen were required to feed the coal fires. On passenger liners and warships, this could

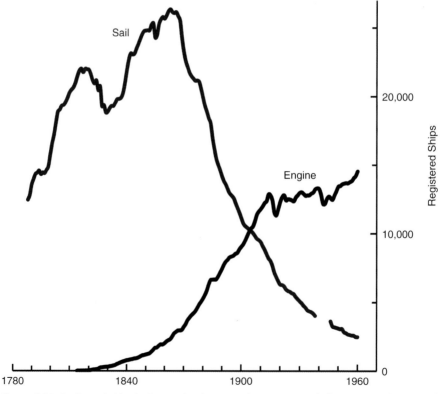

Figure 9.31 Registered ships in the UK. The data come from Brian Mitchell, 2011, *British Historical Statistics*, Cambridge University Press.

be two hundred men. On the other hand, sailing vessels have essentially unlimited range. Every four years, the French department of Vendée holds a solo sailing race where the boats depart the port of Les Sables-d'Olonne and sail 40,000 km around the world, keeping Antarctica off the starboard side, without assistance of any kind. The 2016 Vendée Globe was won in 74 days by Frenchman Armel Le Cléac'h. However, sailing vessels could not compete with powered ships in cargo volume. Figure 9.32 shows the cargo volume of British ships over time. By 1905, the volume for powered ships was five times that of sailing ships.

9.5.2 Ocean Freight

Marine shipping has become the dominant mode of freight transport. Ships are not restricted to specific directions as rail lines are. Freighters can change course in minutes. Ships can carry much more freight than trains. Intermodal freight trains pulling more than 300 containers are inconveniently long for stations, sidings, and for vehicles waiting at grade crossings. Container ship capacity is measured

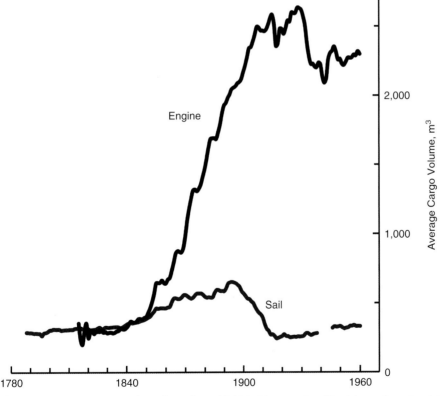

Figure 9.32 Average cargo volume for registered British ships over time. The data are for net register tonnage for all shipping and come from Brian Mitchell, 2011, *British Historical Statistics*, Cambridge University Press. For shipping, tonnage is not a mass unit, but rather a unit of volume equal to 100 cubic feet, or 2.83 m^3.

in twenty-foot equivalent units (TEUs). The larger container ships built during the 2010s have a capacity of 20,000 TEUs. The most common container length is forty feet, or 12.2 m, so it is appropriate to think of these ships as having a capacity of 10,000 containers in practice.

The mass limit for a freighter, including the cargo and the fuel and everything else that is not actually part of the structure of the freighter, is called the *dead-weight*. The deadweight for the largest container ships is 200 kt. The length is 400 m and the beam (width) is just under 60 m. These ships are typically propelled by very large diesel engines with capacities approaching 100,000 hp. The engines may run at 80 rpm and they are directly coupled to the propeller. The typical ship velocity is 30 km/h. At this speed, it takes 30 days to sail from Shanghai to Rotterdam in the Netherlands, 22,000 km. Figure 9.33 shows shipping container traffic over time. There has been a steady increase of 33 million TEUs per year.

Container ports have evolved with the ships. It is a challenging logistics problem to unload the large container ships efficiently. At one container per minute, moving 6,000 containers takes four days. This is 3,000 km of sailing at 30 km/h.

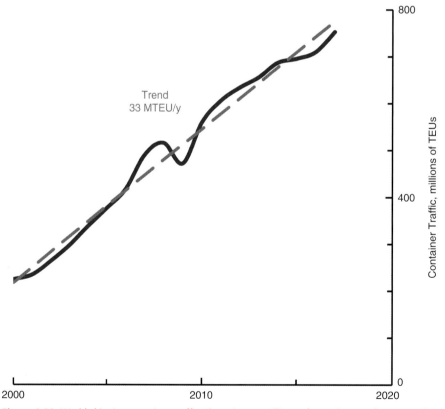

Figure 9.33 World shipping container traffic. The units are millions of TEUs (twenty-foot-equivalent units). The typical container is twice this length, so the actual number of containers is a bit more than half of these values. The data come from the World Bank.

Table 9.10 The major regional container port volumes in 2015. The numbers are in millions of TEUs.

Shenzhen/Hong Kong/Guangzhou (China)	61
Shanghai/Ningbo (China)	57
Singapore	31
Rotterdam (Netherlands)/Antwerp (Belgium)	22
Busan (South Korea)	19
Qingdao (China)	17
Dubai Ports	16
Los Angeles/Long Beach (US)	15
Tianjin (China)	14
Port Kelang (Malaysia)	12

From the American Association of Port Authorities, 2015, "World Port Rankings."

Table 9.10 shows the container shipping traffic for the world's major ports, grouped regionally. A substantial fraction of the container traffic is *transshipment*, where a container is moved from one ship to another. Transshipment counts double in the table compared with a container that is moved from a train to a ship. The largest transshipment ports are Singapore, Hong Kong, and Shanghai. Nothing demonstrates better the industrial might of China than its dominant position in this table.

9.6 By Air

One of the challenges that oceanic birds have is that their feeding areas can be a long way from their nesting islands. The problem is that most of the ocean is biologically barren. The fish and squid that the birds eat are associated with local upwellings of nutrients. For example, the Black-Footed Albatross nests on Midway Island, west of Hawaii, and it often feeds offshore near Monterey, California. The albatross has several physiological adaptations. They can drink sea water, with the salt expelled from nose tubes. The birds produce an oil in their stomachs that can be regurgitated to feed the young or metabolized if necessary. The largest ocean birds are the Wandering Albatross of the Southern Ocean, with a wingspan of 4 m. Round-trip foraging runs during the nesting season of 7,000 km have been recorded for these birds with satellite trackers. Passenger on ships have noted that the Wandering Albatross can cover great distances along the surface of the water without flapping their wings. The birds use a technique called dynamic soaring that depends on the fact that because of friction, the wind speed near the surface of the ocean is lower than it is a few meters above the surface. The birds glide just above the surface, and then pitch up vertically, almost coming to a standstill. This is shown in Figure 9.34. In the process the bird has converted its kinetic energy to gravitational energy. The increased wind at this height blows the bird so that it gains kinetic energy in addition to its gravitational energy. Then it heads down to glide across the surface again. Dynamic soaring has no commercial applications, but radio-controlled gliders have used the technique to reach remarkably high speeds, above 800 km/h.

9.6.1 Flight Range

We abstract the process of flight as a sawtooth pattern of alternating gliding and powered altitude gain shown in Figure 9.35. We let the glide distance be Δr and the glide slope be s. The altitude gain requires energy and is given by $s\Delta r$. Albatross gain energy through the technique of dynamic soaring. Other birds flap their wings to gain altitude. We can use the diagram to estimate the maximum distance d that a bird can fly by flapping its wings without eating. We can relate the altitude gain $s\Delta r$ and the metabolized mass Δm as

$$\eta E\Delta m = -mgs\Delta r \qquad\qquad 9.2$$

where η is an effective flight efficiency in converting food energy to altitude gain, E is the average energy density of the metabolized mass, m is the mass of the bird, and g is the acceleration due to gravity, 9.8 m/s². We can solve for Δr as

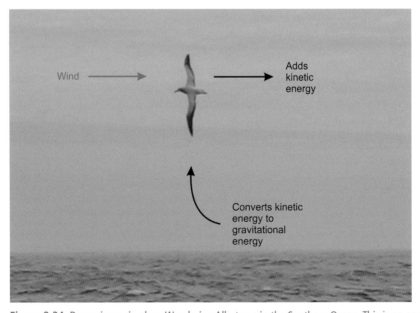

Figure 9.34 Dynamic soaring by a Wandering Albatross in the Southern Ocean. This is an annotated still from a remarkable 2007 video that is courtesy of travel blogger Macgellan. The video shows the bird traveling a kilometer without flapping its wings. A link to the video is given in the references.

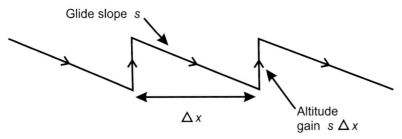

Figure 9.35 Flight abstracted as the alternation of gliding and powered altitude gain.

Figure 9.36 The Bar-Tailed Godwit (*Limosa lapponica*) in flight. Credit: Maciej Olszewski/Shutterstock.

$$x = -\frac{\eta E}{sg}\ \frac{m}{m}\ .$$

9.3

We interpret Δr and Δm as differentials and integrate to find the maximum distance d as

$$d = \left(\frac{\eta E}{sg}\right)\ \ln(r)$$

9.4

where r is the ratio of the mass of the bird at the beginning of the flight to the mass at the end.

We will use the Bar-Tailed Godwit as an example (Figure 9.36). This bird nests in the northern latitude summer along the west coast of Alaska. It spends the southern latitude summer in New Zealand. Its fall migration recorded by satellite trackers is a week-long ocean flight of 11,000 km. One advantage of this direct

Table 9.11 Flight parameters for the Bar-Tailed Godwit.

	Bar-Tailed Godwit	737
Glide slope s	0.07	0.07
Mass ratio r	2.3	1.4
Effective flapping efficiency η	25%	30%
Energy density of fat (9 Cal/g) E	38 kJ/g	43 kJ/g

The glide slope, mass ratio, and the estimate for the effective flight efficiency for the godwit are taken from Henk Tennekes, 2009, *The Simple Science of Flight*, 2nd Edition, MIT Press. For comparison, representative numbers are shown for the Boeing 737 aircraft.

migration is that the bird avoids predators. Godwits are wading birds that feed in marshes and on mudflats. They do not eat on the open ocean and they do not drink sea water. During the flight, the bird metabolizes over half of its mass. The flight parameters are given in Table 9.11. When we plug these numbers into Equation 9.4, we get the migration distance

$$d = \frac{25\% \cdot 38\,\text{kJ/g}}{0.07 \cdot 9.8\,\text{m/s}^2}\ \ln(2.3) = 11{,}500\,\text{km}. \qquad 9.5$$

The effective glide slope s would be reduced somewhat by tail winds. Offsetting this factor is that the birds metabolize some protein, which has a lower energy density than fat.

The effect of none of the parameters in Equation 9.4 is a surprise. Higher efficiency, energy density, and mass ratio give longer range, as does a smaller glide slope. With appropriate re-interpretation of the parameters, the equation can also be applied to aircraft. In this context, it is called Breguet's Equation. Representative numbers for a Boeing 737 aircraft are also shown in the table for comparison. Remarkably, none are strongly affected by scaling from a bird with a mass of 300 g to a 50-t plane. Commercial aircraft burn kerosene, which has somewhat higher energy density than fat, and the efficiency of a jet engine is better than a flapping godwit. However, the godwit makes up for it with a larger mass ratio. In principle, the same equation could also be applied to a train, with the rolling resistance coefficient C_r replacing the glide slope s. The fact that C_r is two orders of magnitude smaller than s gives the train its fundamental efficiency advantage over an airplane for freight.

9.6.2   Airline Passenger Travel

Airplanes have been extremely successful for passenger travel. Figure 9.37 shows the history for the United States, France, and China on a km per person basis. Air traffic developed earlier in the United States than in Europe. The first American low-cost airline, Southwest, was founded in 1967. However, low-cost airlines are now also well established in Europe and the gap between French and American travel has narrowed. In 1989, a French person traveled 33% as much as an American, but by 2009, it was up to 61%. Chinese flying is small compared with the US and France, but it has grown rapidly, 15%/y annualized from 1999 to 2009. Compared with rail and car travel, much of air travel is discretionary. There was a significant drop in American travel after the attacks on September 11, 2001, and in 2009 as a result of the Great Recession.

Figure 9.38 shows the world history in terms of the number of passengers. World air travel has grown at 5%/y for fifty years. The number of passengers in 2017 corresponds to a round trip flight by every person in the world in four years. This curve shows no sign of saturation.

Airplanes do not need a highway or rail network connecting to destinations. This greatly reduces the land footprint for this travel mode. The facilities scale easily with the demand. A major international terminal like Japan's Narita Airport is a sophisticated installation with integrated rail and highway access. On the other hand, a village may be served by a strip of asphalt, a few parking spaces

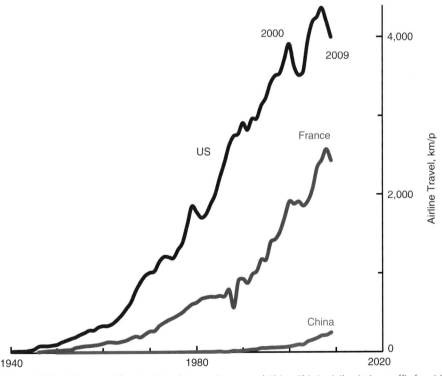

Figure 9.37 Airline travel for the United States, France, and China. This is civil aviation traffic for airlines registered in these countries. The passenger travel data come from Brian Mitchell, 2013, *International Historical Statistics*, Palgrave MacMillan. The population data come from the United Nations Population Division.

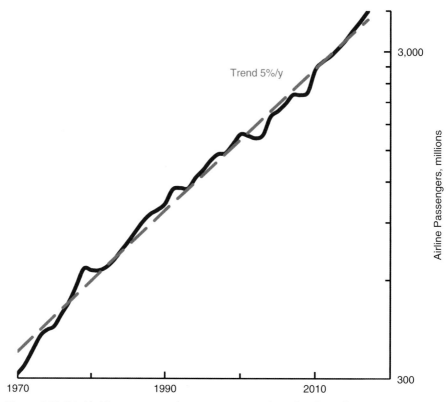

Figure 9.38 World airline passengers from 1970 to 2017. The scale is logarithmic. The data come from the World Bank.

with U-bolt tie downs, and a chain-link fence link with a combination lock. Some Inuit communities in the Far North only have air access; there are no roads to their towns. Aircraft energy efficiency has improved steadily, and commercial aircraft are now more energy efficient than personal cars (Figure 9.11), much safer (Table 9.3), and of course, faster. The steady increase in world air travel, far beyond population growth, is an index of the improving state of humanity.

9.7   Prospects for Self-Driving Cars

Today there is an enormous effort to develop driverless vehicles. This began with a Grand Challenge contest in 2004 that was sponsored by the US Defense Advanced Research Projects Agency (DARPA). This was the same agency that developed the Internet. DARPA offered a million-dollar prize for a self-driving vehicle that could navigate 230 km across the desert from Barstow, California to Primm, Nevada, in less than ten hours. The agency was interested in developing autonomous convoys that would allow military bases to be supplied without casualties. Fifteen vehicles started the competition that year. None finished and the farthest any vehicle drove was 12 km. The next year DARPA sponsored another contest with a prize of two million dollars. This course was 210 km across the Southern Nevada desert. The Stanford University team car Stanley won the prize, completing the route in 6 hours and 53 minutes. The team leader was Professor Sebastian Thrun. DARPA sponsored one more Grand Challenge in 2007, with a 100-km course around George Air Force Base in Victorville, California. This time the prize was won by Carnegie Mellon University's car, Boss. Following the Grand Challenges, the Google Company set up a research program in driverless vehicles, with Thrun leading the effort. A Google subsidiary, Waymo, started a driverless ride service in Phoenix, Arizona in 2018, albeit with a backup driver present in the car.

A major advantage of the self-driving technology would be the potential for reducing the number of accidents. The computer would not drift off to sleep, and lidar and radar sensors can detect obstacles in fog and dust. On the other hand, driving is a complex skill that adults acquire over many years. Moreover, driving conditions are different in different countries. Figure 9.39 shows an intersection in India. The photograph gives a sense of the challenge to self-driving vehicles in that country.

It is premature to speculate on a time frame for a transition to self-driving vehicles. However, we can discuss the energy consumption impacts. A driverless ride service should be cheaper because there would be no paid driver. Lowering the cost of a service would tend to increase the demand for the service and the associated energy use. A driverless vehicle would not need a steering wheel and this would free up space and encourage new car designs. If driverless vehicles became as

Figure 9.39 Traffic in Mumbai. Credit: sladkozaponi/Shutterstock.

safe as buses and trains, the passengers would not wear seat belts and they could work, eat, play games and sleep. The interior of the car could be personalized in the way recreational vehicles (RVs) and boats are. Self-driving technology would allow vehicles to act like micro land cruise ships, driving at night and then stopping during the day for passengers to sightsee. Owning a driverless vehicle would be like having a chauffeured car without having to pay a chauffeur. We could imagine the day of a family of four, where the car first drops off the mother at the office building where she works and then the father at a warehouse where he works. After the car returns home, the children would be taken to their schools. In the afternoon, the trips would repeat in reverse, picking up the kids at school first and then the parents from their jobs. In between, the car would take the children to sports practice and to music lessons. After dinner there would be trips to take the parents to a play and the children to visit their friends. Finally, the car would bring everyone home. In this scenario, there would be many trips without passengers in the car. The trips with no passengers would increase energy consumption. Autonomous cars should make driving safer and more pleasant, and the time could be used productively. It would be a boon for the disabled and the elderly. But all of this would encourage more trips by car and with it higher energy consumption.

Concepts to Review

- The Otto and Diesel engines
- Henry Ford and the Model T
- Transportation energy demand
- Electric vehicle progress
- Transportation oil shares
- Commuting trends
- Passenger travel efficiency
- Passenger safety
- Freight shares
- Freight efficiency
- Road construction
- Interstate Highways
- Evolution of fuel efficiency and horsepower
- Shinkansen high-speed passenger trains
- The 1980 Staggers Act that deregulated American transportation
- Malcom McLean and the shipping container
- Classification yards
- Intermodal shipping
- The Lac Mégantic oil train disaster

- Los Angeles smog and the catalytic converter
- The Volkswagen diesel scandal
- The transition from sail to power in ships
- The growth of container shipping
- Dynamic soaring
- Flight range
- Passenger air travel trends
- The impact of self-driving cars on energy demand

Problems

Problem 9.1   Aircraft range

Calculate the range for the Boeing 737 aircraft using the representative parameters in Table 9.11.

Problem 9.2   Trains

The major rail freight route in California passes through Barstow and San Bernardino on its way to the ports of Los Angeles and Long Beach. The line is operated by the BNSF Railway, but the Union Pacific Railroad has trackage rights. The mileposts and elevations for the route are given in the Figure 9.40.

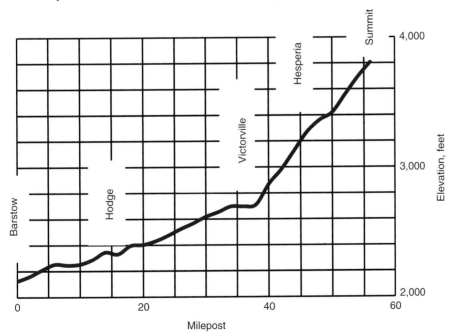

Figure 9.40 Mileposts and elevations for the BNSF (Burlington-Northern Santa Fe) from San Bernardino to Barstow over the Cajon Pass. This figure comes from a 2006 Master's Thesis by Travis Painter at the University of Illinois, "Recovering Railroad Diesel-Electric Locomotive Dynamic Brake Efficiency."

Assume a train hauling grain for export leaves Barstow, headed west toward Summit. There are 60 full grain cars. Each car weighs 125 t, including 100 t of grain. There are four locomotives, each weighing 188 t.

a. Calculate the change in the gravitational energy of the train in GJ in moving the train from Barstow up to Summit.

b. Assuming that the rolling-resistance force is 0.001 times the gravitational force, calculate the energy consumed by the rolling resistance in GJ in moving the train from Barstow to Summit.

c. Calculate the total diesel fuel required in metric tons, to move the train from Barstow to Summit, assuming that the overall efficiency of the diesel-electric motors is 30%, and that the energy density of diesel fuel is 42 GJ/t. You may neglect aerodynamic drag and you may assume that the velocity in Summit is the same as the velocity in Barstow, so that you do not have to consider changes in the kinetic energy.

d. Currently the best-selling locomotives are GE's Evolution Series. There is a version, ES44AC with AC motors and a version, ES44DC, with DC motors. AC motors are better than DC for sustained high traction at low speeds, and this makes them suitable for coal and grain trains. The DC version is cheaper, and it is often used for the trains that haul containers, which are lighter than coal and grain cars. The number 44 refers to the nominal horsepower, 4,400 hp. Some of this power is consumed in other parts of the locomotive than the motors, and some is consumed in the conversions for the electric motors. We will assume that the available power for traction for each locomotive is 3,000 hp. The maximum grade on the route from Barstow to Summit is 1.6%. Calculate the maximum speed in km/h that could be maintained at this grade, taking the gravitational force and the rolling resistance into account.

Problem 9.3   Passenger ships

The most common ship engines, by far, are diesels. These may have mechanical transmissions, or they may be configured like locomotives with electric generators and electric motors that turn the propellers. For this problem you may take the density of diesel fuel to be 0.84 kg/l and the energy content to be 42 MJ/kg. In practice small boats use distillate diesel fuels with a slightly higher energy content and large ships use residual diesel fuels with a somewhat lower energy content.

a. The Oasis-class cruise ships were the largest passenger vessels ever built at the time they were constructed. They carry 5,400 passengers and displace 100 kt. The Oasis of the Seas entered service in 2009. Their propulsion is from three 20 MW electrical thrusters that can be oriented at any azimuth angle. Assuming an overall efficiency of 25% for the diesel engines and the thrusters, calculate the passenger efficiency in passenger·km/kgoe at its rated speed of 23 knots (nautical miles per hour).

b. The Nord Star 24-foot trawler-style boat has been used as a water taxi in Scandinavian countries. It displaces 3 t and it would typically carry 4 passengers. With a Volvo Penta D3 170-hp diesel engine at 3,000 rpm, its speed is 22 knots. Locate a technical data sheet for this engine online and find the fuel consumption at this engine speed. Calculate the passenger efficiency in passenger·km/kgoe.

Problem 9.4   Tankers

Oil tankers are quite efficient in transporting oil and this helps keep the price of oil before taxes and subsidies relatively constant in different parts of the world.

a. A tanker is 314 m long and 46 m wide. The draft, or distance between the surface of the water and the bottom of the ship is 18 m when loaded with oil. Taking the density of sea water to be 1.025 t/m^3, calculate the mass of the water displaced. Treat the tanker as a rectangular box.

b. Taking the force of the propeller moving the ship forward to be 2.4 MN, calculate the distance travelled in the first minute after the engine starts up, neglecting drag.

c. There are several components of drag, including wave creation, dynamic pressure, drag due to the shape of the hull, and the friction of the hull with water. Friction dominates for tankers. The drag coefficient varies somewhat with velocity and the size of the ship, but for tankers a representative value is $C_d = 0.0025$. The density to apply in Equation 7.34 is the density of sea water and the area is the area of the hull that gets wet. Calculate this area.

d. Calculate the maximum velocity of the ship in m/s.

e. If the mass of the ship empty is 50 kt, calculate the mass of the crude oil cargo.

f. Calculate the freight efficiency in t·km/kgoe at its maximum speed. You may assume an overall efficiency of the diesel engine and the propeller of 20%.

Further Reading

- Spencer Lisenby, 2018, "Fastest RC airplane in the world! Transonic DP – 545mph!!" This is a video of a glider that uses dynamic soaring at Bird Spring Pass, California. Available at www.youtube.com/watch?v=MoaWlKC3wIM
- Macgellan, 2007, video of a Wandering Albatross in the Southern Ocean is available at www.youtube.com/watch?v=buuxFP–Ezo
- John McPhee, 2006, *Uncommon Carriers*, Farrar, Straus & Giroux. This book is a collection of essays on the author's experiences riding with tank truck drivers, coal train engineers, barge operators, and freighter captains. It is both informative and great literature.

- Vaclav Smil, 2010, *Prime Movers of Globalization: The History and Impact of Diesel Engines and Gas Turbines*, MIT Press. A good discussion of the role of different engines.
- Henk Tennekes, 2009, *The Simple Science of Flight*, 2nd Edition, MIT Press. A wonderful introduction to the flight of both birds and airplanes.
- The Transportation Safety Board of Canada, 2014, Railway Investigation Report R13D54, *Runaway and Main-Track Derailment, Montreal, Maine, and Atlantic Railway, Freight Train MMA-002, Mile 0.23, Sherbrooke Subdivision, Lac Mégantic, Quebec, 06 July 2013*. Available at www.tsb.gc.ca/eng/rapports-reports/rail/2013/r13d0054/r13d0054.pdf

10 Climate Change

Right now, the land and the ocean are taking up almost half of the carbon dioxide we add to the atmosphere by burning fossil fuels, but the future is fundamentally unknown.

Paul Wennberg, Professor of Atmospheric Chemistry and Environmental Science and Engineering, Caltech, and Science Team member for the Orbiting Carbon Observatory 2

10.1 Introduction

The earth loses heat through infrared radiation to space. This heat loss balances the energy absorbed from the sun. Some atmospheric gases, like water vapor, carbon dioxide, and methane absorb in the infrared. Because the temperature of the atmosphere falls with altitude the atmospheric gases radiate less to space than the surface of the earth would in their absence. The result of this shift in the radiation balance is a warmer earth. The Swedish chemist and Nobel Laureate Svante Arrhenius was the first to quantify the effect in his 1896 paper, "On the influence of Carbonic Acid in the Air Upon the Temperature of the Ground." He wrote "Now it will be shown in the sequel that a variation of the carbonic acid [carbon dioxide] of the atmosphere in the same proportion produces nearly the same thermal effect independently of its absolute magnitude." This can be written mathematically as

$$T_{CO_2} \propto \log\left(c / c_o\right) \tag{10.1}$$

where T_{CO_2} is the change in temperature in kelvins, c is the CO_2 concentration, and c_o is a reference concentration. The symbol $\propto$ denotes a proportional relation. Arrhenius deduced that the change for a doubling of the concentration of carbon dioxide would be 6 K.

10.1.1 The Callendar Effect

After Arrhenius, the next significant research came from an English mechanical engineer named Guy Callendar (Figure 10.1). Callendar did not have a permanent laboratory position. His primary work, supported by grants from the British

Figure 10.1 Guy Callendar (1898–1964). Callendar was an English mechanical engineer whose pioneering work on global warming was done in his spare time. Credit: G. S. Callendar Archive/University of East Anglia.

Electrical and Allied Industries Research Association (BEAIRA), was to make measurements to develop tables of the thermodynamic properties of steam. In 1938, he wrote a paper, "The Artificial Production of Carbon Dioxide and Its Influence on Temperature." Figure 10.2 shows a graph from Callendar's paper plotting the temperature change against the carbon dioxide concentration. He deduced his curve from a study of the infrared radiation from atmospheric carbon dioxide back to earth. Digitizing the figure gives a temperature rise of 1.55 K for a doubling of the CO_2 level from 200 ppm to 400 ppm.

Callendar ended his paper with

In conclusion it may be said that the combustion of fossil fuel, whether it be peat from the surface or oil from 10,000 feet below, is likely to prove beneficial to mankind in several ways, besides the provision of heat and power. For instance, the above mentioned small increases of mean temperature would be important at the northern margin of cultivation, and the growth of favorably situated plants is directly proportional to the carbon dioxide pressure (Brown and Escombe, 1905). In any case, the return of the deadly glaciers should be delayed indefinitely. As regards the reserves of fuel these would be sufficient to give at least ten times as much carbon dioxide as there is in the air at present.

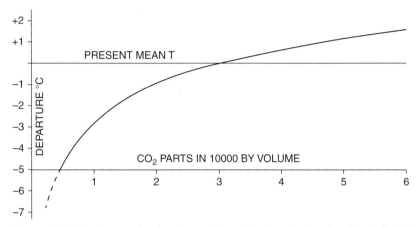

Figure 10.2 "The change of surface temperature with atmospheric carbon dioxide." This is Figure 2 in Guy Callendar, 1938, "The artificial production of carbon dioxide and its influence on temperature," *Quarterly Journal of the Royal Meteorological Society*, vol. 64, pp. 223–240.

At that time world coal reserves were twenty times larger than they are now, so the sentence about reserves is no longer correct. In Callendar's day, journals published a transcript of comments on the paper and the author's responses. In his paper and in the exchange, many threads of the modern discussion on climate change appear: atmospheric dynamics, the calculation of temperature changes from carbon-dioxide levels, the trend in CO_2 levels, temperature indexes, urban heat islands, the role of CO_2 in the ice ages, the movement of climate zones towards the poles, and CO_2 fertilization of plants. His theory was called the *Callendar effect*. However, a cooling period set in during the 1950s and Callendar died in 1964. Computer technology began to develop in the 1960s and 1970s, and researchers wrote computer climate models that took into account many more factors than Calendar did. The effect came to be called *global warming*. In 1979, an American Academy of Sciences study group headed by MIT professor Jule Charney surveyed the computer models and suggested a range for the temperature change from a doubling of the carbon-dioxide level of 1.5 to 4.5 K. Today climate modelers define two different temperature responses. The first is the equilibrium climate sensitivity (ECS), which is the steady-state response to a step-change doubling of the carbon-dioxide level. In the 2013 5th Assessment Report of the United Nations Intergovernmental Panel on Climate Change (IPCC) the range of the ECS in the computer models is 2.1 to 4.7 K, comparable to the Charney range. The second is the transient climate response (TCR), where the CO_2 levels double over a 70-year period in 1% annual steps. In Section 5.7, the projection for the 10% to 90% lifetime t_l for the fossil-fuel era was 107 years. This makes the TCR more relevant than the ECS, where the equilibration time is of the order of a thousand years. The range of the TCR in the computer models in the 5th Assessment Report is 1.1 to 2.6 K.

The discussion has expanded from global warming to a wide range of climatic impacts. In this chapter, we will discuss the trends for three different components: carbon-dioxide levels, sea level rise, and the temperature indexes. In addition, we will make projections by linking the trends to the fossil-fuel projections in Section 5.7. This will allow us to estimate the effects of climate-change policies that limit fossil-fuel burning.

10.1.2   California Forest Fires

There are areas where fossil-fuel policies will have limited effect. California Governor Jerry Brown repeatedly blamed climate change for California forest fires. From the perspective of evolutionary biology, California forests are defined by fire. The chaparral country of Southern California routinely burns at the end of the dry season. This is the time when inland high-pressure systems drive hot *santana* winds westward towards the coast.[1] The Sequoia trees in the Sierras are

Figure 10.3 The Camp Fire in California, November 8, 2018 at 10:45am. This image has a false-color infrared overlay to indicate the active fire. Credit: NASA Landsat 8 Operational Land Imager.

---

[1] The actor Humphrey Bogart was a skilled sailor. He named his boat *Santana*. Nowadays the weather personalities call the winds Santa Annas.

products of fire. They are among the largest trees in the world, reaching a height of eighty meters and a diameter of eight meters. Their seed cones only open after a fire. To grow, the seeds require full sun, which is only available after a fire. People who live in these forests put themselves in harm's way because the fires move so quickly that they cut off escape. The 2018 Camp Fire began on the morning of November 8, and by midday the fire had extended 20 km to the west (Figure 10.3). The fire killed 86 people and obliterated the town of Paradise. The Pacific Gas and Electric Company was blamed for the fire because of a downed power line and declared bankruptcy in 2019. Recall that PG&E also declared bankruptcy in 2001 in the wake of the California Electricity Crisis (Section 7.9.2). However, the only way to eliminate the risk is to clear the trees and brush within a large distance of every building. But then the people would no longer be living in the forest.

10.2   Carbon Dioxide

The tight relationship between atmospheric CO_2 levels and cumulative fossil-fuel production in Figure 1.30 leaves no doubt that fossil-fuel burning is responsible for the rise in CO_2 levels. The carbon dioxide measurements shown in the graph are annual averages. We can get additional information from daily measurements. Figure 10.4 shows data from La Jolla, California. The measurements, like the ones

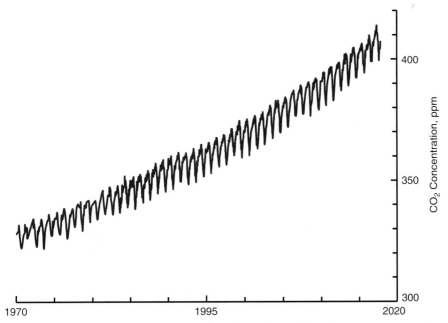

Figure 10.4 Carbon dioxide concentrations in La Jolla, California. These data represent over a thousand measurements of daily average atmospheric CO_2 concentrations derived from flask air samples. The data set is maintained by the Scripps CO_2 Program, R. F. Keeling, S. C. Piper, A. F. Bollenbacher and S. J. Walker. The data are available at http://scrippsco2.ucsd.edu/data/atmospheric_co2/ljo

in Hawaii, were started by Charles Keeling. They continue today under a team led by his son Ralph. The curve has a large ripple, with a drop during the northern summer. The drop is primarily from photosynthesis by land plants that are pulling carbon dioxide out of the air to produce glucose. The ripple is reduced at stations in the middle of the ocean and at stations in the Southern Hemisphere, where the land fraction is much smaller.

Figure 10.5 shows the data for a single year. The green circles indicate the data that were taken during the growing season from May 15 to September 15. The first green circle is 361 ppm, taken on May 19. The last is 349 ppm, taken on September 9. After the growing season, the CO_2 levels recover. The dominant process then is aerobic respiration, which converts glucose and oxygen to carbon dioxide and water (Equation 6.4). The respiration process is also ongoing during the growing season, so the effect we see then is photosynthesis net of respiration.

Figure 10.6 shows how the growing season dip has evolved from 1970 to 2017. It is a linear relationship with an r^2 of 0.92. We use the endpoints of the regression line shown on the graph to evaluate the rise. The rise is 31% over this period. This tells us that somewhere there are a lot of happy plants. Recall that we also saw a rise in US timber density, 48% over the years from 1953 to 2007 (Figure 2.22). Possible factors that contribute to both of these are carbon dioxide fertilization, reactive nitrogen compounds from power plants entrained in rain, and longer

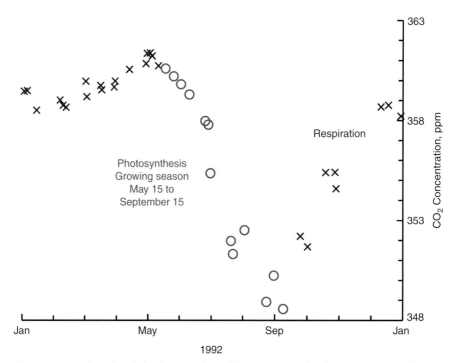

Figure 10.5 Carbon-dioxide levels in La Jolla, California in 1992. The data are a portion of the set plotted in Figure 10.4.

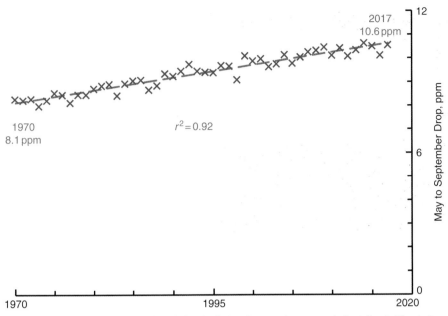

Figure 10.6 The drop in carbon-dioxide levels during the growing season in La Jolla, California from 1970 to 2017. The data points are the difference between the May average and the September average atmospheric CO_2 concentrations.

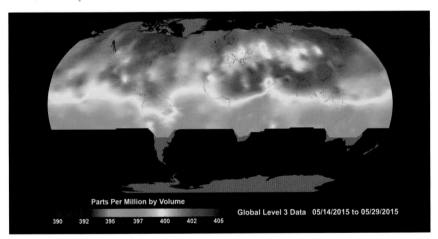

Figure 10.7 Atmospheric carbon dioxide concentrations measured by the Orbiting Carbon Observatory 2 from May 14, 2015 to May 29, 2015. Credit: NASA.

growing seasons. It also suggests that corn farmers should give some credit for their bumper crops to environmental factors.

In 2014, the Jet Propulsion Laboratory in Pasadena, California, launched the Orbiting Carbon Observatory 2 (OCO-2) to map carbon dioxide concentrations.[2] Figure 10.7 and Figure 10.8 show the results early and late in the northern growing

---

[2] An earlier mission, Orbiting Carbon Observatory (OCO) suffered a rocket failure. Hence OCO-2.

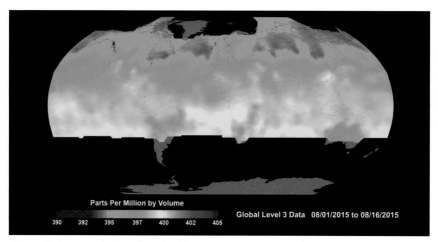

Figure 10.8 Atmospheric carbon dioxide concentrations measured by the Orbiting Carbon Observatory 2 from August 1, 2015 to August 16, 2015. Credit: NASA.

season. In May there are red swirls for high concentrations. By August, there are blue areas showing the photosynthesis drops. What is striking about the maps is the large activity in the northern latitudes. In addition, there is a great deal of regional structure.

Alas, these maps do not help us explain the La Jolla measurements. However, as Southern California sailors know, the dominant winds at La Jolla in summer are produced by the North Pacific High. This is a sustained high-pressure system with winds that push air southward along the Pacific coast in the summer. We might look for increased timber density in Alaska, British Columbia, Washington, and Oregon to explain the La Jolla measurements.

10.3   Sea Level

We would expect sea level to rise in an era of global warming because high temperatures melt glaciers. In addition, water expands when it warms. This also operates in reverse. During the last Ice Age, sea level was a hundred meters lower than it is today. Around 18,000 years ago, temperatures began to rise and deglaciation began. The temperatures continued to increase for ten thousand years. The average sea level rise rate during this time was ten millimeters a year, or one meter per century. There have been fears that because of global warming, this high rise rate could return. It is easy to measure the sea level rise rate relative to the land with tide gauges. In recent years, satellites have used radars to measure sea level heights. However, the interpretation of these measurements depends on complex geophysical models that are still evolving. We will work with tide-gauge data series that are long enough to make comparisons before and after the hydrocarbon transition in 1956.

Aside from sea level rise, there are other processes that affect a coastline. Ostia was the port for the city of Rome during the time of the Roman empire two thousand years ago. Because of silting by the Tiber River, the ancient port is now three kilometers from the sea. During the deglaciation from the Ice Ages, coral reefs grew to offset the sea level rise. In Figure 5.32, we saw that the city of Boston, Massachusetts has expanded by landfill. We also saw in Figure 5.29 that land subsidence from oil production raised the relative sea level in Long Beach, California.

Figure 10.9 shows the tide-gauge measurements for Juneau, Alaska. The tide-gauge data are collected by the Permanent Service for Mean Sea Level (PSMSL). To make comparisons easier, all of the tide-gauge plots we show will use the 2017 measurement as the zero reference, and a scale from –1 m to +1 m. The relative sea level at Juneau has been dropping since the tide gauges were set up in 1944 during World War II. The fall since then has been close to a meter. The reason for this is that the area was covered by glaciers during the last ice age and the land is rebounding. The same thing is happening in Scandinavia.

Recall that in Section 6.6 that we could associate the variation in the Pacific Sardine catch with the Pacific Decadal Oscillation (PDO). Sea level rise can also be associated

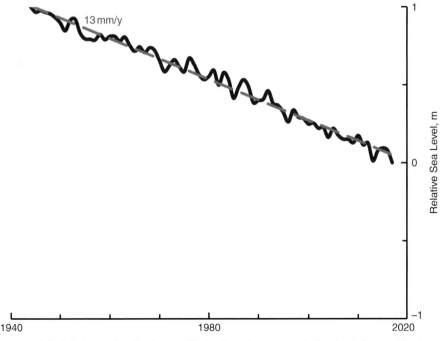

Figure 10.9 Relative sea level at Juneau, Alaska. The reference year is 2017 and the trend line is a regression line. The data come from the Permanent Service for Mean Sea Level (PSMSL).

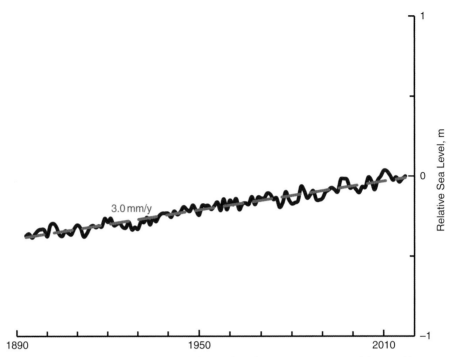

Figure 10.10 Relative sea level at New York City. The reference year is 2017 and the trend line is a regression line. The data come from the Permanent Service for Mean Sea Level (PSMSL).

with natural variations. Figure 10.10 shows the tide-gauge record for New York City. This series starts in 1893. A regression line through the data has a slope of 3.0 mm/y.

In Figure 10.11, the data are replotted, this time with the rise rate plotted as the slope of a 31-year running regression. This is a conventional time frame for defining climates. This gives a sine wave with a period of sixty years like the PDO. The minimum rise rate is 1.3 mm/y and the maximum is 4.9 mm/y. For comparison, the Atlantic Multidecadal Oscillation (AMO) temperature index is plotted alongside the sea level rise. The AMO is also plotted as a running 31-year average. The AMO is computed as a temporal component of sea surface temperatures, so the temperature unit, K, has been retained. There is a clear relationship. There is another problem with the New York tide-gauge data. The gauge is not on stable land. The group SONEL (Système d'Observation du Niveau des Eaux Littorales) measures land movement with GPS receivers at tide-gauge stations. They found that from 2011 to 2014, the land dropped at a rate 2.1±0.6 mm/y. This means that the New York tide-gauge data are contaminated both by ground movement and by the AMO. We will look for a tide gauge on stable land in the Pacific.

The city of Honolulu, Hawaii, has a superb tide-gauge series that starts in 1905. It is one of the few sets at PSMSL that have no missing years. The SONEL GPS measurement for the tide-gauge station there shows the land dropping 0.23±0.18 mm/y from 1999 through 2013. This is small enough that we can neglect it. Figure 10.12

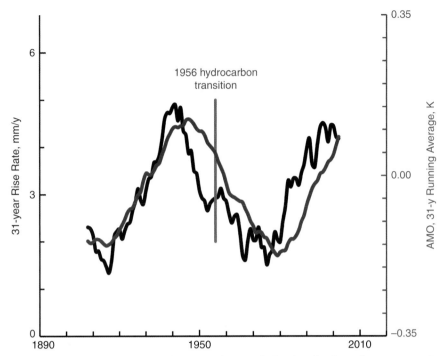

Figure 10.11 The 31-year rise rate for New York City, calculated as the slope of a 31-year running regression. This graph uses the same data as Figure 10.10. The AMO data come from NOAA and are available at www.esrl.noaa.gov/psd/data/timeseries/AMO/

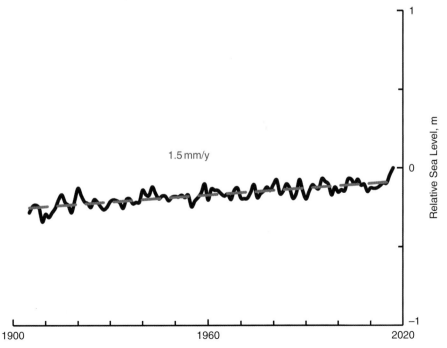

Figure 10.12 Relative sea level at Honolulu, Hawaii. The reference year is 2017 and the trend line is a regression line. The data come from the Permanent Service for Mean Sea Level (PSMSL).

shows the tide-gauge measurements. The regression line from 1905 to 2017 has a slope of 1.5 mm/y.

Figure 10.13 shows the running 31-year rise rate. It is interesting that even though the station is in the Pacific Ocean, it does not show contamination by the Pacific Decadal Oscillation (PDO) that we associated with changes in the Pacific Sardine catch. The PDO is a spatial climate pattern and it turns out that the Hawaiian Islands are on the its node lines of temperature and pressure.[3] By comparing the rise rates before and after the hydrocarbon transition in 1956, we have a natural experiment to test the effect of burning fossil fuels. Recall from Section 1.4 that 87% of all fossil-fuel burning occurred from 1956 on. The figure gives no indication that the rise rate changed after 1956.

Even if the rise rate changes significantly in the future, coastal cities and towns are resilient. In Section 5.4.2, we saw that the city of Long Beach recovered from a subsidence of 9 m, larger than any conceivable sea level rise this century. Long Beach is a city with money from oil that is literally underfoot, but smaller towns can adjust also. Figure 10.14 shows Robin Hood's Bay, a beautiful village on the

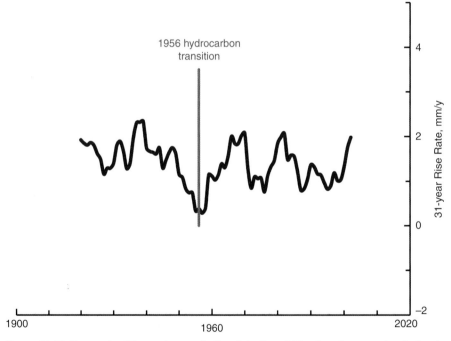

Figure 10.13 The running 31-year rise rate for Honolulu, Hawaii. The slope for a year is calculated as the slope of the regression line for a 31-year period centered on that year. This graph uses the same data as Figure 10.12.

[3] The spatial patterns are shown in Figure 2b in the paper that defined the PDO, Nathan Mantua et al., 1997, "A Pacific interdecadal climate oscillation with impacts on salmon production," *Bulletin of the American Meteorological Society*, pp. 1069–1079.

Figure 10.14 Robin Hood's Bay in North Yorkshire, England, with its 12-m seawall. From freeimageslive/7988.

English North Sea coast. In earlier times its main business was smuggling, but now it is tourism. Robin Hood's Bay is full of twisting narrow lanes and medieval buildings. However, the village had also been damaged by storms over the years. In 1975 it completed an enormous sea wall 12 m high.

10.4   Temperature Indexes

Temperature records are more complex than tide-gauge data. Sea level has a single surface that is coupled by the gravitational potential around the world. On the other hand, air temperature varies strongly with altitude and sea water temperature varies with depth. The tradition has been to combine land thermometer records taken at a height of 1.25 m with sea water temperature measurements. It is conventional to form an index calculated as a weighted average of temperature measurements from thermometers in different places. This is like an inflation index in economics that is calculated as a weighted average of price rises. The choice of weighting matters. For example, it is common in calculating an inflation index to exclude food and energy. This allows an economist to focus on factors that may be amenable to policy. However, for the consumer, inflation in food and energy may be the major concern. Similarly, for temperature there are many questions. A farmer may find an increase in winter daily minima of little concern, but an increase in summer maxima might require changing crops. Often the land measurements are readings from maximum–minimum thermometers at standard times of the day. What does an average of these two numbers mean? There are some places that are not stable enough for thermometers, like pack ice. There are also some places where almost no one lives, like the interior of the Sahara Desert.

The advantage of the land thermometers is that they are located where people live. That is also their weakness. People produce heat when they run air conditioners and heaters and this affects the measured temperatures. In addition, streets and buildings absorb radiation from the sun during the day and release heat at night. In the scientific literature, these are called *urban heat islands*. More than half of the thermometers in the world climatology network are at airports where there are large asphalt areas that absorb solar radiation. Figure 10.15 shows a spectacular example at Providence, Rhode Island. The airport is 12 K above the forest that surrounds the city. An urban heat island can locally swamp the changes in temperature indexes.

Since 1979 temperature measurements of the upper atmosphere have also been made from satellites. These use microwave receivers called *radiometers* to measure the radiated power from molecular oxygen. The advantage of the satellite radiometers is that the same instrument covers the entire earth. In addition, the radiometers make measurements of the upper atmosphere where the effect of increased carbon dioxide concentrations is predicted to occur. However, the satellite orbits change over time and the satellites themselves last only a few years. The mathematical procedures for compensating for the change in orbits and for recalibrating for the next satellite are complex.

The sea covers 70% of the earth's surface and its heat capacity is thousands of times that of the atmosphere. The ocean has its own temperature fluctuations that are not connected in an obvious way to fossil-fuel burning and carbon dioxide. For example, the Pacific Decadal Oscillation (PDO) shuts down the Pacific sardine fishery at an interval of 60 years. The historical data for ocean temperatures consist primarily of ship records. In the early days, sailors dipped a bucket in the water and brought it up for a thermometer measurement. Credit should be given to the British Royal Navy for collecting millions of these bucket measurements. Around the time of World War II, temperature readings for engine cooling water began to be used instead. More recently buoys were set out that transmit the temperature and wave heights by radio. Each of these approaches has limits. Ship measurements are limited to freighter tracks and areas of interest to navies. The thermal properties of the buckets changed over time. Early buckets were made of wood, while later buckets were canvas. Engine cooling water inlets are below the depth that a bucket would collect water. As ships grew bigger the depth of the inlets

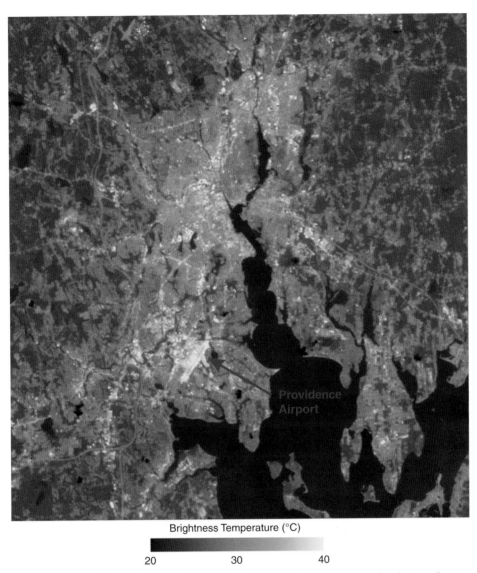

Figure 10.15 Satellite infrared image of Providence, Rhode Island. Credit: NASA. The photograph was shown by Ping Zhang, 2010 "Potential Drivers of Urban Heat Islands in the Northeast, USA," at the 2010 Fall meeting of the American Geophysical Union.

increased. Most of the buoys are along the coasts, very few are in the middle of the ocean.

Starting in the 2000s, thousands of robot divers have been deployed that measure temperature and salinity with depth as part of the Argo Project (Figure 10.16). Conceptually Argo is a most remarkable scientific venture. The robots hang out most of the time at a depth of 1,000 m. Every ten days they dive to 2,000 m and then rise to the surface, measuring temperature and salinity along the way.

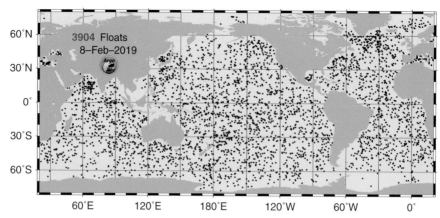

Figure 10.16 The Argo Project. The black dots are the positions of the robot divers that reported temperature and salinity profiles in the 30 days leading up to February 8, 2019. Available at www.argo.ucsd.edu/

At the surface they transmit the data profile and then they return to the deep. In the future, it is likely that the Argo network and the satellite radiometer measurements will provide the keys to understanding the thermal dynamics of the atmosphere and the ocean. However, for our discussion we will use the surface thermometer series because the satellite and Argo series are not long enough to compare before and after the hydrocarbon transition in 1956.

10.4.1   The Central England Temperature Index

Figure 10.17 shows the longest running temperature series. This is the Central England Temperature (CET) index developed by Gordon Manly in 1952 and continued to the present by the Hadley Centre of the British Met Office. The CET starts in 1659. One complication is that early thermometer measurements were taken inside unheated buildings. These readings would be affected by the thermal time constant of the buildings. Manly made adjustments to try to account for this effect. While this series does not include marine measurements, English weather is heavily influenced by the ocean. There are several things to notice about the plot. Generally, the temperature has been rising. However, the annual fluctuations dominate. There are years before 1700 that are warmer than years after 2000. The three coldest years in the record are 1695, 1740, and 1879. These were all associated with famines, 1695 in Finland, and 1740 and 1879 in Ireland. The 1695 and 1740 famines caused great loss of life. However, by time of the 1879 famine, steamships were available to send food to keep people from dying.

In Figure 10.18, the CET is replotted as a 31-y running average. There is a noticeable dip in the late 1600s. There was a well-documented lack of sunspots at the time called the Maunder Minimum, named for the British solar astronomers Annie and Walter Maunder. Sunspots are an index of solar activity and scientists

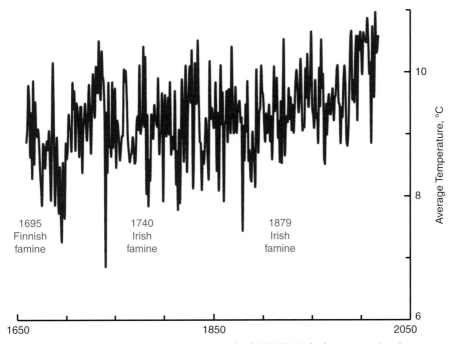

Figure 10.17 The Central England Temperature index (CET). This is the longest running thermometer index, starting in 1659. The data shown here are annual averages. The data come from the British Met Office.

Figure 10.18 The CET plotted as a 31-year running average. The underlying data are the same as in Figure 10.17. The retreat of the Lower Grindelwald Glacier, Switzerland, is calculated from drawings and paintings. The data are from Daniel Steiner et al., 2008, "Sensitivity of European glaciers to precipitation and temperature – two case studies," *Climatic Change*, vol. 90, pp. 413–431.

have long been curious as to whether there is a connection between a sunspot minimum and cold temperatures. We would need another sunspot minimum to test the idea, as there is not a good understanding of how variations in solar activity affect temperatures. The curve shows a rise in the twentieth century of one kelvin in two distinct steps that are separated by a shallow decline in the 1950s and 1960s. Looking at the curve, one can see why Guy Callendar's theory of global warming was attractive when he published his paper in 1938. The temperature was rising. The ending of the Little Ice Age is well documented by the retreat of European glaciers. The figure also shows the retreat for the Lower Grindelwald Glacier in Switzerland. There is not a good alignment between the retreat of the glaciers and the jump in the Central England Temperature index. The CET starts rising at the beginning of the twentieth century, well after the retreat of glaciers in the middle of the nineteenth century.

10.4.2   The HadCRUT4 Index

Figure 10.19 shows the most prominent world temperature index. It is called HadCRUT4. This index is a collaboration between the Hadley Centre of the British Met Office and the Climate Research Unit of the University of East Anglia. The Hadley Centre is responsible for the sea surface measurements and the Climate

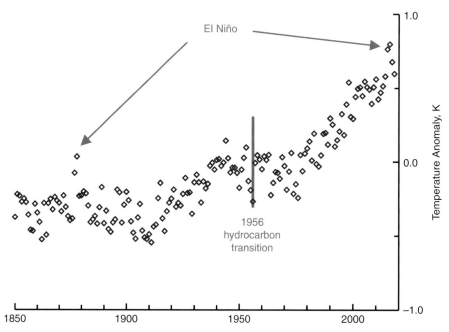

Figure 10.19 The HadCrut4 world annual temperature index from 1850 to 2018. This index is described in C. P. Morice, J. J. Kennedy, N. A. Rayner, and P. D. Jones, 2012, "Quantifying uncertainties in global and regional temperature change using an ensemble of observational estimates: The HadCRUT4 dataset," *Journal of Geophysical Research*, vol. 117, D08101, doi:10.1029/2011JD017187.

Research Unit maintains the land data. One difference with the CET is that the absolute temperatures are not plotted in HadCRUT4. In HadCRUT4, the temperature anomaly is plotted. The anomaly is defined for a particular location as the temperature relative to the temperature averaged over a reference period. The anomalies for each location are then spatially averaged over the earth.

There are outliers in HadCRUT4 that stick out above the rest of the points: 1877–1878 and 2015–2016. These were years with a prominent El Niño Southern Oscillation. Its most important characteristic is a warming of the water in the eastern tropical Pacific Ocean. This reduces the upwelling off the coast of Peru, often causing a collapse of the Peruvian anchovy fishery, normally the largest in the world. In the 1878 El Niño, there was great loss of life from starvation. The effects of the El Niño Southern Oscillation are seen all over the world. The effects can be beneficial. For example, average rainfall is higher in Southern California during El Niño years.

A major difference between the HadCRUT4 index and the CET index is that averaging over the globe greatly reduces the temperature fluctuations. Recall that in the CET, annual fluctuations are comparable to the rise in temperature over the three hundred years of the series. This is not true for HadCRUT4, and we do not need to calculate a 31-year average to reveal the structure. However, it should be kept in mind that the year-to-year fluctuations that a person experiences are more like the CET than HadCRUT4. As with the CET, there are two steps separated by a small decline in the 1950s and 1960s.

10.4.3   Radiative Forcing and HadCRUT4

In recent years, researchers have developed methods for estimating the transient temperature response to carbon-dioxide level changes from the measured temperature indexes. The methods fit well with the approach we have been using to make projections for ultimate production from linearized plots and we will follow their lead. The results fall within the TCR range of the IPCC 5th Assessment Report computer models. It is conventional to define intermediate quantities in Equation 10.1, writing

$$T_{CO_2} = R\,F_{CO_2} \qquad\qquad 10.2$$

where F_{CO_2} is the *forcing*. The forcing is the shift in the global radiation balance expressed as a power density with units of W/m^2. The R factor is a thermal resistance with units of $K/(W/m^2)$. In the 5th Assessment Report, the forcing for CO_2 is calculated as

$$F_{CO_2} = 3.66 \times \log_2\left(c\,/\,c_o\right) W/m^2. \qquad\qquad 10.3$$

The coefficient of 3.66 is derived from infrared spectroscopic measurements. It is conventional to use base 2 logarithms that increase by 1 when the concentration doubles. Figure 10.20 shows the forcing estimates from the 5th Assessment Report. The points plotted in the figure include all forcing components except the short-lived volcanic gases. The contribution of the components aside from carbon dioxide was estimated at $500\,mW/m^2$ in 2011. This contribution is uncertain because of aerosols, which are small particles suspended in the atmosphere. The scattering from aerosols is thought to partly offset the forcing from greenhouse gases. The aerosol forcing component has been revised in the past and it will likely be revised in the future. There is a knee in the plot around the 1956 hydrocarbon transition, which reflects the increased production of oil and gas.

The comparison between radiative forcing and temperature is shown in Figure 10.21. We interpret the slope of the regression line as the thermal resistance factor R. The value of $370\,mK/(W/m^2)$ is comparable to the R-factor of a double-glazed window.[4] We can use this slope and the intercept to make a linear radiation forcing model for temperature. This is shown in Figure 10.22. There is a reasonable correspondence between the model and the data. However, the model does not pick up the step before the hydrocarbon transition in 1956, nor does it get the decline in the decades around the hydrocarbon transition.

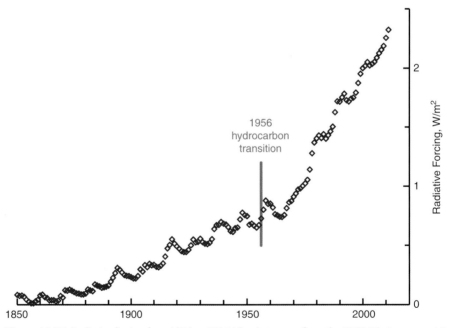

Figure 10.20 Radiative forcing from 1850 to 2011. The data come from the 2013 5th Assessment Report of the UN Intergovernmental Panel on Climate Change (IPCC). The reference year is 1750.

---

[4] American readers will be used to thinking of R-factors for windows and insulation in British units. Double-glazed windows with an R-factor in British units of $2\,°F/(Btu/h/ft^2)$ have an R-factor in SI units of $350\,mK/(W/m^2)$.

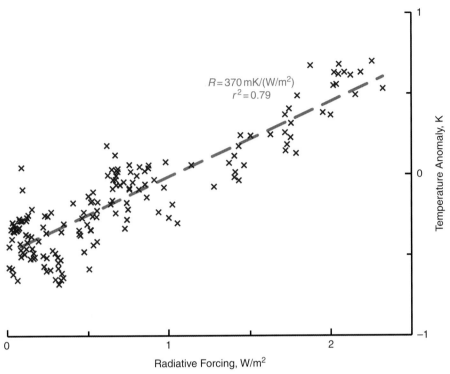

Figure 10.21 The HadCRUT4 temperature index (Figure 10.19) versus the radiative forcing (Figure 10.20). The beginning of the range is 1850, set by HadCRUT4. The end of the range is 2011, set by the 5th Assessment Report. The trend line is a regression line.

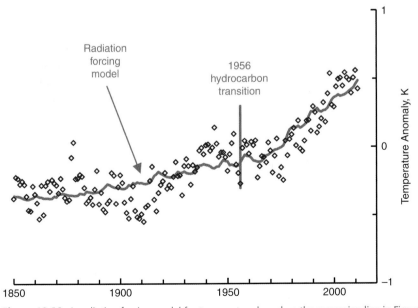

Figure 10.22 A radiation forcing model for temperature based on the regression line in Figure 10.21.

To understand the problem better, we plot the temperature residuals, which are the differences between the measurements and the model. These are shown in Figure 10.23. For comparison, we have replotted the Atlantic Multidecadal Oscillation (AMO) that we met in analyzing the New York City tide-gauge data. The figure shows a 300-mK imprint of the AMO in the residuals that is embedded in the world temperature index.

10.4.4   Summer Maximums and Winter Minimums in the United States

One of the concerns climate scientists have is that temperatures may become more extreme in the future. The United States has developed a large network of volunteer observers that record temperatures. These measurements go back more than a hundred years. Figure 10.24 and Figure 10.25 give results from the program. The figures show the temperature for the country on the hottest and coldest days of the year. The trend for summer maximums is positive, with a slope of 0.4°/century. However, the variation from year to year overwhelms the trend; r^2 is only 0.05. The year with the hottest day for the country was in 1936 at 31.1 °C, before the hydrocarbon transition in 1956. Among the 48 contiguous states, ten of the all-time state record highs were set in 1936, and for two other states the highs that year tied other years for the record. In contrast, so far in the twenty-first century,

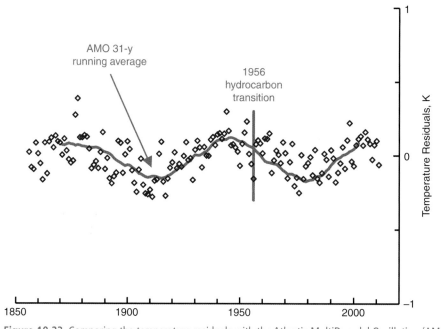

Figure 10.23 Comparing the temperature residuals with the Atlantic MultiDecadal Oscillation (AMO) temperature index. The AMO data are replotted from Figure 10.11.

only one state record has been set outright and one record from the twentieth century was tied.

The trend for winter minima in Figure 10.25 is positive, at 1.3°/century. This is an indication of milder winters. However, the variation from year to year overwhelms the trend again; r^2 is only 0.15. The coldest winter day was in 1979–1980 at −8.6 °C, after the hydrocarbon transition.

Figure 10.24 and Figure 10.25 urge caution in making conclusions about temperature extremes. In practice people respond to high summer temperatures by installing air conditioners. The city of Phoenix is the capital of the state of Arizona, and the population is over a million people. It typically has a day each summer where the temperature reaches 45 °C. It is clear this is not a problem because many people move to the area when they retire to escape cold winters. Air conditioning is universal in homes and office buildings there. More broadly, farmers also adapt to higher temperatures. The yield for corn, the most important US agricultural crop, was higher in each of the years 2015, 2016, 2017, and 2018 than in any of the years before 2015.

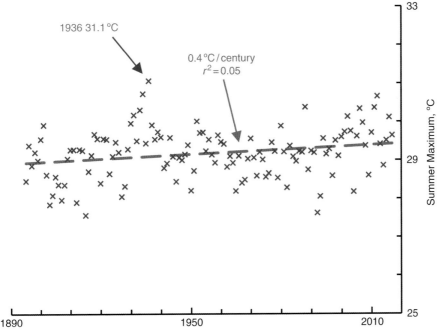

Figure 10.24 The average temperature on the hottest day each summer in the United States. The maxima are calculated from the daily averages for the lower 48 states over the months of June, July, and August. The trend line is a regression line. The data come from the National Oceanic and Atmospheric Administration (NOAA) National Centers for Environmental Information, Climate at a Glance, National Time Series, available at www.ncdc.noaa.gov/cag/

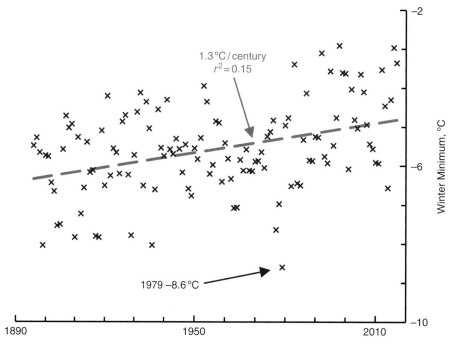

Figure 10.25 The average temperature on the coldest day each winter in the United States. The minima are calculated from the daily averages for the lower 48 states over the months of December, January, and February. The trend line is a regression line. The data are from the National Oceanic and Atmospheric Administration (NOAA) National Centers for Environmental Information, Climate at a Glance, National Time Series, available at www.ncdc.noaa.gov/cag/

10.5 Projections

The United Nation Intergovernmental Panel on Climate Change (IPCC) uses a set of scenarios called *Representative Concentration Pathways* (RCPs) to specify the forcings in climate models. These scenarios cover a wide range of emissions, but the most commonly cited scenario is RCP 8.5, where the forcing in 2100 is $8.5\,W/m^2$. This is a big-coal scenario. The coal consumption given as representative in the paper that defines the scenario is shown in Figure 10.26. Consumption rises rapidly from 1.1 t/p in 2020 to 3.4 t/p in 2100. The cumulative production of coal through 2100 would be 2,100 Gt. Moreover, coal consumption is still rising in 2100. This compares with the original reserves from Table 4.4 of 1,219 Gt. It indicates an ultimate production of a multiple of the original coal reserves. As we saw in Chapter 4, this is contrary to the historical experience that a fraction of original coal reserves is produced. In this section, we will use the projections from Section 5.7 for the calculations.

Economists have attempted to develop models of the economic impacts of climate change through the end of this century. We discussed the limitations of an

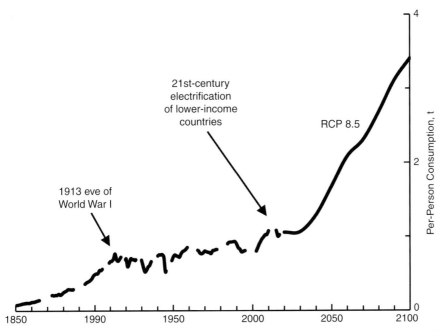

Figure 10.26 World per-person coal consumption in the scenario RCP 8.5. Consumption is taken from Figure 5 in Keywan Riahi et al., 2011, "RCP 8.5 – A scenario of comparatively high greenhouse gas emissions," *Climatic Change*, vol. 109, pp. 33–57. The dashed line shows the historical production from Figure 4.3. The RCP 8.5 data are converted from energy units to production units at the historical average of 2 t/toe. The population data are the medium variant from the United Nations Population Division.

economic model in Section 3.5 that failed in a short-term projection for the labor market. Any GDP changes from climate change would likely be overwhelmed by the ongoing background rise in income. To illustrate, in 2017 the world GDP (PPP) was $17,000 per person. From 2007 to 2017, a period that includes the Great Recession, the inflation-adjusted annualized growth was 2%/y. At this growth rate, the per-person GDP in 2100 would be $90,000. The per-person GDP at the end of the century could be a multiple of what it was in 2017. Such a world would be resilient far beyond current capabilities. For these reasons, we will not attempt economic projections.

Table 10.1 restates the projections for ultimate production U in Table 5.3 in terms of the carbon content. The total is 1,012 GtC, of which 58% is hydrocarbons. We found in Figure 1.30 that there has been a precise linear relation between the cumulative production of fossil fuels and the atmospheric CO_2 levels. This allows us to use U to project a maximum CO_2 level by extending the trend line in the figure. This is shown in Figure 10.27. For an ultimate production U of 1,012 GtC, the projected maximum CO_2 level c is given by

$$c = 298 + 0.254\,U \text{ ppm} = 555\,\text{ppm} \qquad 10.4$$

This is 36% higher than the 407 ppm level in 2017.

Table 10.1 Summary of the projections for world hydrocarbons and coal. These are the same as those given in Table 5.3, but expressed in GtC using the carbon coefficients in Table 1.6.

	Cumulative production q, GtC (2017)	Ultimate production projection U, GtC	t_{90}
Hydrocarbons	229	587	2082
Coal	203	425	2065
Fossil fuels	432	1,012	2075

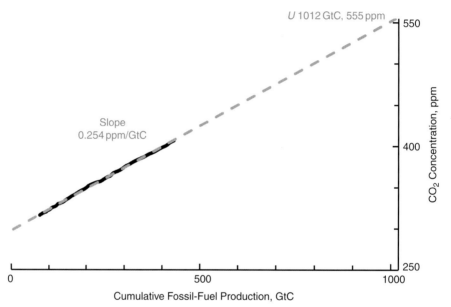

Figure 10.27 Atmospheric CO_2 levels versus cumulative fossil-fuel production. This is Figure 1.30, but redrawn with a different horizontal axis to accommodate U from Table 10.1.

We calculate the radiative forcing F_{CO_2} associated with future fossil-fuel burning from Equation 10.3 as

$$F_{CO_2} = 3.66 \times \log_2\left(\frac{555}{407}\right) W/m^2 = 1.6 \, W/m^2 \qquad 10.5$$

where 407 ppm was the CO_2 level in 2017. We can now use Equation 10.2 to project the long-run temperature rise T_{CO_2} associated with future fossil-fuel burning as

$$T_{CO_2} = R \, F_{CO_2} = 370 \times 1.6 \, mK = 600 \, mK \qquad 10.6$$

where the R-factor of 370 mK/(W/m²) is the slope of the regression line in Figure 10.21. Given a t_{90} of 2075, "long run" means the last quarter of this

Table 10.2 Summary of the projections.

Ultimate production projection U	Maximum CO_2 concentration c	Future fossil-fuel burning F_{CO_2}	Long-run temperature rise T_{CO_2}	Temperature sensitivity $\dfrac{dT_{CO_2}}{dU}$
1,012 GtC	550 ppm	1.6 W/m²	600 mK	900 nK/MtC

The forcing and the long-run temperature change are those associated with carbon dioxide from future fossil-fuel burning only. They do not include other radiative forcing components or fossil-fuel emissions through 2017.

century. The long-run temperature rise is 600 mK from carbon dioxide from future fossil-fuel burning. It does not include other radiative forcing components. With this qualification, it could be compared with the historical 1.0-K rise in the HadCRUT4 temperature index in Figure 10.19. The projections are summarized in Table 10.2.

We can write the temperature sensitivity to changes in ultimate production by differentiating Equation 10.6 with respect to U. We get

$$\frac{dT_{CO_2}}{dU} = \frac{370 \cdot 3.66 \cdot 0.254}{\ln(2)\,550} \text{ mK/GtC} = 900 \text{ nK/MtC.} \tag{10.7}$$

When scaled by a representative policy emissions reduction, this sensitivity could be compared to the 300-mK imprint of the Atlantic Multidecadal Oscillation (AMO) in the HadCRUT4 temperature index.

National and state policy often aim at reducing fossil-fuel consumption. To the extent that a policy is successful we could credit it with a reduction in U and multiply by $\frac{dT_{CO_2}}{dU}$ to calculate a temperature change for the policy. As an example, we consider the fossil-fuels policy of the state of California. By comparison with the federal government and other states, California has had an aggressive policy on renewables and strong support for electric vehicles. Figure 10.28 shows the history of carbon-dioxide emissions for California and for the rest of the states. For both California and the rest of the states, the largest drop, 10%, occurred during the Great Recession. In 2016, California was 3.3% below the rest of the states. For the purpose of this discussion we credit state policy with reducing U by 3.3% of the 2000 reference emissions, which were 113 MtC. The resulting credit is 3.7 MtC. We multiply by $\frac{dT_{CO_2}}{dU}$ to get the long-run temperature change, 3 μK. One thing that limited California was that it shut down the San Onofre nuclear plant in 2012. When a nuclear power plant goes down, that generation must be replaced by alternative sources or carbon-dioxide emissions will go up. The state plans to shut down its remaining nuclear plant, Diablo Canyon, in 2025, so this is likely to happen again.

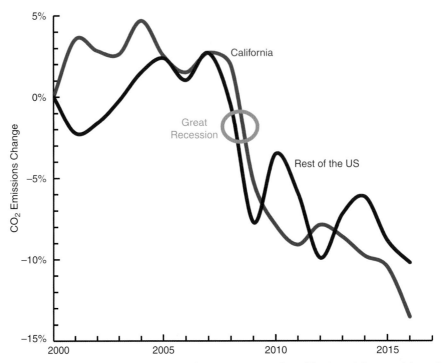

Figure 10.28 Carbon-dioxide emissions changes since 2000 for California and the rest of the United States. The US carbon-dioxide emissions come from the BP *Statistical Review*. The California emissions come from the State of California Greenhouse Gas Emission Inventory.

10.6   Summary

It would be good to keep in mind the warning by Caltech professor Paul Wennberg in the epigram at the beginning of the chapter. We do not know what the future evolution of the CO_2 concentrations will be. The complexity of the CO_2 distributions in the maps from the Orbiting Carbon Observatory 2 gives a sense of the modeling challenge. The linear relationship with cumulative fossil-fuel production has been extremely tight so far, and that is why we have assumed that the relationship will continue. It is possible that as fossil-fuel production slows, the sinks could catch up with the emissions. It is also conceivable that efforts to capture and sequester carbon dioxide will be successful. Either of these would make the projection for the maximum CO_2 level too high. On the other hand, if the CO_2 sinks saturate, the projection would be too low.

The Honolulu tide gauge does not give evidence that we are returning to the meter per century rise of the exit from the last ice age. The measurements do show a rise of 1.5 mm/y from 1905 to 2017 that we could ascribe to melting glaciers. However, 87% of fossil-fuel burning has occurred in the period from the hydrocarbon transition in 1956 to 2017 and Figure 10.13 shows similar rise rates before and after the transition. There is no model that predicted this sea level rise rate for the last century. The behavior of glaciers is simply not well understood.

National and state policy on fossil fuels can be evaluated by the sensitivity of the long-run temperature to the ultimate fossil-fuel production. The projection for this sensitivity is 900 nK/MtC. For coal production units, the sensitivity may be expressed as 500 nK/Mt. To illustrate, in 2019 the Rocky Hill Coal Project in the Hunter Valley of Australia was blocked by a judge who was concerned about meeting Paris Agreement obligations. The mine plan was for an ultimate production of 21 Mt, so we could associate his decision with a 10-µK temperature change. One subtlety is that the Rocky Hill mine would have produced metallurgical coal for making coke to reduce iron ore, rather than steam coal for generating electricity. In this role, electricity from alternative sources is not a substitute. Rather, the traditional alternative substitute for coke in steel production is charcoal made from wood. On the debit side, the mine planned to employ 110 people, and mining is a high-paying job in the Hunter Valley. As with the Oregon loggers (Section 6.5.3), the negative impacts may hit rural areas the hardest. In 2018 French President Emmanuel Macron announced a diesel fuel tax increase to fight climate change. People living in the country would be more affected by this tax than people living in the city because they drive more and they have lower incomes. The announcement triggered demonstrations involving hundreds of thousands of people over many weeks. The protestors wore the yellow safety vests that drivers are required to keep in their vehicles (Figure 10.29).

Figure 10.29 The *Gilets Jaunes* (yellow vests) protest against a carbon tax increase on diesel fuel in 2018. Credit: Alexandros Michailidis/Shutterstock.

In Chapters 4 and 5, we drew graphs that compared the projections for ultimate production with the cumulative production. The temperature sensitivity to U calculated in Table 10.2 gives us a way of thinking about these plots in policy terms, where we interpret shifts in U as an indication of the effects of policy. Figure 10.30 shows the evolution of the ultimate production projections expressed in carbon content. A vertical gray bar with tick marks has been added after the latest projection. Each tick indicates how much policy has to bend the U curve for a temperature change of 100 mK.

Climate-change policy is very different from the air-pollution policy that we discussed in Section 9.4.2. For bad air, there are quick policy solutions: requiring pollution control equipment and fuel switching. The improvements are dramatic. To long-time Los Angeles residents, the end of the smog alerts was a miracle, while young residents would not even know what a smog alert is. In contrast, for climate-change policy, the changes calculated from the temperature sensitivity are for the last quarter of the century, and the Atlantic Multidecadal Oscillation (AMO) may mask 300 mK of the effects.

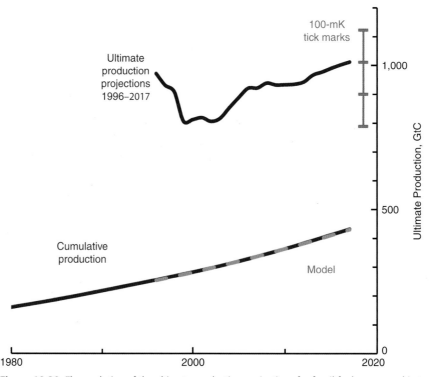

Figure 10.30 The evolution of the ultimate production projections for fossil fuels expressed in terms of the carbon content. These are calculated by converting the projections for coal in Figure 4.46 and the projections for hydrocarbons in Figure 5.47 to their carbon equivalents and adding them. The positions of the 100-mK tick marks are calculated from the long-run temperature sensitivity to U in Table 10.2, 900 nK/MtC.

Concepts to Review

- The logarithmic relationship between CO_2 concentration and temperature change
- The photosynthesis signature in CO_2 records
- The contribution to relative sea level rise from land movement
- The AMO fingerprint on New York City tide-gauge records
- The trend for the Honolulu tide gauge
- Urban heat islands
- The Central England Temperature index
- Evolution of summer maximums and winter minimums in the United States
- The relationship between radiative forcing and the HadCRUT4 index
- The AMO imprint on the temperature residuals
- Projections for carbon-dioxide levels and the associated temperature rise
- Using the long-run temperature sensitivity to U as a way to evaluate policy

Problems

Problem 10.1   The Maassluis tide gauge

Maassluis is a town in the Netherlands on the River Maas 10 kilometers from its mouth (Figure 10.31). The town is named for a nearby lock (*sluis* means lock in Dutch). Maassluis has the world's longest continuous tide-gauge series. It begins in 1848.

Download the annual mean sea level data in millimeters for Maassluis from the Permanent Service for Mean Sea Level (PSMSL) and transfer the data to an Excel spreadsheet.

Figure 10.31 The North Sea port of Maassluis, the Netherlands. This is the site of the longest running tide gauge in the world. Maas refers to the Maas River. In French, this river is known as the Meuse. Credit: Arch/public domain.

a. Find the sea level rise rate in mm/y from 1848 to the latest by linear regression.
b. Plot the rise rate as a running 31-y regression and interpret, noting the 1956 hydrocarbon transition.

Further Reading

- Svante Arrhenius, 1898, "On the influence of carbonic acid in the air upon the temperature of the ground," *The London, Edinburgh, and Dublin Philosophical Magazine and Journal of Science*, vol. 41, pp. 237–276.
- Guy Callendar, 1938, "The artificial production of carbon dioxide and its influence on temperature," *Quarterly Journal of the Royal Meteorological Society*, vol. 64, pp. 223–240.
- Nicholas Lewis, 2018, "Climate sensitivity to cumulative carbon emissions." Lewis is an independent scientist in the tradition of Guy Callendar. His post gives a discussion of the concept of the temperature sensitivity to cumulative carbon-dioxide emissions. It is followed by extensive comments. The link is https://judithcurry.com/2018/12/11/climate-sensitivity-to-cumulative-carbon-emissions/#more-24552
- Justin Ritchie and Hadi Dowlatabadi, 2017, "Why do climate change scenarios return to coal?" *Energy*, vol. 140, pp. 1276–1291. Ritchie and Dowlatabadi were the ones who called attention to coal production problems in RCP 8.5.
- United Nations Intergovernmental Panel on Climate Change (IPCC), Working Group 1, 2013, 5th Assessment Report, *Climate Change 2013, the Physical Science Basis*. Table 9.5 lists the ECS and TCR values in the computer models. Table AII.1.2 lists the radiative forcings in Figure 10.20.

Index